T0328277

Environmental and Health Issues in Unconventional Oil and Gas Development

Environmental and Health Issues in Unconventional Oil and Gas Development

Edited by

Debra Kaden

Ramboll Environ US, Boston, MA, USA

Tracie Rose

RHP Risk Management Inc., Chicago, IL, USA

Amsterdam • Boston • Heidelberg • London • New York • Oxford
Paris • San Diego • San Francisco • Singapore • Sydney • Tokyo

Elsevier
Radarweg 29, PO Box 211, 1000 AE Amsterdam, Netherlands
The Boulevard, Langford Lane, Kidlington, Oxford OX5 1GB, UK
225 Wyman Street, Waltham, MA 02451, USA

British Library Cataloguing-in-Publication Data
A catalogue record for this book is available from the British Library

Library of Congress Cataloging-in-Publication Data
A catalog record for this book is available from the Library of Congress

ISBN: 978-0-12-804111-6

For information on all Elsevier publications
visit our website at http://store.elsevier.com/

Contents

List of Contributors

Carl Adams, Jr
Ramboll Environ Brentwood, TN, USA

Karen J. Anspaugh
Indiana University Robert McKinney School of Law and Surrett & Anspaugh, Traverse City, MI, USA

Uni Blake
American Petroleum Industry (API), Regulatory and Science Affairs, Washington, DC, USA

Katharine Blythe
Ramboll Environ UK Ltd, Edinburgh, United Kingdom

Margaret Cook-Shimanek
University of Colorado at Denver and Health Sciences Center, Denver, CO, USA

Eric J. Esswein
University of the Witswatersrand, School of Public Health, Johannesburg, South Africa; National Institute for Occupational Safety and Health (NIOSH), Western States Division, Denver, CO, USA

Bernard D. Goldstein
Graduate School of Public Health, Pittsburgh, PA, USA, University of Cologne, Cologne, Germany

Anthony R. Holtzman
K&L Gates LLP, Harrisburg, PA, USA

Robert Jeffries
Ramboll Environ UK Ltd, London, United Kingdom

Aleksander S. Jovanovic
Steinbeis Advanced Risk Technologies, Stuttgart, Germany

Debra A. Kaden
Ramboll Environ US, Boston, MA, USA

Bradley King
National Institute for Occupational Safety and Health (NIOSH), Western States Division, Denver, CO, USA

Alan J. Krupnick
Center for Energy and Climate Economics, Resources for the Future (RFF), Washington, DC, USA

Wayne G. Landis
Institute of Environmental Toxicology, Huxley College of the Environment, Western Washington University, Bellingham, Washington, USA

MariAnna K. Lane
Institute of Environmental Toxicology, Huxley College of the Environment, Western Washington University, Bellingham, Washington, USA

David Richard Lyon
Environmental Defense Fund, Austin, TX; University of Arkansas, Environmental Dynamics Program, Fayetteville, AR, USA

Matthew M. Murphy
BatesCarey LLP, Chicago, IL, USA

Daniel J. Price
Ramboll Environ US, St. Louis, MO, USA

Ortwin Renn
SOWI V, Universität Stuttgart, Stuggart, Germany

Elyse Rester
Ramboll Environ USA, Senior Associate, Global Water Practice, USA

Kyla Retzer
National Institute for Occupational Safety and Health (NIOSH), Western States Division, Denver, CO, USA

Tracie Rose
RHP Risk Management Inc., Chicago, IL, USA

Alicia Jaeger Smith
BatesCarey LLP, Chicago, IL, USA

Mark Travers
Ramboll, Copenhagen, Denmark

Scott D. Warner
Ramboll Environ USA, Principal, Water Resources Management, Global Water Practice, USA

Robert Westaway
School of Engineering, University of Glasgow, Glasgow, Scotland, UK

Craig P. Wilson
K&L Gates LLP, Harrisburg, PA, USA

About the Authors

EDITORS

Dr Debra A. Kaden has more than 25 years of experience in toxicology and environmental health sciences, with emphasis in the area of air toxics. She is a senior practitioner in Environmental Health Science practice. Dr Kaden is a member of the International Society for Exposure Science – where she chairs the Strategic Communications and Outreach Committee – as well as a member of the Society of Toxicology (SOT) and the Society for Risk Analysis (SRA), and Past President of the New England chapter of SRA. Dr Kaden has authored more than 25 peer-reviewed publications in toxicology and environmental health sciences. Her expertise includes critical reviews of the toxicological effects on human health from many of the volatile organic compounds potentially associated with emissions from gas-drilling processes, including benzene, acrolein, and formaldehyde. She has evaluated the scientific literature relevant to pollution resulting from activities that might be associated with establishment of well sites, such as traffic. Her experience in the field of exposure-science issues allows an appreciation of spatial and temporal influences on potential community exposures. Dr Kaden holds an SB in biology, and SM and PhD degrees in Toxicology from the Massachusetts Institute of Technology, and has postdoctoral research experience at New York University Medical School and Harvard Medical School. She spent nearly 20 years as the air toxics staff lead at the Health Effects Institute, where she was involved in studies involving mobile source pollution including diesel emissions, benzene, 1,3-butadiene, formaldehyde, and other aldehydes; and criteria pollutants such as ozone and particulate matter.

Ms Tracie Rose is a cultural anthropologist with more than 20 years of experience supporting all aspects of dispute resolution, communication, and perception within topics including personal injury, product liability, toxic tort, bankruptcies, trial arbitration, and both real and threatened litigation. Ms Rose is a principal of RHP Risk Management Inc. in Chicago. She holds an MA in Cultural Anthropology, Northern Illinois University and BA, Anthropology; American Labor History, Northern Illinois University. Her experience includes research, pretrial, trial, and post-trial in mass tort, class action, product liability, premises liability, regulatory, due diligence, environmental, chemical release and exposure, accounting liability, bankruptcy, antitrust, and insurance actions in both state and federal jurisdictions. Ms Rose has worked in-house for insurance counsel, national trial counsel, and large multi-national firms. She works closely with presenters and witnesses in the development, preparation, and presentation of scientific concepts through data and demonstratives. She is experienced in document mining and objective analysis across many topics. Her publications and speaking engagements include topics such as exposure analysis and risk, worker safety and health, state of art of historic liabilities, and warnings. She lives with her husband, Don, in the suburbs of Chicago and has 2 adult children, Jessica and Wyatt, and daughter-in-law, Kate.

ALPHABETICAL LIST AND BRIEF BIOGRAPHIES OF AUTHORS

Dr Carl E. Adams Jr has been a consultant and director to more than 1000 US and foreign industrial wastewater management projects in 32 countries. Dr Adams holds PhD in environmental health engineering at the University of Texas, Austin, and a bachelors of engineering and masters degree from Vanderbilt. Dr Adams has authored over 100 technical publications and coauthored and edited four books and engineering manuals. Dr Adams is considered one of the leading international experts on industrial wastewater management, his special areas of technical expertise include aerobic and anaerobic treatment of high-strength and high-salt industrial wastewaters, nitrification–denitrification, membrane technology, chemical oxidation, source control, and water recycle and reuse. He has served as a visiting adjunct professor at Vanderbilt University for application of advanced wastewater technology. Dr Adams has developed and patented cutting edge technical processes for the management of wastewater including VOC BioTreat™ control technology for facilities with fugitive off-gases proximate to industrial wastewater treatment plants. Dr Adams' patents are installed in over 75 applications in 18 countries. He holds distinctions with the Academy of Distinguished Alumni, Department of Civil and Architectural Engineering at the University of Texas–Austin, and with the Academy of Distinguished Alumni, School of Engineering at Vanderbilt University.

Ms Karen J. Anspaugh has focused her 20 years of practicing law exclusively in oil and gas. Ms Anspaugh holds an undergraduate degree from Oral Roberts University where she graduated Magna cum laude and a Juris Doctorate from the University of Tulsa. She is licensed to practice in Indiana, Michigan and Missouri. Ms Anspaugh has drafted over 1000 title opinions on the topic of oil and gas. Karen has assists client with regulatory compliance, transactions including leasehold and mineral asset acquisitions, and contract and assignment negotiations. Ms Anspaugh is the managing partner of Surrett & Anspaugh, which practices in Oklahoma, Pennsylvania, Illinois, Indiana, Kentucky and Michigan. She previously served "Of Counsel" at Bingham Greenebaum Doll, Indianapolis and Vorys Sater Seymour & Pease, Columbus, Ohio, and as managing partner of Anspaugh Law. Ms Anspaugh teaches oil and gas law at Indiana University Robert McKinney School of Law, and serves on the board of the Indiana Oil and Gas Association and as a member of professional and industry associations including the American Association of Petroleum Landmen, Energy and Mineral Law Foundation, Illinois Oil and Gas Association, Indiana Bar Association, Indiana Coal Council, Indiana Oil and Gas Association, Kentucky Oil and Gas Association,

Michael Late Benedum APL Chapter (Ohio, Pennsylvania, West Virginia), Michigan Association of Professional Landmen, Michigan Bar Association, Michigan Oil and Gas Association, Tri-State Association of Professional Landmen (Illinois, Indiana, Kentucky), and the Women's Energy Network - Appalachian Chapter. Ms Anspaugh is sought after for speaking engagements to peers and stakeholders.

Ms Uni Blake is a toxicologist with over 15 years of work experience in projects related to environmental health, environmental data analysis, risk assessments, and projects relating to the regulation of toxic substances in air, surface, and ground water. Ms Blake holds a Bachelor's Degree in Chemistry from The College of Wooster and a Master's Degree in Toxicology from the American University. Ms Blake is currently a Scientific Advisor at the American Petroleum Institute's Regulatory and Science Affairs Department in Washington, DC. Her work includes a focus on potential human health impacts of oil and gas development. Ms Blake is a member of Society of Petroleum Engineers (SPE), Society of Toxicology (SOT) and the Society for Risk Analysis (SRA).

Dr Katharine Blythe is a Senior Consultant in the Edinburgh office of Ramboll Environ. Dr Blythe holds a PhD in Environmental Plant Physiology from Manchester Metropolitan University, the United Kingdom, an MSc Pollution and Environmental Control from University of Manchester, the United Kingdom, and a BSc Ecology from University of Durham, the United Kingdom. She is a member of AIEMA. Dr Blythe has over 8 years of experience in environmental project management and coordination including 7 years within academic environmental fields. Dr Blythe has been involved in a variety of environmental impact assessments (EIA) and has provided environmental input to other projects throughout the United Kingdom including applications for planning consent, pollution prevention and control (PPC), Controlled Activities Regulations (CAR) and Radioactive Substances Act (RSA) permits. Her experience ranges from the initial consultation and scoping stages of EIA, to preparation and submission of the planning application, environmental statement and other environmental and technical reports working within and managing teams of environmental specialists. She worked with clients in diverse sectors including unconventional gas (coal bed methane (CBM)), renewable energy (wind farms and energy from waste plants), opencast mineral workings, waste developments and leisure, and residential developments.

Dr Margaret K. Cook-Shimanek, MD, MPH, has since 2013, been a Resident in Occupational and Environmental Medicine at the University of Colorado at Denver and Health Sciences Center, Aurora, Colorado. Between 2011 and 2013, Dr Cook-Shimank was a Resident in General Preventive Medicine and Public Health at the University of Colorado Health Sciences Center, and for 1 year (2008–2009), Dr Cook-Shimanek was an Intern in Pediatrics at the Children's Hospital Colorado. Dr Cook-Shimanek holds an MD from the University of North Dakota, School of Medicine and Health Sciences, and a Master's of Public Health in Epidemiology from the Colorado School of Public Health. Dr Cook-Shimanek is a recipient of a 2015 Resident Research Award from the American College of Occupational and Environmental Medicine (ACOEM). Dr Cook-

Shimanek's research has included the examination of Workers Compensation Claims data for the Colorado oil and gas industry.

Mr Eric J. Esswein, MSPH, CIH, CIAQP is a Commissioned Officer, Captain in the US Public Health Service assigned to the Centers for Disease Control and Prevention (CDC), National Institute for Occupational Safety and Health (NIOSH) and a Guest Lecturer at the University of Witswatersrand, School of Public Health. Mr Esswein holds an MSPH in Public Health/Industrial Hygiene from the University of Utah, School of Medicine and a graduate certificate in Emergency Management and Planning from the University of Colorado Denver. For the past 24 years Mr Esswein has been a Senior Industrial Hygienist with CDC–NIOSH Western States Division. Mr Esswein's career also included his service as the Industrial Hygiene Section Chief of the US Public Health Service, and for a year as a Senior Industrial Hygienist for NIOSH in Johannesburg, South Africa, and as part of the NIOSH Health Hazard Evaluation program. Mr Esswein has extensive experience conducting field studies related to occupational exposure risks and controls in industries both nationally and internationally. His fieldwork includes quantitative exposure assessments in upstream aspects of oil and gas extraction, as well as drilling, hydraulic fracturing and well servicing. Mr Esswein is the originator and principle investigator of the NIOSH Field Effort to Assess Chemical Exposures in Oil and Gas Workers. He is the first author of article *"Occupational exposures to respirable crystalline silica in during hydraulic fracturing"* (J. Occup. Environ. Hyg. 10, 347–356, July, 2013). Mr Esswein has significant emergency response experience including on-scene technical team lead (Louisiana) NIOSH Gulf Oil Spill response; industrial hygiene team lead 2001 World Trade Center terrorist attack; first responder, team lead US Capitol bioterrorism attack; responder, SARS pandemic Taiwan; and on-scene responder southeast Asian tsunami, Deputy Commander, Applied Public Health Team 3. He is an award winning inventor of sampling and decontamination technologies and holder of US patents. Mr Esswein is the recipient of US Public Health Service Outstanding Service Medals, Commendation Medal, USPHS Citation, Outstanding Unit Citations, USPHS Unit Commendations, and the American Industrial Hygiene Association, Rocky Mountain Section, Industrial Hygienist of the Year, 2004.

Dr Bernard D. Goldstein, MD is a Faculty Emeritus of the University of Pittsburgh Schools of the Health Sciences. Dr Goldstein is the former dean of the University of Pittsburgh Graduate School of Public Health. He is an environmental toxicologist whose research interests have focused largely on the concept of biological markers in the field of risk assessment and public health. Dr Goldstein has published in the areas of blood toxicity, the formation of cancer-causing substances (free radicals) following exposure to inhalants, various aspects of public health decision-making, and global issues in environmental medicine. His current interests include the public health impact of unconventional natural-gas drilling, specifically in the Marcellus Shale area, and the scientific framework for sustainability related to unconventional natural gas drilling. Prior to his tenure at the University of Pittsburgh, Dr Goldstein was professor and chairman of the Department of Environmental and Community Medicine at the University of Medicine and Dentistry of New Jersey-Robert Wood Johnson Medical School, where he established and directed the largest academic environmental and occupational health program in the United States, the Environmental and Occupational Health Sciences Institute. Further Dr Goldstein served as an officer with the US Public Health Service and was appointed by Ronald Regan as an assistant administrator for research and development at the US Environmental Protection Agency (USEPA). Dr Goldstein received his medical degree from New York University and undergraduate degree from the University of Wisconsin. He is a physician, board certified in Internal Medicine, Hematology, and Toxicology. Dr Goldstein is author of over 150 publications in the peer-reviewed literature, as well as numerous reviews related to environmental health. He is an elected member of the National Academies of Science, Institute of Medicine (IOM) and of the American Society for Clinical Investigation. Dr Goldstein has chaired more than a dozen National Research Council and IOM committees primarily related to environmental health issues. He has been president of the Society for Risk Analysis and has chaired the NIH Toxicology Study Section, EPA's Clean Air Scientific Advisory Committee, the National Board of Public Health Examiners, and the Research Committee of the Health Effects Institute.

Mr Anthony Holtzman is a law partner in K&L Gates Harrisburg office. He holds a Juris Doctorate from Dickinson School of Law, Pennsylvania State, graduating summa cum laude, and was a member of Penn State Law Review and is an award-winning writer. He holds a BA, from Messiah College. Mr Holtzman's practice is focused on environmental law, constitutional law, appellate litigation, gaming law, and commercial litigation. Mr Holtzman is admitted to state and federal courts including the United States Supreme Court. He regularly practices before the US Court of Appeals for the Third Circuit, the US District Court for the Middle District of Pennsylvania, Pennsylvania's appellate and trial courts, and the Pennsylvania Environmental Hearing Board. Mr Holtzman represents clients of energy industry in matters that involve agency enforcement activities, allegations of water and surface contamination, property rights disputes, deed and contract interpretation issues, challenges to permits, and issues that arise under freedom of information laws. Further Mr Holtzman advises energy industry clients on Pennsylvania and federal regulatory compliance issues, including stream crossing, spill remediation, waste processing, storm water discharge, mechanical integrity, sensitive species, permitting, and reporting issues. In addition to his environmental law activities, Mr Holtzman regularly advises public and private sector clients on issues of Pennsylvania and federal constitutional law. He has significant experience with the Pennsylvania Constitution's Bill of Rights, along with its provisions regarding the structure and powers of the Pennsylvania General Assembly and requirements for the enactment of legislation. On the federal side, he has represented clients with regard to the First Amendment, Second Amendment, Tenth Amendment, Eleventh Amendment, dormant Commerce Clause, Takings Clause, Equal Protection Clause, and Speech or Debate Clause issues, among others. Beyond practicing law, Anthony serves as an adjunct instructor at Messiah College.

Mr Robert Jeffries is a Principal in Ramboll Environ's London office. Mr Jeffries holds a BA MAppSc. Mr Jefferies has 25 years of environmental consulting experience, specializing in the assessment and remediation of contaminated land, management of multidisciplinary teams and implementation of multisite and international due diligence and environmental assessment programs. During his career, he has been based in locations around the world including the United Kingdom, Australia, USA and southeast Asia and has held various leadership positions in operations, project, and client management. Mr Jeffries has extensive experience with projects involving oil and gas, manufacturing, mining, chemical, and pharmaceutical sectors. Further Mr Jeffries' expertise in the unconventional oil and gas sector (including fracking) includes downstream and upstream projects. Mr Jeffries has provided technical support for clients in the development of shale gas hydrofracturing projects in northwest England. This includes the evaluation of potential impacts upon groundwater and surface water resources from shale gas processes and the compilation of an Environmental Statement as part of planning applications.

Dr Aleksandar S. Jovanovic is a professor and project director European Union at the University of Stuttgart, Center for Interdisciplinary Risk and Innovation (ZIRIUS). Dr Jovanovic has worked since 1977 internationally (European Union, Japan, the United States) in research and industry. Since 1987, Dr Jovanovic has been affiliated with the University of Stuttgart, where he has lead over 30 large projects for the European Union and industry focused in the areas of emerging industrial risk issues, best practices, asset integrity management, decision making and support systems, intelligent software systems, and cost–benefit analysis. Further Dr Jovanovic has acted on assignments to the European Union as "Seconded National Expert" and "Contractual Agent" in the Directorate General (DG) Research and DG Enterprise. Currently Dr Jovanovic leads the European Virtual Institute for Integrated Risk Management EU-VRi cofounded by the University of Stuttgart. He has been a lecturer and professor at universities in Milan (Politecnico), Stuttgart, La Jolla, California (the United States), Paris (École Polytechnique), Zagreb and Novi Sad. Dr Jovanovic is the chairman of CEN–CWA Committee "Emerging Risks" and the liaison-member of the ISO-Committee PC262 (ISO 31000 Standard "Risk Management"). Dr Jovanovic is a Steinbeis medal winner in 2013. His writings include emerging risks and corporate responsibilities, risk-based decision making, failure risk analysis, and risk management.

Mr Bradley King, MSHP, CIH, is a certified industrial hygienist at the National Institute for Occupational Safety and Health (NIOSH) in Denver, Colorado. Mr King came to NIOSH as a fellow in Association of Teachers of Preventive Medicine (ATPM) fellow in 1999, and has since remained in Hazard Evaluations and Technical Assistance Branch (HETAB) as an industrial hygienist. He holds a Master's of Public Health degree in Environmental and Occupational Health from Saint Louis University and is a PhD candidate for a degree in Environmental Health Science at Johns Hopkins University. Mr King has conducted health-hazard evaluations in a wide variety of worksites throughout the United States. and currently conducts exposure assessment research activities in the oil and gas extraction industry. Mr King holds the rank of Commander in the US Public Health Service. Is the 2015 president elect of the AIHA local chapter, Rocky Mountains Section.

Dr Alan Krupnick is a consultant to state governments, federal agencies, private corporations, the Canadian government, the European Union, the World Health Organization, and the World Bank. Dr Krupnick holds a PhD in economics from the University of Maryland. Dr Krupnick has held leadership seats on advisory committees to the US Environmental Protection Agency, as a senior economist on presidential councils advising on environmental and natural resource policy issues, and is a regular member of expert committees from agencies including the National Academy of Sciences and the US Environmental Protection Agency (EPA). He is the Co-Director of Resources for the Future's (RFF) Center for Energy and Climate Economics (CECE) and a Senior Fellow at RFF. Alan's own research focuses on analyzing environmental and energy issues, in particular, the benefits, costs and design of pollution and energy policies, both in the United States and in developing countries. Dr Krupnick is the lead author of *Toward a New National Energy Policy: Assessing the Options study*, examining the costs and cost-effectiveness of a range of federal energy policy choices in both the transportation and electricity sectors. Dr Krupnick's focuses his research in the development and analysis of preference surveys, as well as undertaken research on natural gas supply and impact on energy prices and policies; the costs and benefits of converting the US heavy-duty truck fleet to run on liquefied natural gas; and the costs and benefits of expanded regulation around deep-water oil drilling.

Dr Wayne Landis is a professor and director of the Institute of Environmental Toxicology, Huxley College of the Environment, and Western Washington University. Dr Landis holds a PhD in zoology from Indiana University, an MA in biology from Indiana University, and a BA in biology from Wake Forest University. Dr Landis' current area of research is ecological risk assessment at large spatial and temporal scales. His research contributions include: creation of the action at a distance hypothesis for landscape toxicology, the application of complex systems theory to risk assessment, and development of the relative-risk model for multiple stressor and regional-scale risk assessment and specialized methods for calculating risk due to invasive species and emergent diseases. Dr Landis holds patents and has written on the use of enzymes and organisms for the degradation of chemical weapons. Dr Landis has authored over 130 peer-reviewed publications and government technical reports, made over 300 scientific presentations, edited four books, and wrote the textbook, *Introduction to Environmental Toxicology*, now in its fourth edition. He has consulted for industry; nongovernmental organizations as well as federal (United States and Canada), state, provincial, and local governments. Dr Landis serves on the editorial boards of the journals *Human and Ecological Risk Assessment* and *Integrated Environmental Assessment and Management*, and is the ecological risk area editor for *Risk Analysis*. He is a member of the Society of Environmental Toxicology and Chemistry (SETAC) and served on the SETAC Board of Directors from 2000 to 2003. Dr Landis was named to the Science Panel for the Puget Sound Partnership, a state agency that focuses on the restoration of Puget Sound. In 2007 he was named a Fellow of the Society for Risk Analysis. In 2012 Dr Landis was presented the Lifetime Achievement Award of the Annual International Conference on Soils, Sediments, Water, and Energy.

Ms MariAnna Lane holds a BS in Environmental Science, with an Environmental Toxicology emphasis, and a Chemistry minor from Western Washington University, graduating cum laude. Ms Lane's particular areas of interest include the interaction of science and policy, science communication, environmental toxicology, ecological risk assessment, risk communication, environmental chemistry, chemical fate and transport, botany, and mycology.

Mr David Lyons is a scientist for the Environmental Defense Fund (EDF). He holds an MS in Forestry from the University of Kentucky and a BA in Biology from Hendrix College. As a scientist with EDF, Mr Lyon works on studies that quantify methane emissions from the natural-gas value chain. His efforts include analyzing emissions data and researching technologies and policies to reduce natural gas leakage and minimize the climate impacts of natural gas development. Prior to EDF, Mr Lyons worked for the Arkansas Department of Environmental Quality (ADEQ) as coordinator of the state's air pollution emissions inventory program. While at ADEQ, Mr Lyon's was the project manager of a US Environmental Protection Agency (USEPA)–funded study to evaluate emissions and air quality impacts of natural gas development in the Fayetteville Shale. Further, Mr Lyon managed a half-million dollar project to develop and implement a web-based emissions-inventory reporting system for a multistate consortium of environmental agencies. While in Arkansas, Mr Lyon was a leader in the Sierra Club and advocated for effective environmental regulations. He has experience in teaching about air pollution and performing research in biogeochemistry and restoration ecology.

Mr Matthew M. Murphy is a founding partner of BatesCarey LLP, Chicago, Illinois whose practice focuses on complex insurance matters, including transportation and construction risks, reinsurance transactions and arbitrations, organization formulation and dissolution, coverage litigation, broker-company relations, claim oversight and examinations, policy and contract preparation, and regulatory affairs, in addition to general corporate governance and litigation. Mr Murphy holds a Juris Doctorate from IIT/Chicago-Kent College of Law where he graduated with honors, and a BA degree in finance from the University of Illinois. He is admitted to practice in the state courts of Illinois and Wisconsin, in the US District Court for the Northern District of Illinois, the Eastern and Western Districts of Wisconsin, and US Court of Appeals for the Seventh Circuit. Through his experience Mr Murphy has represents a number of corporations outside of the insurance industry, providing litigation services and general legal counsel. Before focusing on private legal practice, Mr Murphy held several executive positions with Aon Corporation subsidiaries, including Senior Vice President and Director of Aon Risk Consultants and Virginia Surety Company, Dearborn Insurance Company, and Vice President of Aon Reinsurance Agency.

Prior to that, he was with Crum & Forster Managers Corporation in various insurance claim, regulatory and ceded reinsurance positions. Matthew began his insurance career with Sentry Insurance, A Mutual Company. He has been recognized by clients and fellow lawyers as one of the Best Lawyers in US News, 2015; and Leading Lawyers, 2014.

Mr Daniel J. Price, RG is a Principal Consultant in the St. Louis, Missouri office of Ramboll Environ. Mr Price has for more than 25 years provided technical expertise and environmental site characterization, remediation, and liability management to industry, commercial, and investment entities. Mr Price holds a BS in Geology, Missouri State University, and is a Registered Geologist. He has managed a variety of projects, from property-transaction-site assessments to full-scale RCRA facility investigations, Comprehensive Environmental Response, Compensation, and Liability Act (CERCLA) remedial investigations, interim response actions and remediation. Through his practice, Mr Price has provided expertise in the assessment of environmental liabilities related to owned and legacy properties. His experiences include working as a team and project manager conducting an investigation and implementing remediation efforts involving a release of petroleum solvents at a facility in Canada. Supporting remedial efforts of soils of *in situ* chemical sources. Technical support entailing the reduction of hydrocarbon concentrations in groundwater on a property of concern. Technical oversight and opinions evaluating environmental and business risks associated with injection of nonhazardous oilfield wastes (NOW) including naturally occurring radioactive material (NORM) (Class II wells) and nonhazardous industrial solid wastes (NID) (Class I wells) into salt caverns and the caprock above a salt dome. Mr Price has provided technical support to assess the environmental liabilities associated with the acquisition of a natural gas company providing service to residential and commercial customers. Environmental liabilities identified were related to the historic operation of manufactured gas plants (MGPs). Mr Price has evaluated the environmental and liability mid-stream assets of an oil and gas production company, the assets are a part of the Eagleford Shale play in West Texas and include 11 gas gathering systems (GGSs) each consisting of one or more compressor stations and 619 miles of pipeline.

Dr Ortwin Renn is a Professor of Environmental Sociology and Technology Assessment, as Dean of the Economic and Social Sciences Department, and as director of the Stuttgart Research Center for Interdisciplinary Risk and Innovation Studies at the University of Stuttgart. Dr Renn holds a Doctoral degree in Social Psychology from the University of Cologne. Dr Renn serves as Adjunct Professor for "Integrated Risk Analysis" at Stavanger University (Norway) and as Affiliate Professor for "Risk Governance" at Beijing Normal University. His career further includes teaching and research positions at the Juelich Nuclear Research Center, Clark University (Worcester, Massachusetts), the Swiss Institute of Technology (Zurich) and the Center of Technology Assessment (Stuttgart). Dr Renn codirects the German Helmholtz-Alliance: "*Future infrastructures for meeting energy demands. Towards sustainability and social compatibility.*" Dr Renn directs the nonprofit research institute for the investigation of communication and participation processes in environmental policy making (DIALOGIK). He is a member of the Scientific and Technical Council of the International Risk Governance Council (IRGC) in Lausanne; the National Academy of Disaster Reduction and Emergency Management of the People's Republic of China; and the State Sustainability Council in his home state of Baden-Württemberg. Dr Renn has served on panels including the "Public Participation in Environmental Assessment and Decision Making" of the US National Academy of Sciences in Washington, DC (2005–2007), the German Federal Government's "Commission on Energy Ethics after Fukushima" (2011), and on the Scientific Advisory Board of EU President Barroso (2012–2014). Dr Renn is a member of the Senate of the Berlin-Brandenburg Academy of Sciences (Berlin) and of the Board of Directors of the German National Academy of Technology and Engineering (Acatech). In 2012 he was elected President of the International Society for Risk Analysis (SRA) for the 2013–2014 term. His honors include an honorary doctorate from the Swiss Institute of Technology (ETH Zurich), an honorary affiliate professorship at the Technical University Munich, the "Distinguished Achievement Award" of the Society for Risk Analysis (SRA) and several best publication awards. In 2012 the German Federal Government awarded him the National Cross of Merit Order in recognition of his outstanding academic performance. Dr Renn is primarily interested in risk governance, political participation as well as technical and social change towards sustainability. Dr Renn has published more than 30 books and 250 articles, most prominently the monograph *Risk Governance* (Earthscan: London 2008). His publications, white papers, monographies, and articles focus on risk, social conflict, uncertainty, growth, risk versus hazard, risk communication, decision making, and perception. His research on risk, society, and governance has included topics such as oil and gas operations, nanotechnology, food safety, and emerging technology.

Ms Elyse Rester is a Senior Associate in the Emeryville, California office of Ramboll Environ. Ms Rester has 4 years of experience in environmental consulting, with particular expertise in water treatment and regulatory management. Before joining Ramboll Environ Ms Rester graduated from the Georgia Institute of Technology with a Bachelors of Science in Environmental Engineering in 2010 and a Masters of Science in Environmental Engineering in 2011. Ms Rester is licensed in the State of Tennessee as a Professional Engineer. Currently, Elyse provides consulting services specific to wastewater, stormwater, and related compliance activities under the Clean Water Act (CWA), National Pollutant Discharge Elimination System (NPDES), Resource Conservation and Recovery Act (RCRA), and specific state and local regulatory programs.

Ms Kyla Retzer, MPH is an epidemiologist and Safety Researcher with the Center for Disease Control and National Institute for Occupational Safety and Health (CDC–NIOSH) Western States Division since 2010. Ms Retzer holds a Master's of Public Health from University of North Texas Health Science Center, Fort Worth.

Ms Alicia Jaeger Smith is a Special Counsel to BatesCarey LLP, Chicago, Illinois where she focuses her practice on insurance coverage law. Ms Smith holds a

Juris Doctorate from Washington University in St. Louis and was a member of the Washington University Law Quarterly. She holds a BS in civil engineering from Northwestern University. Ms Smith has handled complex coverage cases arising under general liability policies, excess and umbrella policies, errors and omissions policies, and professional liability policies for insurers throughout North America, Bermuda, and London. Ms Smith has counseled clients with respect to claims handling practices, policy analysis, and claims resolution. She has represented clients in litigation, mediation and arbitration proceedings in the United States and London. Before joining BatesCarey, Ms Smith practiced law at Hanson Peters Nye where she handled primarily directors' and officers' liability and employment practices liability claims, and counseling clients regarding policy drafting. As part of her experience, Ms Smith was seconded to Limit Syndicate 2000 (part of QBE) in London in 2003, advising claims and underwriting personnel regarding claims and reserving analysis, and providing in-house training on various aspects of US law. Ms Smith is admitted to practice in Illinois State Courts and in US District Court for the Northern and Central Districts of Illinois.

Mr Mark Travers, PG is an Executive Vice President, and part of Global Practice Development at Ramboll. Mr Travers is currently based in Copenhagen, Denmark. Prior to his Danish move, Mr Travers spent 2 years working at Environ Singapore, but as spent the majority of his career based out of offices in the United States. Mr Travers has completed graduate studies in business administration, and earned an MS in Engineering Geology from the Purdue University and a BS in Geology from Illinois State University. Mr Travers has more than 30 years of experience in applied science and engineering, with particular emphasis in multimedia site assessment and remediation; municipal and hazardous waste management; environmental and geotechnical engineering, contaminated sediment assessment and remediation; construction engineering; mine and ore-processing site development and operation; and natural resources restoration. Mr Travers has a career that includes projects for both private corporations and government organizations globally. Mr Travers has provided support or management on projects focused on oil and gas resources in the United States, the European Union, and Australia.

Mr Scott Warner, PG is a Principal in the Emeryville, California office of Ramboll Environ US. Mr Warner has more than 25 years of consulting experience in hydrogeology, design/assessment of groundwater remediation strategies, groundwater modeling, and water resources. Mr Warner earned an MS in geology from Indiana University, and a BS in engineering geology from UCLA. He is a registered Professional Geologist, Certified Hydrogeologist and Certified Engineering Geologist in California. He has been accepted as an expert in matters including hydrogeochemistry, groundwater remediation, and remediation cost recovery, and has worked on projects in North America, South America, and Europe. Mr Warner has specific expertise in designing and assessing *in situ* groundwater remediation measures, including enhanced bioremediation, permeable reactive barriers (PRBs) and geochemical manipulation for chlorinated hydrocarbons, petroleum hydrocarbons, metals and radioactive constituents. Mr Warner has authored or coauthored more than 50 publications on groundwater remediation, hydraulics, and geochemistry for both peer-reviewed journals and conference proceedings, and currently serves as co-editor of an Oxford University Press book on dense nonaqueous-phase liquid (DNAPL) characterization and remediation. Mr Warner is the codeveloper/coinstructor of national and international courses for the Remediation Technology Development Forum (RTDF), USEPA and the Interstate Technology Regulatory Council (ITRC). Mr Warner has served as the President of the Bay Planning Coalition, a San Francisco Bay Area organization advocating balanced use and regulation of the San Francisco Bay and Delta resources.

Dr Robert Westaway is a Senior Research Fellow (Systems Power and Energy) at the School of Engineering, University of Glasgow. Dr Westaway has a PhD from Cambridge University. He has worked at several British and overseas universities and ran his own consultancy company before joining the University of Glasgow in 2012. His research interests in geothermics include the development of techniques to correct raw heat flow data for effects of palaeoclimate and topography and investigation of effects of erosion, sedimentation and magmatism on the thermal state of the Earth's crust. With multidisciplinary research interests, Dr Westaway has authored almost 200 publications. His work includes the development of conceptual models to facilitate the understanding of high-conductive heat flow or hot groundwater in poorly understood geothermal fields. Dr Westaway is involved in the commercial development of a geothermal field in Slovakia, and has also recently acted as a consultant regarding the assessment of geothermal energy resources in Britain.

Mr Craig Wilson is a law partner in K&L Gates' Harrisburg office and practice group coordinator for the firm's global Environment, Land and Natural Resources Practice Group. Mr Wilson holds a Juris Doctorate and MSEL from Vermont Law School, graduating summa cum laude and was Managing Editor of the Vermont Law Review. He holds a BA from Dartmouth College. Mr Wilson has been associated with K&L Gates since 1992 and a partner with the firm since 2001. He is admitted to state and federal courts including the Supreme Court of the United States. Mr Wilson concentrates his practice in the areas of energy, environment and natural resources, primarily in environmental, land and natural resources and secondarily in energy, liquefied natural gas, oil and gas, and real estate land use, planning and zoning. Mr Wilson has counseled clients exploring for producing and transporting natural gas; developing and operating energy projects and other commercial and industrial projects; seeking environmental permits, zoning, and land development approvals from government agencies; and clients who are parties to business transactions or litigation involving potential environmental liabilities. Mr Wilson is often requested as a speaker on his expertise to audiences including peers, stakeholders, and interest groups.

Acknowledgment

We would like to thank Elsevier and specifically Marisa LaFleur for their patience and encouragement on our idea for the collective approach we desired for this book on a very big and contentious subject. We wish to thank all our authors who generously gave their time, their expertise, and their creativity. We wish to thank our respective firms, Ramboll Environ and RHP Risk Management Inc. for their willingness to provide us the space to pursue the opportunity to bring this brain trust together.

We would also like to thank Amnon Bar-Ilan, Gretchen Greene, Jonathan Kremsky, Ken Mundt, Don Rose, and Suzanne Persyn for their suggestions and support throughout this process.

Introduction

Debra A. Kaden, Tracie L. Rose

Unconventional oil and gas (O&G) development is a complicated and sophisticated process and industry; it is also quite controversial. There are few individuals who are neutral in their opinions about hydraulic fracturing, as compared to many who are quite vocal about (both for and against) the process. There are many editorials, blogs, letters, and opinion pieces written about both sides of the debate. Therefore, when we were approached about organizing a book about unconventional O&G development, we thought it important to cover many of the issues of controversy in a balanced manner.

Like its conventional counterpart, unconventional O&G development presents many potential issues, along with both benefits and risks. The industrial operation is complex; public understanding is incomplete, and research findings are inconsistent. However, by better understanding the process and its issues, we believe that regulators, industry, and the public can make informed decisions about the relative benefits, safety, and risks of unconventional O&G development, as well as how those benefits, safety, and risks compare to other sources of energy. Clearly, the need for energy is not going away. Instead, it is our obligation to understand the facts and form knowledgeable opinions to guide policy.

We have laid out the areas we believe to be of the highest concerns today, as well as those that were least understood. We further thought about who we most wanted to hear from and began approaching our authors. Finally, we listened to the opinions of prospective authors we contacted as to topics we may have left out. The chapters in this book reflect our findings of the topics of principal interest namely air, water, waste, chemical use, transportation, worker safety and health, public health, perceptions of risk, state and federal regulations, economics, and sustainability. In order to reflect the different leanings of individual authors, we strived to approach leaders and experts in the fields of interest from a variety of "stakeholder" groups, to participate in a nonpartisan approach presented through the book.

The result of this process is a collection of chapters covering a range of issues by a group of diverse, intelligent, and authoritative writers who rank as some of the top experts in their respective fields. Their discussions and writings are instructive and insightful. We are incredibly indebted to them for the generosity of their time, their insights, and their creativity.

- *Chapter 1: The Unconventional Oil and Gas Process.* In the introductory chapter of the book, Ms Uni Blake (American Petroleum Institute [API]) provides background about the history and technology of the unconventional O&G development process. It examines the history of unconventional O&G, which primarily focuses on the earliest events and conditions in the United States. This is relevant in global forecasting as the United States has been on the forefront of development and technology for these resources (see discussions in the International Perspectives in Chapter 13). Evaluating the background associated with the processes in unconventional O&G allows a critical approach to the discussions in the chapters that follow, and allows subsequent author chapters to focus on their individual topics. Further, Chapter 1 takes the reader on a broad introductory look at the O&G industry in general, and addresses the reasonableness of environmental and public health effects by talking about possible exposure pathways.
- *Chapter 2: Hydraulic Fracturing for Shale Gas: Economic Rewards and Risks.* Authored by Dr Alan Krupnick (Co-Director of Resources for the Futures' (RFF) Center for Energy and Climate Economics (CECE) and a Senior Fellow at RFF), Chapter 2 lays out what is known, uncertain, and unknown about the economic consequences of unconventional O&G development. It examines impacts, environmental health, and commerce of the interchange between risk and value on a national scale, on a regional scale, and on an industry scale.
- *Chapter 3: Methane Emissions from the Natural Gas Supply Chain,* written by Mr David Richard Lyon of the Environmental Defence Fund (EDF), provides insights into one of the highly contentious issues surrounding unconventional O&G, namely air emissions. Although the replacement of coal as an energy source with natural gas is believed to have long-term climate benefits due to reduced levels of carbon dioxide (CO_2) per unit of energy generated, it has the potential to increase methane (CH_4) emissions, resulting in a powerful greenhouse gas (GHG) chemical. Comparisons of the climate impact of natural gas relative to more carbon-intensive fossil fuels is discussed, including the contribution of methane emissions to this formula. Only by understanding potential methane effects can there be a move towards control technologies to reduce methane emissions, thus minimizing potential climate impacts of natural gas development and leading to a stronger overall climate benefits.
- *Chapter 4: A Review of Drinking Water Contamination Associated with Hydraulic Fracturing,* is authored by Ms Elyse Rester and Mr Scott D. Warner (Ramboll Environ US) Chapter 4 focuses on surface and groundwater located within areas where hydraulic fracturing for petroleum recovery is used – water, which has historically, presently, or may in the future serve as a drinking water resource. There are drinking water exposure pathways throughout the hydraulic fracturing water cycle. Two case studies, one in Dimock, PA and one in Pavillion, WY, shed light on the process of determining water contamination as a result of hydraulic fracturing.

Ground-and surface water contamination has been associated with fracking activities; although the contamination often is related to operational issues, human error, rather than subsurface hydraulic fracturing itself. Continued monitoring of groundwater, both pre- and posthydraulic fracturing as well as groundwater sampling events covering longer durations, will be key to identifying and preventing drinking source water contamination in the future.

- In addition to the issue of environmental contamination, the use of large-scale volumes of water presents issues related to sources of water, abundance, and disposal of the associated waste. *Chapter 5: Water Use and Wastewater Management – Interrelated Issues with Unique Problems*, by Mr Daniel Price (Ramboll Environ) and Dr Carl Adams, Jr (retired) presents water use and wastewater management, including the recent introduction of recycling of flowback water as an untapped source of water for use in the process of fracturing. Various treatment technologies are discussed, including membranes, electro-coagulation and evaporation/crystallization, and reusable waters from waste and offsite disposal. An innovative new technique, Zero Discharge Systems (ZDS) is laid out by the authors.
- As anyone following the media on unconventional O&G knows, water issues including concerns over legal rights and contamination is front and center. In *Chapter 6: A Primer on Litigation That Involves Alleged Water Well Contamination From Hydraulic Fracturing*, Mr Craig P. Wilson and Mr Anthony Holtzman (K&L Gates) review the litigation history involving allegations of contamination of residential water wells by industrial activities surrounding unconventional O&G development. The factual allegations, causes of action, requests for relief, and evidentiary issues involved in these cases are described and the ongoing contested legal and factual issues are explored.
- In *Chapter 7: Worker Safety and Health*, Mr Eric J. Esswein, Ms Kyla Retzer, Mr Bradley King (NIOSH), and Dr Margaret Cook-Shimanek retrospectively look at fatalities that are attributed to the O&G industry and their causes, statistically, and also looks forward to evaluate the risks for chemical exposure among workers in the unconventional O&G development industry. The authors present an innovative and collaborative program between government and industry designed to heighten worker safety and health.
- In *Chapter 8: Public Health, Risk Perception & Risk Communication: Unconventional Shale Gas in the United States and the European Union*, Dr Bernard D. Goldstein (University of Pittsburgh), Dr Ortwin Renn (University of Stuttgart), and Dr Aleksander S. Jovanovic (University of Stuttgart) collectively address perspectives and societal appetite for risks associated with unconventional O&G development on public health. Further they examine the publics' *perception* of risks surrounding the industry, both in the United States and in Europe. In this chapter, the authors relate recent regulatory decisions to both these issues: evidence for risk, and perception of risk.

- Of course, once oil or natural gas are recovered from shale plays, there needs to be some way of moving it to where it will be used. *Chapter 9: Transportation of Shale Gas and Oil Resources*, by Ms Alicia Jaeger Smith and Mr Matthew W. Murphy (BatesCarey LLC), review the various modes and routes of transportation of shale gas and oil, including pipeline, tanker, rail, barge, and truck. The authors explore the issues and concerns associated with the movement of shale gas and oil, including environmental and safety concerns, infrastructure and capacity concerns, territorial and property rights and insurance considerations. The chapter concludes with an overview of regulations governing the industry. Although the focus of the chapter is the United States, the issues have global implications.
- The original purpose of *Chapter 10: An Evaluation of the Hydraulic Fracturing Literature for the Determination of Cause–Effect Relationships and the Analysis of Environmental Risk and Sustainability* was to examine the sustainability of unconventional O&G development. However, Ms MariAnna Lane and Dr Wayne G. Landis (Western Washington University) soon concluded that a systematic evaluation of the cause–effect relationships between the potential sources of stressors and impacts to typical endpoints were not yet sufficient to conduct an ecological risk assessment. The controversy surrounding unconventional O&G development is not helped by the many conflicting publications produced. The authors moved forward by exploring the impact of funding sources on conclusions reached, and analyze the published scientific literature to see whether funding sources influence publication findings. The authors then propose a preliminary cause–effect conceptual model to act as a starting point around which to organize future research on the impact and eventual sustainability of unconventional O&G development.
- In *Chapter 11: Induced Seismicity*, Dr Robert Westaway (University of Glasgow) examines historic seismic events and reviews the current knowledge surrounding induced seismicity, drawing on examples from the United States and the United Kingdom. Both the issues related to seismic activity observed associated with disposal of waste water, as well as induced seismicity associated with hydraulic fracturing within certain geological conditions are discussed.
- In *Chapter 12: State and Federal Oil and Gas Regulations*, Ms Karen J. Anspaugh (Indiana University Robert McKinney School of Law and Surrett & Anspaugh) looks at how hydraulic fracturing is looked at by individual states and the federal government in the United States. She discusses how through safe practice there are many benefits to the US economy including afford-ability, abundance, and read availability. The author discusses how fossil fuels impact almost every aspect of the lives of Americans. Examines how the US national energy policy could be most effective. The author identifies the laws of the land and the regulatory bodies on top.
- Finally, in *Chapter 13: International Perspective of Challenges and Constraints in Shale Gas Extraction*, Ms Katharine Blythe, Mr Robert Jeffries (Ramboll

Environ), and Mr Mark Travers (Ramboll) extend the discussion from other chapters, which focus largely on the United States and the United Kingdom, to the worldwide view. The chapter describes the current status of unconventional O&G development in terms of commercial or exploratory production of shale gas. It provides regional summaries of relevant geological and environmental information, and presents legislative and socio-political considerations across the globe.

Clearly, there are very complex issues surrounding unconventional O&G development, and any of these 13 chapters could likely occupy a separate book on their own. Rather than ask our authors to comprehensively review every aspect of their topic, we asked them to highlight their fields and expertise, and provide sufficient reference documentation so that the interested reader could further investigate the topic independently. Each author and author group brings their training, knowledge, experiences, research, and publications into their writings. While we anticipated that as human beings we all have biases, our authors may bring a tilt to their writings, we are pleased at the scholarly balance that each chapter holds, and how the overall mixture of authors brings a balance to this highly controversial topic. We hope that this collection of chapters on such a broad group of issues will provide a jumping-off point for the reader to understand the issues surrounding unconventional O&G and approach the debates in a collegiate manner with good intent to find solutions.

The global need for energy will not go away; neither will the economic and political power that comes with those with energy resources. While there are many different approaches to energy production, they each hold their own risks and benefits. Only by understanding the risks and benefits, not only for unconventional O&G development but also for all energy sources, can we make informed decisions as to a path forward. Furthermore, by accurately identifying potential risks, it may be possible to modify methods to minimize these risks. Fear of the unknown or dreams of economic gain will not mitigate risks associated with any energy resource. Only by examining the issues in a balanced, systematic, and methodological manner can we move forward from the current polarized debate, where decisions are often based on politics or perception as played out in the media, rather than facts and science. In that way, governments, industry, and individuals can make balanced decisions about future options for energy development.

CHAPTER 1

The Unconventional Oil and Gas Process, and an Introduction to Exposure Pathways

Uni Blake
American Petroleum Industry (API), Regulatory and Science Affairs, Washington, DC, USA

Chapter Outline

INTRODUCTION TO THE OIL AND GAS INDUSTRY

The Oil & Gas (O&G) sector is highly dynamic and technical, with a long history: the first commercial gas well was dug in Fredonia, NY in 1821; the first oil well was dug in 1859 in Titusville, PA; the first time rotary drilling technique was applied to an oil well was in 1901 in TX; and the first horizontal well was drilled in 1937 in the Havener Run field in OH (Dickey, 1959; Primer, 2009; Pees, 1989).

Environmental and Health Issues in Unconventional Oil and Gas Development.http://dx.doi.org/10.1016/B978-0-12-804111-6.00001-7

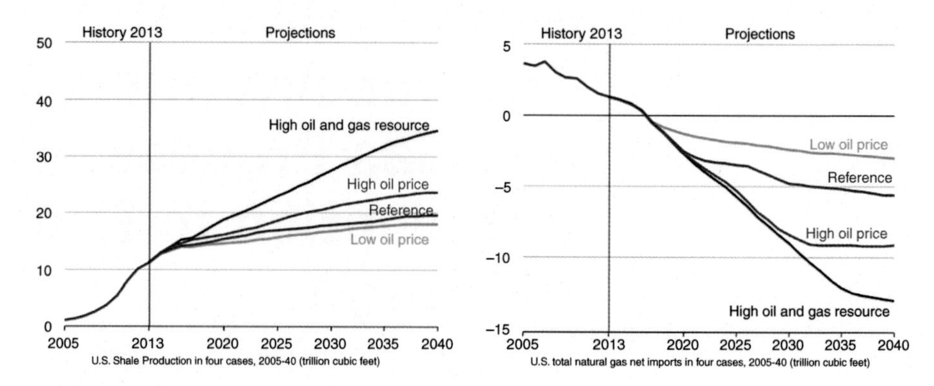

FIGURE 1.1
US net exporter. *Source: US Energy Information Administration (2015).*

As the easily recoverable O&G fields started to mature, focus shifted to accessing the resources from less traditional sources that had been considered uneconomical. The demand for energy rose concurrently, and by the energy crisis in 1973, the US Department of Energy (DOE) was funding programs geared towards developing less traditional O&G resources. By the late 1990s dramatic advancements in the technologies allowed for the development of unconventional resources. The result was a dramatic increase in the pace of domestic O&G production in the United States. It is anticipated that under the current growth scenario, the US is on track to become a net exporter of O&G (Figure 1.1) (see Chapter 2).

Introduction to the Oil and Gas Industry Sector

INDUSTRY SEGMENTS

The O&G industry's value chain is divided into three segments. The upstream segment is responsible for exploration and production activities. It includes the businesses involved in drilling, extraction, and recovery of the resources. The midstream segment is responsible for the development, and management of the infrastructure that connects the production of the resources to the next segment. For natural gas, the midstream segment businesses include: processing, pipeline transportation, rail transportation, and shipments (see Chapter 9). The downstream segment acts as the transformation point for the resources into consumable products. It also includes the refining and marketing of the resources.

INDUSTRY STRUCTURE AND PLAYERS

The industry is composed of many different companies and organizations that contribute to various segments of the value chain: fully integrated companies have business in all segments; independent producers, independent refiners, and independent marketers focus only on a particular segment; pipeline companies transport the resources through a network of pipelines, and compressor stations; and, service companies support the primary functions of the various segments. There are also trade associations, and professional organizations that support the industry, and government agencies that regulate and ensure compliance.

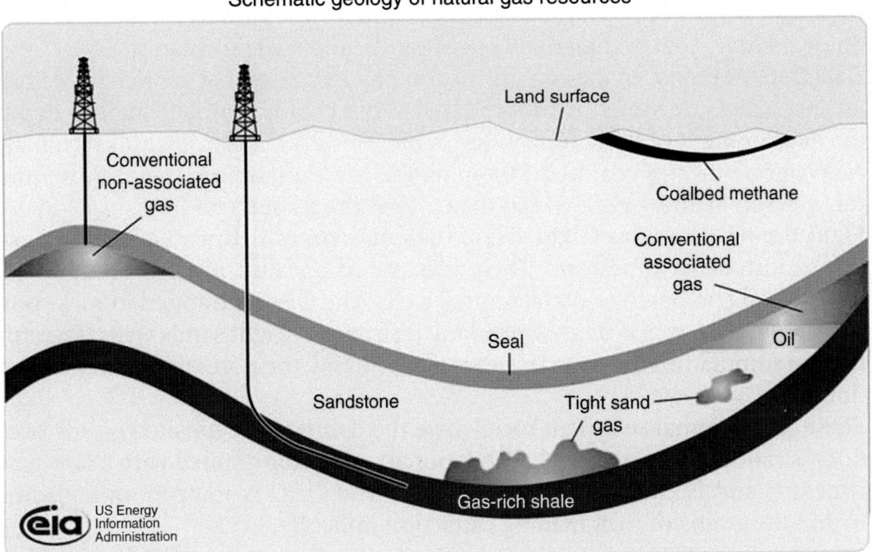

FIGURE 1.2
Different types of formation. *Source: US Energy Information Administration (2010a).*

INDUSTRY ACTIVITIES

The O&G industry supplies 60% of the US energy needs through the extraction and processing of hydrocarbons. The hydrocarbons originally formed when organic debris, together with mud and silt, settled on the ocean floor millions of years ago. The sediment under the pressure of the overlaying rocks, and the high temperatures from within the earth's crust eventually changed to O&G (Passey et al., 2010). Over time, some of the O&G migrated upwards seeping through rock pores and spaces of "reservoir rocks" (e.g., fissured limestone and sandstone). The upward seep eventually became trapped by impermeable rock barriers "cap rocks" (e.g., mudstone). Water, salt, hydrogen sulfide, and carbon dioxide are sometimes also trapped within the reservoir rocks. The upward migration of hydrocarbons lays the foundation of the types of wells, and processes that are used to extract the resource. Figure 1.2 illustrates the various types of O&G formations.

Conventional Wells

Early wells extracted O&G from porous and permeable reservoir rocks. The rocks contained interconnected pores, and once tapped the oil or gas flowed easily into the wellbore, and up to the surface under natural pressures. Hence, the resource is extracted by "conventional" means.

Unconventional Wells

The term "unconventional" has been used in many different contexts. It has been used to describe the type of formation being developed, the scale of operation required to produce the formation, and the cost required to produce a well. For the purpose of this chapter *unconventional* refers to reservoirs that are unfavorable for production using conventional (traditional) recovery methods. These include: coal bed methane, tight sands, and shale formations.

Coal Bed Methane. In the coal formation process, gases that are rich in methane are generated and stored within the coal. Since coal is found at shallow depths, the development of these wells is favorable. However, it is not without challenges. Water permeates coal beds; trapping the gas. Production, therefore, requires the water to be drawn off, which then allows the resource to flow.

Tight Sands. The term "tight sands" describes an impermeable and hard rock like sandstone or limestone. These formations are difficult and costly to produce, when compared to conventional wells. The O&G is trapped in rocks pores that are not connected or are irregularly distributed. Tight sands wells, therefore, require stimulation to create fractures that will link the pores to aid in the extraction of the resource.

Shale. The original sediment rock where the debris was deposited on the ocean floor is called the "source rock." The source rock is fine grained with a laminated structure, and is classified as a shale rock. The O&G is trapped and adsorbed within the grains of rock making extraction difficult.

Onshore Unconventional Resource Development

The industry has known for decades that tight sands and shale contain O&G, but commercial extraction was not feasible due to technology limitations, and unfavorable economics. Some production of unconventional resources occurred despite the challenges, but it was considered marginal due to the low permeability of the formations, and the low incidences of natural fractures. To increase wellbore intersection with natural fracture systems, directional drilling was introduced along with small-scale fracturing in the 1940s (Reed et al., 1982). Figure 1.3 outlines a timeline of the unconventional development.

By 2011, 95% of wells developed were in unconventional formations, and accounted for 67% of natural gas production, and 43% of oil production in the United States (US National Petroleum Council, 2011). The leading basins (plays) in the production of unconventional gas in the United States as of 2015 are the Marcellus in the Appalachian basin, Eagle Ford in Texas, and the Haynesville in Texas and Louisiana (Figure 1.4). The leading oil production basins are the Permian basin, and the Eagle Ford in Texas, and the Bakken in North Dakota (EIA, 2014).

To economically and commercially extract O&G from unconventional resources, a combination of hydraulic fracturing (HF), and horizontal drilling is required. Figure 1.5 outlines the approximate timeline of the development of a hydraulically fractured well.

EXPLORATION PHASE

During the exploration phase, teams of environmental scientists and engineers review data to determine where the shale resources are located. Three- and four-dimensional time-lapse images are used to capture the formations and reservoirs, greatly increasing the ability of exploration teams to accurately pinpoint

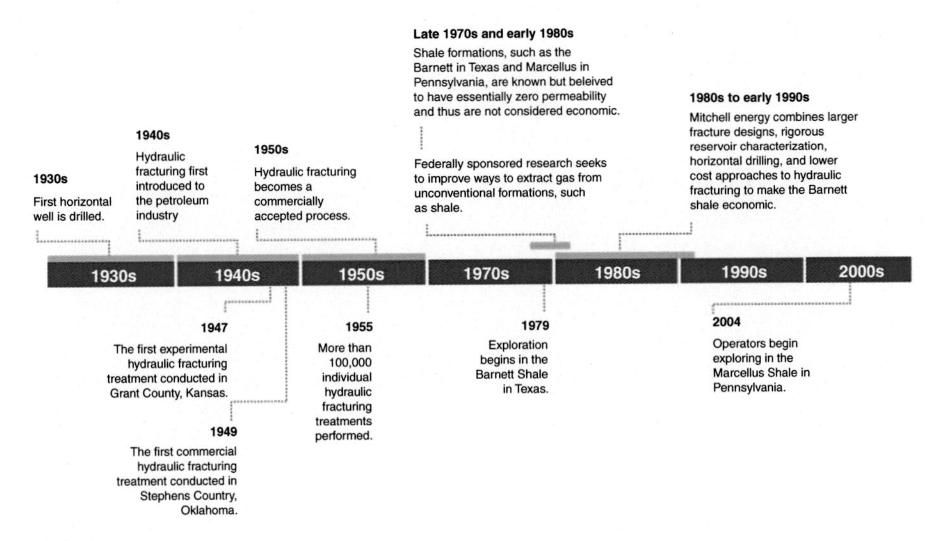

FIGURE 1.3
History of HF in the United States. *Source: US Government Accountability Office (2012).*

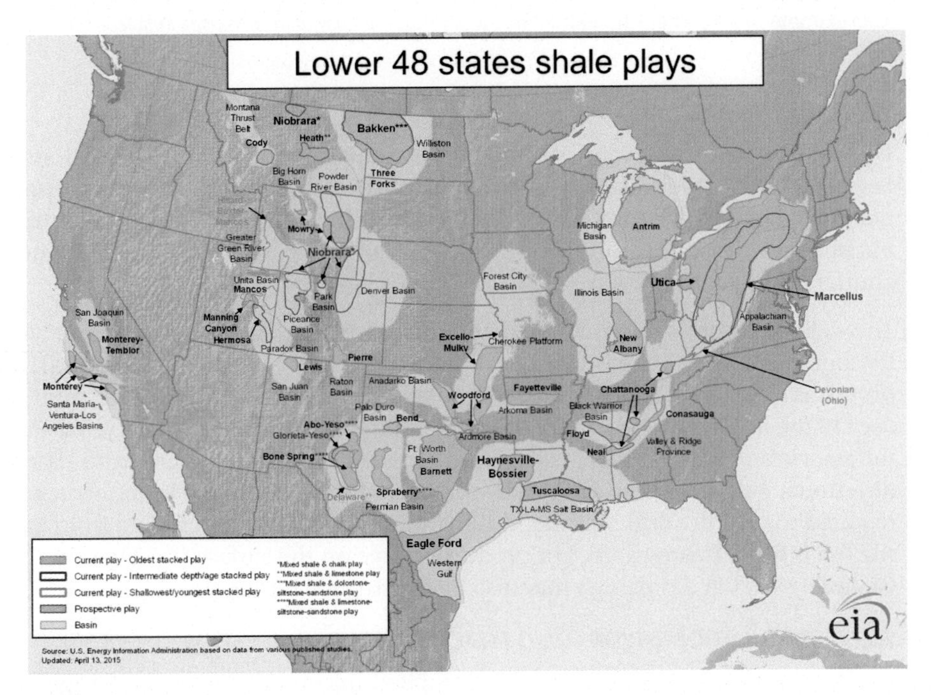

FIGURE 1.4
Shale basins (plays) in the United States. *Source: US Energy Information Administration (2010b).*

the production zones. This phase is represented by the first two steps in Figure 1.5. Companies may sign leases with landowners, and undertake more detailed local evaluations that include seismic exploration, and the drilling of exploratory wells.

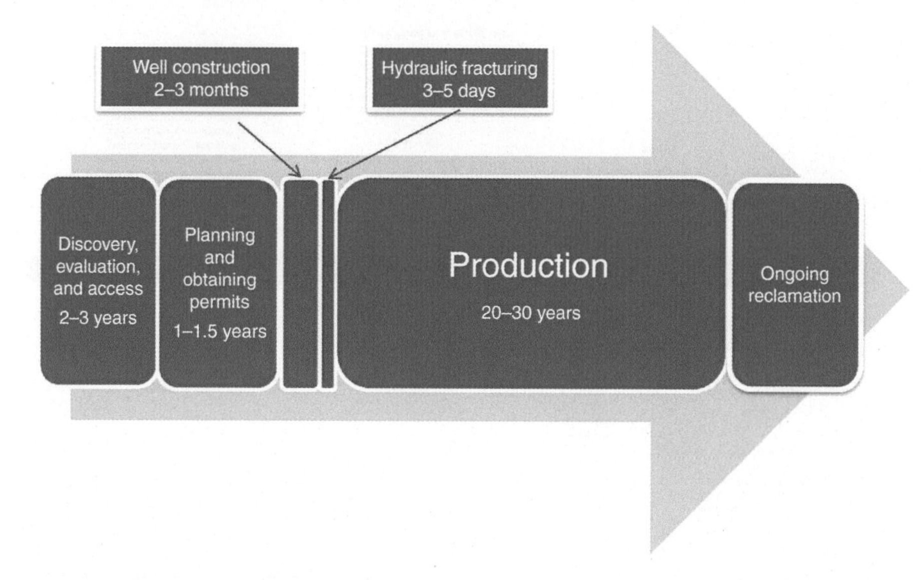

FIGURE 1.5
Approximate development timeline (*not to scale). *Courtesy of the American Petroleum Institute, Washington, DC.*

WELL CONSTRUCTION

During this phase, relevant permits and leaseholds are secured. The pad area is cleared and a workover rig is brought onsite. The well construction engineering plan includes information about drilling mud considerations; drill bit selection; casings; pore pressures, well control concerns, drinking water aquifers, and economics.

The wellbore is first drilled to a predetermined depth below the deepest fresh water source, and the casing is inserted into the hole, and cemented into place. The process is repeated at different predetermined depths (Figure 1.6). Drilling mud is used in the drilling process. Once the mud returns to the surface, it is recycled and the associated rocks cuttings are disposed of based on permit requirements. For a directional wellbore, the use of specialized bottom hole assembly components are required to turn the drill bit. After drilling, the workover rig is removed from the site leaving a "Christmas tree" or production tree on the surface (see Figure 1.6). Gathering lines and pipelines may also be constructed during this phase.

WELL STIMULATION (HYDRAULIC FRACTURING)

During this phase the formation is stimulated into production. The design of the stimulation process is key to the economics and the production rate of a well. The most commonly used stimulation fluids are water-based fracturing fluids, and as such the process is called "HF" or "slick-water fracturing." In formations that are sensitive to water, foam fracturing can be used. There is also an option for waterless stimulation that utilizes propane (see Chapter 5). HF lasts 3–5 days, and is the shortest phase of the operation.

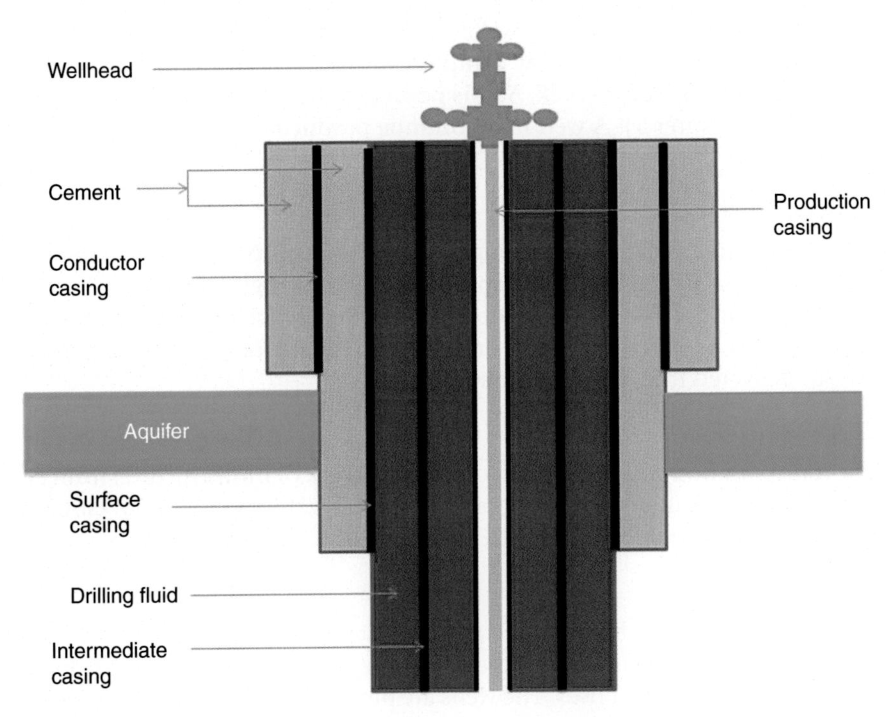

FIGURE 1.6
Well head and casing (*not to scale). *Courtesy of the American Petroleum Institute, Washington, DC.*

HF Fluids and Stages

Information regarding the characteristics of the reservoir is first gathered to optimize the design of the "frac job." Diagnostic information is reviewed prior to, during, and posttreatment. Frac jobs are similar across the formations; however, the design or the "blend" of the chemicals is dependent on the formation characteristics.

The first stage of HF is the acidization stage. Several thousand gallons of water are mixed with dilute acid, and pumped into the wellbore to clear debris, and the cement that was used to place the casing. During the second stage (pad stage), a mixture of biocide, scale inhibitor, iron control/stabilizing agent, friction reducing agent, gelling agent, and sometimes a cross-linking agent are mixed with thousands of gallons of water to create what is referred to as "*slickwater*." The slickwater is pumped down the wellbore under pressure to open pre-existing fractures within the formation. Next, the proppant material (i.e., fine mesh sand or ceramic material) is combined with the slickwater and pumped down the wellbore to prop open the fractures. Finally, the wellbore is flushed with water to remove the excess proppant from the wellbore.

The excess water that flows back from the formation to the surface is called "flowback." It contains a diluted amount of the chemicals used in the HF process, and also chemicals and substances that are contained within the formation. As the fluids continue to flow up the wellbore, its chemical composition changes. It eventually retains the chemical composition of the water that was originally trapped within the source rock. At this point the fluid is referred to as "produced water" (see Chapter 5). A well may continue producing water, but in diminishing quantities during its lifetime.

PRODUCTION

The production phase is the longest phase of the processes, lasting some 20–30 years. Initially the well may produce O&G at a high rate, and then decline slowly overtime. Once the well is producing, the resource is delivered in gathering lines to a processing plant, or a compressor station where it is processed and/or delivered into the pipeline system.

TRANSPORTATION

The transportation of oil and gas requires a network of infrastructure; from gathering lines to ocean going vessels (see Chapter 9). The industry's investment in infrastructure has increased by 60% between 2010 and 2012.

RECLAMATION

After an oil or gas field has matured and production is no longer economical, the well can be shut down or repurposed. The site is then restored to near-original, or transformed for other uses. The wells are plugged below the ground level, and the surface is restored with little or no visible evidence of the well remaining.

PERMITTING, REGULATIONS, AND OTHER PROGRAMS

In the past three decades, upstream O&G regulations have evolved in response to concerns from landowners, farmers, and municipal officers, and also in response to federal environmental laws (Groundwater Protect Council, 2009). Rapid changes in technology, location of development, and industry management practices, guidance documentation, and standards have kept the rule-making process dynamic with both the federal and state level regulations being continuously upgraded and improved upon (see Chapter 12).

In 2010, the Ground Water Protection Council (GWPC) and the Interstate Oil and Gas Commission (IOGCC) set up FracFocus. FracFocus is a HF chemical registry that provides public access to the chemicals disclosed by operators (Hochheiser, 2014). As of 2015, approximately 100,000 disclosures have been made public (see Chapter 4).

EXPOSURE PATHWAY EVALUATION OF UNCONVENTIONAL OIL AND GAS ACTIVITIES

An exposure pathway evaluation is utilized to evaluate the likelihood, or the extent of an exposure that has occurred. The assessment evaluates the level of connectivity between the source of contaminants and the population by: assessing the concentrations of the contaminant at the source; modeling its pathway with consideration to fate and transport; and understanding the population's

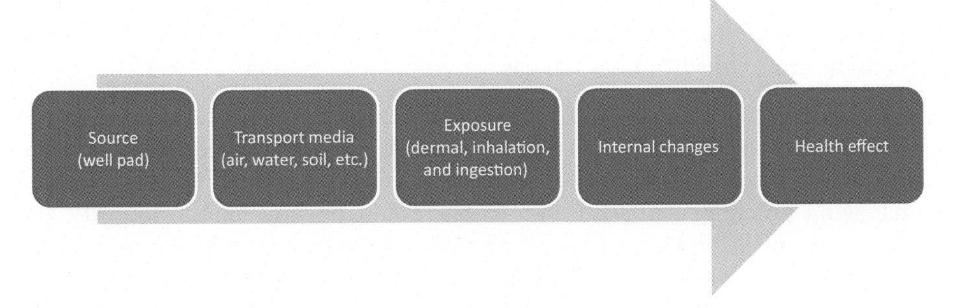

FIGURE 1.7
Conceptual model for a source–outcome–continuum.

exposure factors (Moya et al., 2011). It also takes into consideration the exposure response health outcome. This process is described in Figure 1.7.

Exposure Pathway Analysis

As onshore unconventional O&G development requires the management of large volumes of fluids and air emission; it is therefore crucial to understand the exposures, to ensure that public health is protected. A pathway analysis can shed light on potentially viable pathways of exposure, and monitoring data can help identify pathways (see Chapter 3). A pathway analysis requires the identification of the following five elements:

1. contaminant source or release;
2. environmental fate and transport. The contaminant has to enter into the environment and moves through a media (air, water, soil, etc.);
3. exposure point or area;
4. exposure route. There exists a route by which the exposed population can come in contact with the contaminants (i.e. dermal contact, inhalation, or ingestion); and
5. potentially exposed population.

For a completed exposure pathway, all the five elements must be present. In a potential exposure pathway, one or more elements are not met; and in an eliminated exposure pathway, one or more elements are missing. It is important to note, that even with a completed pathway, a health effect may not be present (Figure 1.7).

Potential Exposure Pathways

AIR

Air emissions may occur via equipment leaks, process venting, and evaporation (see Chapter 3).

Methane emissions: concerns about methane emissions stem from methane being considered a potent greenhouse gas (see Chapter 3). However, it is difficult to apportion methane sources. In response, the US Federal government set a goal of lowering methane emissions from the O&G industry by 40% over the next 10 years. Ongoing research continues to narrow the uncertainties, and also define better estimates for the industry's methane emissions. In the meantime, a combination of industry innovation, and regulations has led to a decrease in methane emissions from natural gas activities (see Chapters 3 and 12).

Volatile organic compounds (VOCs) and other hazardous air pollutants: exposure to VOCs from industry activities is a concern, both from the health and GHG perspective (Federal Register, 2011). Some VOCs are considered carcinogenic, and managing exposure pathways and exposure is a priority for the industry and regulators. Institutional controls are also utilized (see Chapter 3). Air monitoring data suggest that this class of chemicals can be found in ambient air samples, but not at concentrations that exceed human health-based threshold values (Bunch et al., 2014; Zielinska et al., 2014). Other studies, conducted at the community level, detect polycyclic aromatic hydrocarbons; however, these studies do not take into consideration exposure duration or the exposure paradigm (Chapter 8).

WATER

Water is an integral part of the HF process (see Chapters 4 and 5). Potential exposure pathways exist from surface spills, compromised well-integrity, and improper management of wastewater (Lutz et al., 2013; Vidic et al., 2013; Warner et al., 2013) (see Chapter 6).

Pathway Identification

Specific information can be collected about exposure pathways from studies and monitoring activities. However, because of the evolving narrative of the industry's activities, these assessments can only offer a snapshot in time.

Epidemiological studies: Epidemiological studies investigate the relationship between the activity (source of contaminant) and the potential health consequences of the exposure. However, challenges exist in using epidemiological studies to inform industry-wide assessments: (1) populations living near development sites are usually small, limiting the outcomes that can be studied and the size of the effect. (2) Exposure pathways and potential releases are site-specific. (3) Potential health end points are unspecific and vague, and can be the result of unrelated exposures. Due to the significant cost associated with epidemiology studies, pathways assessments tend to rely on environmental measurements and modeling (Nieuwenhuijsen, 2006).

Biomonitoring: Measuring the internal dose of the contaminant or of an associated metabolite can help characterize or rule out exposure (see Chapter 7). However, this method is limited as some biomarkers reflect only recent exposure, and in some cases the source of the exposure cannot be identified (Texas Department of State Health Services, 2010; Braatveit et al., 2007).

Environmental testing and monitoring: The data collected during monitoring can help provide the information needed to better understand exposure pathways (see Chapters 3 and 4). Monitoring is usually either regional or focused on local sites where O&G activities are ongoing.

- *Ambient air monitoring*: modeling methods have greatly improved the manner in which air ambient data can be utilized to characterize exposures (see Chapter 3) (Paulik et al., 2015; Zielinska et al., 2014; Modern Geosciences, 2014).

- *Water quality monitoring*: water quality monitoring typically focuses at locations where the public may come in contact with water (see Chapters 4 and 6). Obtaining baseline water quality samples can help assess the status of exposure pathways (Rahm & Riha, 2014).

Social, Economic, and Cultural Considerations

Social impacts can be direct, indirect, intended, unintended, negative, or positive (see Chapter 8). They include changes in demographics, social infrastructure (healthcare and education system adequacy, emergency services) and changes to transportation and roads (see Chapter 9).

With the increase in development opportunities (26 basins in 28 states) the economic benefits are widespread across the United States. The manufacturing sector is growing and it is anticipated that earnings, and employer-provided benefits from unconventional O&G will surpass $278 billion by 2025. Currently, the O&G industry supports 9.2 million domestic jobs, and contributes more than 7.7% to the US economy (see Chapter 2).

CONCLUSIONS

The location, scale, and pace of unconventional O&G development have afforded the United States economic opportunities, and the prospect of becoming a net exporter of O&G. Meanwhile, the industry, along with local, state and federal governments, and communities are working together to maximize the benefits, and finding ways to manage risks. The landscape is constantly evolving, with innovations that improve efficiency, and effectiveness occurring throughout the industry's value chain. Emerging markets will look to the United States to be a leader in the management of environmental and health challenges of the resource development.

References

Braatveit, M., Kirkeleit, J., Hollund, B.E., Moen, B.E., 2007. Biological monitoring of benzene exposure for process operators during ordinary activity in the upstream petroleum industry. Ann. Occup. Hyg. 51 (5), 487–494.

Bunch, A.G., Perry, C.S., Abraham, L., Wikoff, D.S., Tachovsky, J.A., Hixon, J.G., Urban, J.D., Harris, M.A., Haws, L.C., 2014. Evaluation of impact of shale gas operations in the Barnett Shale region on volatile organic compounds in air and potential human health risks. Sci. Total Environ. 468, 832–842.

Dickey, P.A., 1959. The first oil well. J. Petrol. Technol. 11 (1), 14–26.

DOE, US. (2009). Modern Shale Gas Development in the United States: A Primer. Office of Fossil Energy and National Energy Technology Laboratory, United States Department of Energy.

EIA, 2014. US Energy Information Administration (EIA). Drilling Productivity Report.

Federal Register, 2011. Oil and natural gas sector: new source performance standards and National emission standards for hazardous air pollutants reviews; Proposed rule.

Groundwater Protect Council. State Oil and Natural Gas Regulations Designed to Protect Water Resources, Department of Energy (DOE), Office of Fossil Energy, Oil and Natural Gas Program and the National Energy Technology Laboratory (NETL), DOE Award No. DE-FC26-04NT15455), May 2009.

Hochheiser, H.W., Arthur, R., Layne, M.A., Arthur, J.D., 2014. Overview of FracFocus and analysis of hydraulic fracturing chemical disclosure data, SPE International Conference on Health Safety and Environment.

Lutz, B.D., Lewis, A.N., Doyle, M.W., 2013. Generation, transport, and disposal of wastewater associated with Marcellus Shale gas development. Water Resour. Res. 49 (2), 647–656.

Modern Geosciences, 2014. Air Monitoring Report SE Mansfield Padsite, Mansfield, Texas, December 29.

Moya, J., Phillips, L., Schuda, L., Wood, P., Diaz, A., Lee, R., Clickner, R., et al., 2011. Exposure Factors Handbook, 2011 ed US Environmental Protection Agency, Washington, DC.

Nieuwenhuijsen, M., Paustenbach, D., Duarte-Davidson, R., 2006. New developments in exposure assessment: the impact on the practice of health risk assessment and epidemiological studies. Environ. Int. 32 (8), 996–1009.

Passey, Q.R., Bohacs, K., Esch, W.L., Klimentidis, R., Sinha, S. 2010. From oil-prone source rock to gas-producing shale reservoir-geologic and petrophysical characterization of unconventional shale gas reservoirs. In: International Oil and Gas Conference and Exhibition in China. Society of Petroleum Engineers.

Paulik, L.B., Donald, C.E., Smith, B.W., Tidwell, L.G., Hobbie, K.A., Kincl, L., Haynes, E.N., Anderson, K.A., 2015. Impact of natural gas extraction on PAH levels in ambient air. Environ. Sci. Technol. 49 (8), 5203–5210.

Pee, S., Greenawalt, J., Burgchardt, C., 1989. Petroleum mining and horizontal wells. History of the Petroleum Industry Symposium, American Association of Petroleum Geologists, first ed., September 17–20, p. 10.

Primer, A., 2009. Modern Shale Gas Development in the United States.

Rahm, B.G., Riha, S.J., 2014. Evolving shale gas management: water resource risks, impacts, and lessons learned. Environ. Sci. Processes Impacts 16 (6), 1400–1412.

Reed, R.M., Etnier, E.L., Kroodsma, R.L., Schweitzer, M., Mulholland, P.J., Switek, J., Oakes, K.M., Vaughan, N.D., et al., 1982. Preparation of environmental analyses for synfuel and unconventional gas technologies. ORNL 59 (1), 1.

Texas Department of State Health Services (TDSHS), 2010. DISH, Texas Exposure Investigation: DISH, Denton County, Texas.

U.S. Energy Information Administration (2015) U.S. Annual energy outlook 2015. US Energy Information Administration, Washington, DC.

U.S. Energy Information Administration, 2010a. Schematic geology of natural gas resources. US Energy Information Administration, Washington, DC. Available from: http://www.eia.gov/oil_gas/natural_gas/special/ngresources/ngresources.html (accessed 21.11.2015.)

U.S. Energy Information Administration, 2010b. Lower 48 shale plays. US Energy Information Administration, Washington, DC. Available from: http://www.eia.gov/oil_gas/rpd/shale_gas.jpg (accessed 21.11.2015.)

U.S. Government Accountability Office, 2012. Report to Congressional Requesters, "Oil and Gas: Information on Shale Resources, Development, and Environmental and Public Health Risks." GAO-12-732, p. 7.

US National Petroleum Council, 2011. Prudent development: realizing the potential of North America's abundant natural gas and oil resources. US National Petroleum Council.

Vidic, R.D., Brantley, S.L., Vandenbossche, J.M., Yoxtheimer, D., Abad, J.D., 2013. Impact of shale gas development on regional water quality. Science 340 (6134), 1235009.

Warner, N.R., Christie, C.A., Jackson, R.B., Vengosh, A., 2013. Impacts of shale gas wastewater disposal on water quality in western Pennsylvania. Environ. Sci. Technol. 47 (20), 11849–11857.

Zielinska, B., Campbell, D., Samburova, V., 2014. Impact of emissions from natural gas production facilities on ambient air quality in the Barnett Shale area: a pilot study. J. Air Waste Manage. Assoc. 64 (12), 1369–1383.

CHAPTER 2
Hydraulic Fracturing for Shale Gas: Economic Rewards and Risks

Alan J. Krupnick
Center for Energy and Climate Economics, Resources for the Future (RFF), Washington, DC, USA

Chapter Outline

INTRODUCTION

Despite the many debates and political impasses surrounding global economic growth, environmental protection, climate change, and energy security, there is significant agreement among policymakers, the research community, private sector leaders, and other stakeholders that natural gas can be a game changer for our energy future. More specifically, the ability to cost-effectively develop vast, globally dispersed deposits of natural gas in deep geological shale using advanced horizontal drilling, 3-D seismic imaging, and hydraulic fracturing techniques could well represent new opportunities for domestic and global economic growth and deficit reduction. Further, natural gas compares very favorably to coal and oil in its lifecycle environmental footprint (assuming fugitive methane emissions are kept low). Still, the speed of the shale gas revolution and the lack of data and research about the environmental risks and economic issues leave the extent of the revolution and its sustainability uncertain.

Environmental and Health Issues in Unconventional Oil and Gashttp://dx.doi.org/10.1016/B978-0-12-804111-6.00002-9

In this chapter, we take stock of what is known, uncertain, and unknown about the economic consequences of this revolution, leaving to later chapters, the discussion of the environmental consequences. Our purpose here is to examine the literature in three broad areas: national impacts, regional impacts, and company impacts. Our focus is primarily on socioeconomic research, centering on economics.

THE NATIONAL PICTURE

Before we examine the literature on economic impacts, it is helpful to understand how natural gas fits into the economy. Natural gas is the most versatile of fuels, being used in high proportions in all sectors of the economy except transport (EIA, 2015d). Oil is primarily used in transport, although it has significant industrial use as a feedstock, for example, to make plastics. Coal is used primarily in the electricity sector and nuclear is used entirely to generate electricity. Renewable biofuels are used in transportation (mostly as ethanol) and hydropower, wind, and solar to make electricity (see Chapter 1).

Production of natural gas has been rising rapidly, from 18.1 trillion cubic feet (Tcf) in 2005 to 24.4 Tcf in 2013 (EIA, 2015b). According to the US Energy Information Administration (EIA), these figures will continue to increase to projected levels of 35.5 Tcf in 2040, in their reference case. The most dramatic turnaround for natural gas is that the United States (US) went from expecting to import large quantities of natural gas to expecting to be a net exporter by 2017 (EIA, 2015b, p. 21). The fracking revolution is largely responsible for this.

In the initial stages of the boom, gas prices were in the $10 per thousand cubic feet ($/mcf) range, which, with the breakeven costs of supplying gas using fracking technologies under $5 per mcf, created a tremendous incentive to drill. This production resulted in dramatically lower prices, currently around $2.50 per mcf.

Another way of gauging the market impact of the boom is to use models to answer what prices would have been had the fracking revolution not taken place. The Rice World Gas Model shows a 44% lower natural gas price with the presence of shale during the period 2011–2020; a 43% lower price during the period 2021–2039; and a 51% lower price during the period 2031–2040 (Medlock, 2012). These results mirror results from Brown and Krupnick (2010) using the National Energy Modeling System. The basic story: fracking reduced gas prices by around $2 per mcf – a significant fraction of an average historical natural gas price of nearly $6 per mcf in 2004, prior to the fracking boom.

How have these lower prices affected, and how are they projected to affect, the economy? I start with the national economy, and then examine the effects on the electricity and industrial sectors.

National Impacts. Few experts dispute that the US economy is benefiting, and will continue to benefit, from the innovations that made shale gas more accessible. Many studies classify benefits as more jobs and economic growth (as measured by Gross Domestic Product [GDP]). We focus on the latter because it is so difficult to identify new jobs from shifted jobs.

Studies differ in their findings regarding the duration and size of foreseeable economic effects – Boston Consulting Group on the high side, for example, (Plumer, 2013) and the Stanford Energy Modeling Forum's (EMF) multimodel study (EMF, 2013) on the low side. Key factors explaining these differences include: estimates of when the US economy will reach effective full employment, assumptions about future oil prices, what the supply curve for natural gas looks like at its upper reaches, whether the study takes an *ex ante* or *ex post* perspective, and the type of model used (input–output vs. general equilibrium models, for instance).

The series of model results compiled by Resources for the Future from EMF data sheds light on the GDP impact (Figure 2.1). These model results were generated under similar assumptions to allow a comparison of the results. In the EMF26 report (EMF, 2013), the comparisons were set up to cover a reference case and up to seven other cases, including low shale gas and high shale gas resource cases. To represent the impact of abundant gas supply on the economy, one must compare results for the low shale gas case (with cumulative production 50% below the projections in the EIA's 2012 Annual Energy Outlook (AEO2012) (EIA, 2012)) and the high shale gas case (with cumulative production 50%

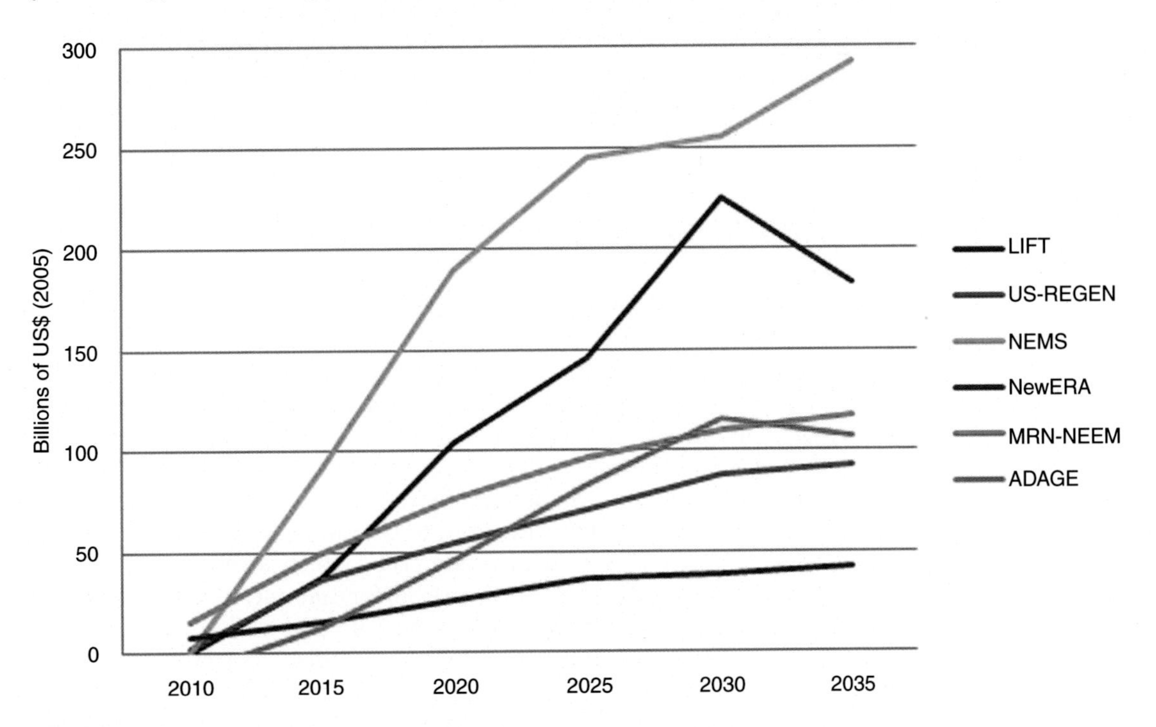

FIGURE 2.1
Change in real GDP in high shale gas case versus low shale gas case in various models.
Source: Krupnick et al. (2014) based on EMF (2013).

above the AEO2012 projections) and report on the models' forecast differences between the two. Six of the EMF models can produce these results, and they all show that the difference between the two cases on GDP is modest in percentage terms: an increase of less than 1.4% in any year. In absolute terms, the benefit is highest according to the EIA's National Energy Modeling System model at about $300 billion in 2035; the other five models cluster between $50 billion and $180 billion in 2035.

Although most models look to effects on GDP as a measure of economic benefit, economists use another measure that better reflects the economic benefits to society – changes in consumer and producer surplus, which are basically the value to consumers of cheaper gas and the products it is used to make (such as electricity) and higher profits. In a back-of-the-envelope analysis, Mason et al. (2014) found that consumer surplus increased $4.4 billion from the natural gas revolution over a 7-year period and producer surplus increased $9.6 billion.

As Houser and Mohan (2014) point out, the United States should not expect enormous labor productivity and innovation impacts from the shale gas revolution. It is unlike the Internet and computer revolutions, which fundamentally changed our way of life, were the catalyst for numerous ancillary industries, and spurred enormous productivity improvements in the way goods are manufactured, distributed, and sold. It is even unlike previous revolutions in the energy sector, such as the switch from wood to coal energy because "the current oil and gas boom does not yield a different and more convenient form of energy; it provides more of the same energy at a lower price" (Houser and Mohan, 2014, p. 55). The economic impacts of shale development are, according to most economic models, large and positive, but do not have transformative economic effects.

Sectoral Effects: Electricity

The electric utility sector is probably the most studied of those that use gas. Numerous studies have examined the effect of lower gas prices on electricity prices, generation mix, shares of generation fuel, and plant retirements and investments. Several of these studies (Brown and Krupnick, 2010; Burtraw et al., 2012; Logan et al., 2013) agree in their estimations that more abundant gas has led to lower electricity prices, ranging from 2% to 7% lower. There is less consensus, however, on the magnitude of the change in natural gas consumption in the power sector. In April 2015, natural gas for the first time overtook coal (briefly) in electricity generation and it is expected to take an increasing share of generation through 2040, mainly at the expense of coal, which is anticipated to fall absolutely (as well as relatively) (EIA, 2015b). So, for example, in the year 2000, the majority (52%) of electricity generation came from coal while 16% of electricity generation came from natural gas, 9% from renewables, 20% from nuclear, and 3% from petroleum and other liquids. In 2013, dependence on coal was reduced to 39% of electricity generation, with natural gas rising to 27% and renewables rising to 13%; nuclear (19%), and petroleum/other (1%) remained similar to earlier percentages. Looking forward to 2040, the percentage of electricity generation from coal is expected to decline further (34%), with a corresponding increase in natural gas (31%) and renewables (18%), while nuclear is projected to decrease (16%). Note, however, that these projections do not account for climate change policies, most importantly the Clean Power Plan (CPP), which could probably advantage gas even more (see Burtraw et al., 2015, Table 3). In a separate analysis conducted specifically on the CPP, the EIA (EIA, 2015a), for its CPP policy case relative to baseline projections found little change to natural gas production and prices compared with its AEO2015 estimates, but there was initially a substantial increase in natural gas generation, which then decreases as renewable generating capacity is added. With alternative baselines (such as higher or lower natural gas production costs) or policy extensions, the role of natural gas in generation and generating capacity can be correspondingly smaller or greater. See later on climate change policy risks to the industry.

Sectoral Effects: Manufacturing

Low natural gas prices have created a significant, if narrow, renaissance in manufacturing sectors feeding the Oil & Gas (O&G) industry and other sectors which use natural gas as a feedstock. These advantages can be seen in foreign companies expanding in the United States, in greater competitiveness of US exports abroad, and in increases in domestic production. A case in point is the plastics industry. Three-quarters of US plastics production uses natural gas as a feedstock, while in Europe and China oil is the primary feedstock. This has given US producers an advantage on the world stage, which shows up as a 12% increase in employment in the plastics sector since 2009 (admittedly a recession-driven down year) and industry estimates of creating 127,000 jobs over the next decade and a tripling of exports by 2030 (Tankersley, 2015; ACC, 2015).

Another example is the American Chemistry Council (ACC, 2011), which finds that a 25% increase in the ethane supply (a natural gas liquid) would generate 17,000 new jobs in the chemical sector; 395,000 additional jobs outside the chemical industry; $4.4 billion annually in additional federal, state, and local tax revenue; and $132.4 billion in US economic output. Other reports include those of PricewaterhouseCoopers (PWC), 2012, which is also on chemicals; Deloitte Center for Energy Solutions (2013) and Ecology and Environment (2011), on the natural gas value chain and indirect sectoral effects; and IHS Global Insight (2011), on general predictions about how shale gas development would impact various industries, such as chemicals, cement, steel, and aluminum.

SUBNATIONAL IMPACTS

Local/State Impacts

Much controversy has surrounded the actual size of local economic benefits. An example concerns a series of studies on impacts in the Marcellus play (Considine et al., 2010, 2011; Considine, 2010) in New York, Pennsylvania, and West Virginia, based on the IMPLAN input–output model. One of these studies, sponsored by the Marcellus Shale Coalition, attributes 44,000 new jobs to shale gas development in Pennsylvania and $3.9 billion in value added, with tax revenues increased by $398 million, all in 2009 (Considine et al., 2010).

Kelsey et al. (2011), using the same model but supplementing it with surveys of businesses, landowners, and local government officials as well as a GIS analysis of landownership, finds gains of only 24,000 new jobs and $3 billion of value-added revenues in Pennsylvania during 2009. Kelsey et al. (2011) attributes these lower economic gains to a consideration of the associated leakage of benefits outside the state.

Two reports address the potential and hypothetical economic impacts in New York if the state were to lift the current moratorium on horizontal drilling (see discussion on state moratoriums in Chapter 12). Though the job estimates in the two reports are somewhat similar, the conclusions are different. Considine et al. (2011) estimate that between 15,000 and 18,000 jobs could be created by allowing drilling in the Marcellus shale, and an additional 75,000 to 90,000 new jobs if drilling in the Utica shale commences. They conclude that this would cause large economic output gains in the state and large increases in tax revenues.

Similarly, a report conducted for the New York State Department of Environmental Conservation concludes that natural gas operations would produce between 12,491 and 90,510 direct and indirect jobs. However, this report implies that these additional jobs are not very significant because they account for only between 0.1 and 0.8% of New York State's 2010 total labor force (Ecology and Environment, 2011). This question of leakage leads to questioning state-level estimates of benefits that do not account for this factor. Thus, the national level analyses are probably superior.

What is unassailable is that state revenues have been up (at least until very recently) due to payments of severance taxes. Richardson et al. (2013) show how these severance tax rates vary by state and that two different formulas are used – one is a tax amount per unit of gas extracted and the other is a percentage of the value of gas extracted (see Figure 2.2a). Figure 2.2b shows the implications of these different methods for a natural gas price of $2.46 per mcf (around the current price as of July, 2015).

According to US Census data, total severance tax state revenue was $13.4 billion in 2009 (almost 2% of the total state tax revenue). In a comparison of state severance tax rates, Ozpehriz (2010) finds that, as expected, for states with more gas production and higher tax rates, a natural gas severance tax accounted for a higher percentage of their total state tax revenue. Alaska severance taxes made up 77% of the state's revenues in 2009, whereas the next top states were Wyoming (43%) and North Dakota (34%).

In a recent analysis of severance tax revenues, the National Conference of State Legislatures found that most states allocate a portion of their revenues to the state's general fund or toward O&G development and regulation. Fewer states allocate their funds to local governments and school- and transportation-related purposes (Brown, 2013). Information on Pennsylvania's impact fee system

(Pifer, 2013) indicates that local governments receive only about 47% of the revenues collected; given the structure of this system, these local distributions may have little or no relationship to local impacts.

Raimi and Newell (2014) have been two of the few researchers studying the impact of the shale boom (both gas and oil) on local public finances. Figure 2.3 shows qualitative results for many municipalities and counties over eight states on the front line of the boom (note, not the bust, which is a recent phenomenon). The overall conclusion is that finances have been improved, but in towns without a diversified economic base and very rapid growth, the demand for services, such as roads, housing, schools, drinking water, and sewage disposal, has outstripped revenues coming in. This is particularly true in states where localities lack much taxation authority and are therefore dependent on the generosity of the state legislature.

The recent and precipitous drop in oil prices does not yet show up in the Newell and Raimi work. However, our analysis shows that some undiversified

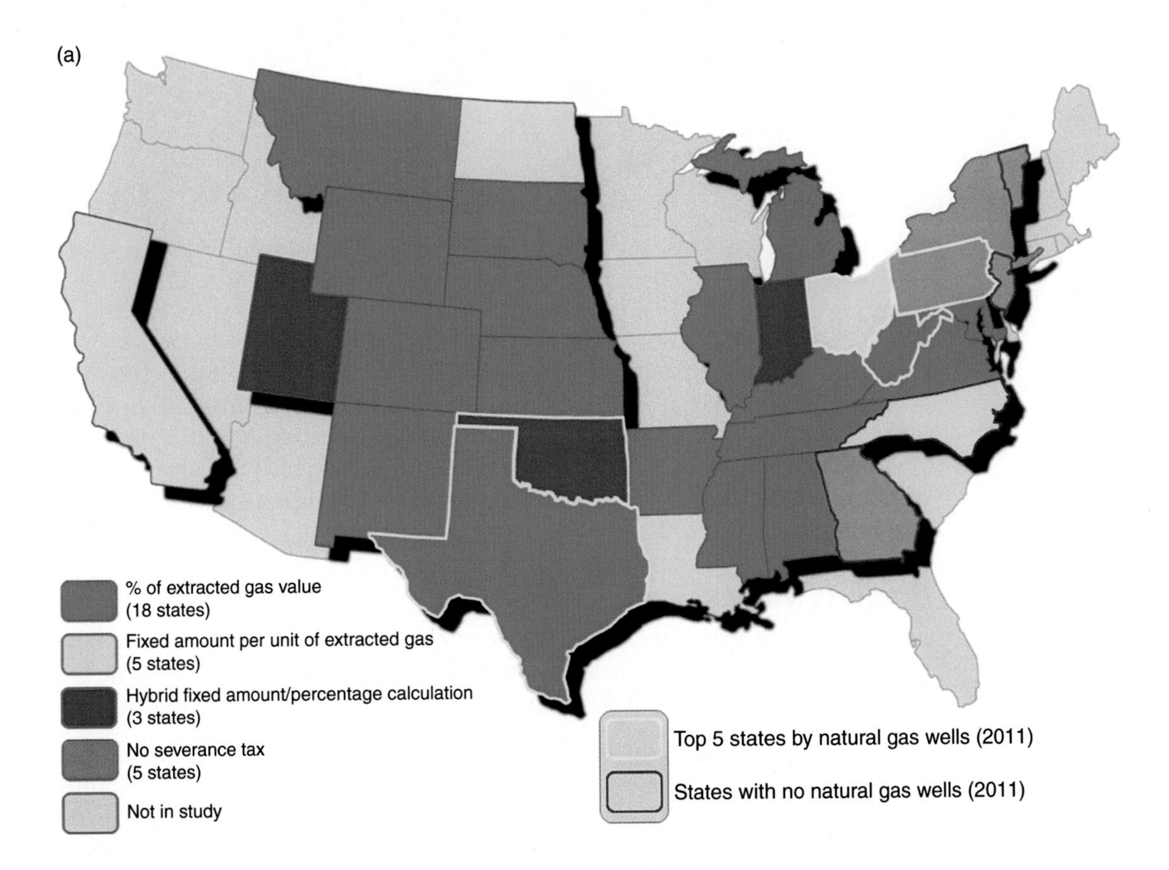

FIGURE 2.2
(a) Severance tax calculation method. (b) Severance taxes at $2.46 per mcf gas price.
Source: Richardson et al. (2013). (Continued)

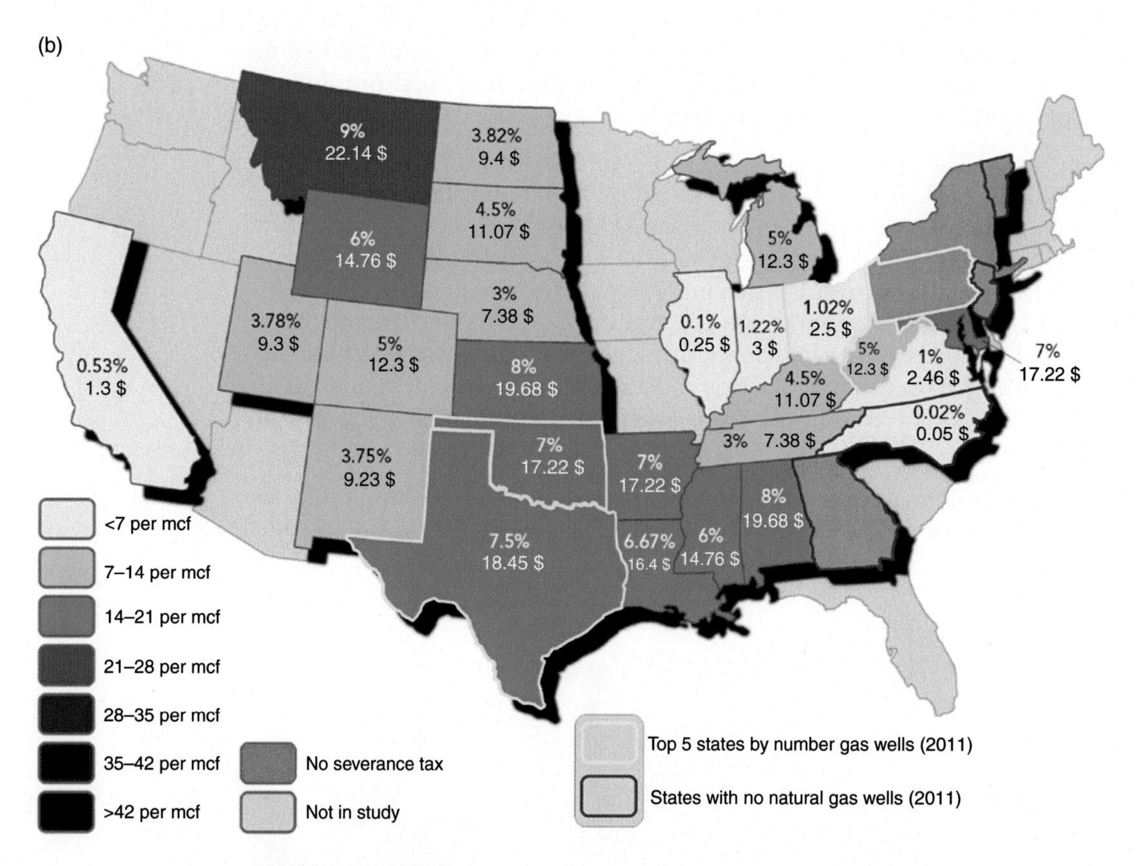

FIGURE 2.2 ***(cont.)***

municipalities are in trouble, but that states are directing more funds their way. Some – like the recent 'surge' bill funding transportation projects in North Dakota – are one-time injections of money. (On February 20, 2015 the North Dakota State House passed the oil patch 'surge' funding bill). A longer standing question is how local–state revenue sharing might be calibrated to the local risks and burdens borne by shale gas development (Figure 2.3).

Property Values

Another way to gauge local impacts is to examine how local property values have changed. Such values reflect both the perceived benefits and costs (including environmental risks) of the boom as evaluated by both buyers and sellers.

Two similar and broadly reinforcing studies have examined the property value impacts of shale gas development, both in Washington County, Pennsylvania. Gopalakrishnan and Klaiber (2014) find that, generally, property values decreased with development, but these decreases were largely transitory and depended on the proximity and intensity of shale gas activity. They also found that the impacts are heterogeneous and that negative effects disproportionately fall on "households that rely on well water, are located

State	Counties	Municipalities
Arkansas	Medium to large net positive	Small to medium net positive
Colorado	Small negative to large net positive	Small to medium net positive
Louisiana	Medium to large net positive	[insufficient data]
Montana	[insufficient data]	Roughly neutral to large net negative
North Dakota	Small to medium net negative	Medium to large net negative
Pennsylvania	Small to large net positive	Small to large net positive
Texas	Roughly neutral to large net positive	Roughly neutral to large net positive
Wyoming	Large net positive	Roughly neutral to small net positive

FIGURE 2.3
Net financial impact on local governments. *Source: Raimi and Newell (2014).*

close to major highways, or are located in more agricultural areas" (Gopalakrishnan and Klaiber, 2014, p. 44).

Muehlenbachs et al. (2015) also analyzed property values in Pennsylvania and New York, from January 1995 to April 2012 at various distances from drilling sites. They found that homes located near shale gas wells experienced an increase in property value if they had access to piped water compared with similar homes farther away, whereas groundwater-dependent homes that were in close proximity (0.63–0.93 miles) to a natural gas well experienced a 10–22% decrease in value compared with similar homes farther away. Overall, then, negative perceptions of environmental risks to groundwater caused close-in housing prices to drop.

Looking at property tax assessments, rather than home sales, Kelsey et al. (2012) found that shale gas development had minor impacts on the real property tax base.

Other Impacts

Many critics of studies showing positive economic effects at the local and state levels say that such studies ignore public costs and costs to the recreation and tourism sectors as well as boomtown effects, such as increased crime and crowded classrooms and the effects of the bust following the boom. Barth (2010) details, but doesn't quantify, many of these impacts. Christopherson and Rightor (2011) posit that

> to fully understand the economic impacts that shale gas drilling has on communities, one needs to look at long-term consequences (that is, economic development) and cumulative impacts (for example, new demands on government services, traffic congestion, noise, and social disruption), which input–output models ignore. Looking backward to past boom–bust cycles related to oil and gas development might help.

One example of increased government costs is crime impacts, in terms of enforcement and prevention. Kowalski and Zajac (2012) did not find "consistent increases" in Pennsylvania State Police incidents or calls for service or Uniform Crime Report arrests in the top Marcellus-active counties, but calls dropped in other counties over the same time period. In a comparison of disorderly conduct arrests in heavily "fracked" rural counties and in unfracked rural counties, Food & Water Watch (2015) found a 4% increase in arrests in fracked counties. Another advocacy group, FracTracker Alliance, has found slightly higher crime rates in fracked counties in Ohio, although these are primarily for traffic or other automotive issues (Auch, 2014).

Boomtown Economics

The dynamic nature of economic impacts has generated a small literature specific to shale gas that adds to a very large literature on the boom-and-bust cycle in communities from all types of resource development.

Specific to shale gas, Christopherson and Rightor (2011) found that the pace and scale of drilling determine the duration of the boom and projected that, although the Marcellus region as a whole will experience a boom for years to come, individual counties and municipalities within the region will experience short-term booms and busts as production rapidly decreases in those areas. This cycle is displayed in terms of royalty payments in Figure 2.4. The drilling of shale gas often requires an out-of-state work force, which causes increases in costs for local governments. Christopherson and Rightor (2011) note that with sudden community population increases comes the need for more policing, more emergency response preparedness, more teachers, and larger schools. Though the increase in population causes temporary booms in certain sectors, such as hotels and restaurants, other local businesses that usually serve more traditional clientele may not share in the growth.

Taking a somewhat wider view (i.e., O&G development and its effect on communities), Brown and Yücel (2013) found that communities or even states

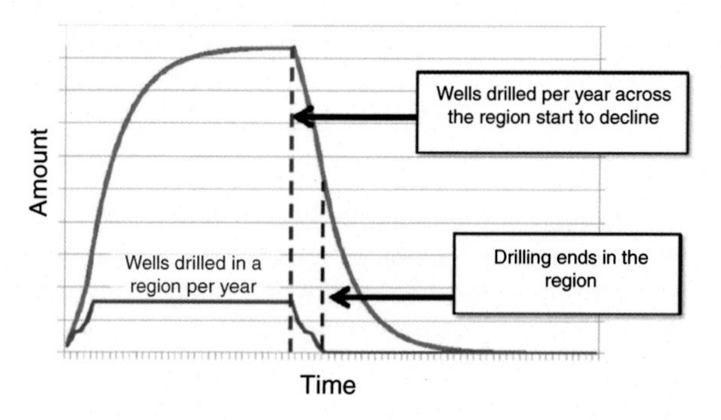

FIGURE 2.4
Royalties over time in a boom–bust cycle. *Source: Christopherson and Rightor (2011).*

that are highly dependent on such developments for their revenues and gross regional product can hurt when economic activity associated with a boom subsides.

In a recent statistical analysis of the boom phenomenon, using county-level census and other data for the US, Allcott and Keniston (2014) find that O&G booms increase growth rates 60–80% over nonproducer counties, with local wages increasing modestly (0.3–0.5% per year). In contrast to fears that such rising wages will cause manufacturing sectors to contract during a boom, manufacturing sector employment and output showed a rise (while falling along with decreased production during "busts").

Other county-level studies include Haggerty et al. (2013), Weber (2012), and Jacobsen and Parker (2014), all focusing on the West. Haggerty et al. found that although a boom increases county incomes, the longer a county is highly dependent on O&G development (the percentage of county income from this sector), the more per capita income growth erodes relative to counties with a lesser and shorter dependence. Quality of life indicators, such as crime, are also correlated with this dependence (Haggerty et al., 2013). Conducting an *ex post* analysis in Colorado, Texas, and Wyoming, Weber found that each million dollars in gas production created 2.35 local jobs in counties developing shale gas, a figure much smaller than economic input–output models projected (Weber, 2012). Jacobsen and Parker (2014) conducted a retrospective analysis of the boom of the 1970s and 1980s and found that boomtowns experienced positive short-term effects – with local incomes increasing by 10% over preboom levels at the height of the boom, and decreased unemployment – but negative long-term effects, with increased unemployment benefits and incomes 6% lower than they would be if the boom had not occurred.

RISKS TO COMPANIES

"Upstream" companies who invest in leases, the infrastructure needed to drill, frack, and move the gas to market; their suppliers and the midstream and downstream portions of the O&G sector; and manufacturing and electric utility demanders, are all, to one degree or another making bets about the future economics of natural gas production, and the course of environmental and economic regulation. In this section, we consider the downside risks to these sectors.

Economic Risk

Unless nearly all the prognosticators are wrong, economic risks to the sector are minimal. Figure 2.5 is one of a class of such figures showing the natural gas needed to break even by play and for an assumed target internal rate of return. This particular figure, from Devon Energy Corporation, is better than most because it gives some idea of the distribution of costs in a given play rather than an average over all the possible wells in a play. Thus, one of the cheapest plays is

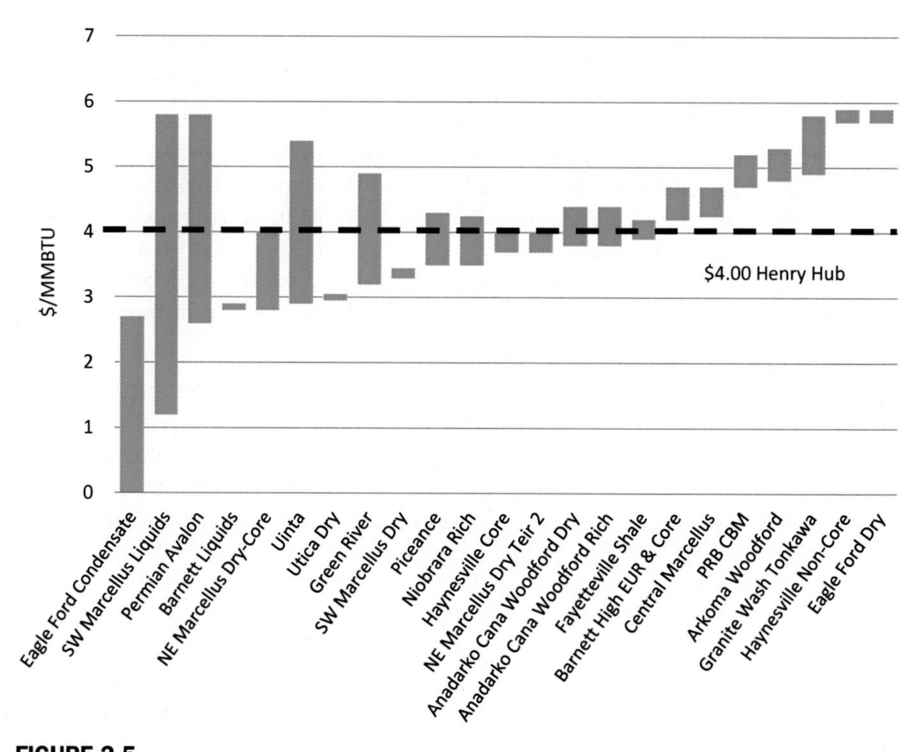

FIGURE 2.5
Gas breakeven prices across plays. *Source: Devon Energy Corporation.*

from the liquids rich portion of the Marcellus play. Liquids bring the cost down because they sell for prices governed by the world oil price – until very recently, trading at far higher prices per Btu than natural gas. While the figure neglects to mention the assumed internal rate of return target, it often is 10% or 15%. The higher the return desired, the higher the gas price has to be to break even.

Figure 2.5 is basically a thick supply curve for the gas industry organized by the lowest cost parts of each play. This curve is fairly flat at the range of prices being predicted by models of natural gas supply and demand, meaning that there is ample supply available at relatively low cost to meet growing natural gas demands. With current prices around $2.50 per million Btu it would seem that many of these plays are uneconomic. But this figure dates from 2011. Phenomenal technological change has been moving these bars downward, so many of parts of some of the plays are economic even at the low current price.

A related concern is whether rising demand will push up prices rapidly. This would be a boon to the O&G industry but could cause problems for manufacturing, the electricity sector, and the liquified natural gas (LNG) export sector. In fact, a host of studies examined this issue in light of new demands coming for natural gas exports and, in our view; the most credible studies suggested that effects on prices would be small (Ditzel et al., 2013). What's more, the same factors that lead US natural gas exporters to be bullish about the future – high liquefaction and transport costs to move gas across oceans – insulate the US market from non-North American competition.

Of course, low gas prices create risks for the O&G sector. One might be somewhat concerned about the energy reforms underway in Mexico adding more supply. However, the government plans to import US gas for at least 5 years and the gas available is, in part, from the same play (Eagle Ford) that US operators exploit. Canadian gas may be more of a low-cost threat, but it is hard to imagine that the Marcellus and Utica plays in the eastern United States will be out competed by Albertan gas piped into the eastern United States.

This is not to say that LNG exporters are not taking some risk. Currently, the price differentials between US natural gas and Asian and European markets are large enough to make it profitable to export to Asia and India. Long-term contracts with exporters and national governments further reduce risks. In addition, no observers expect gas to be produced quickly or at low cost from countries with large reserves, such as China and Brazil. On the other hand, Australia has large resources and a locational advantage over the US in Asia and Russia and Algeria, as well as Middle Eastern suppliers have locational advantages in supplying to Europe, particularly if pipelines can be used (see Chapter 13 for further discussion on the International market). Additionally, it doesn't take much of a price rise in the US, with export costs around $5 per mcf, to wipe out profits associated with these multibillion dollar liquefaction plants.

Social License to Operate: Regulatory Risks and Bans

The rapidity of the shale gas revolution and the community disruption it causes (see eg. Chapter 8) has led to the creation of many community groups opposed to fracking. Adding to this opposition are strong concerns coming from national environmental NGOs over CO_2 emissions from burning fossil fuels, as well as from leaking natural gas, which is a much more powerful greenhouse gas (see eg. Chapter 3). For its part, industry, at a minimum, has problems communicating its message of expertise at O&G development and a "best practice" environmental ethic. As shown through a survey by Siikamäki and Krupnick (2013), industry narratives about shale gas risk are just as likely to backfire, as they are to help, while environmentalist messages are far more successful.

The result of the perceptions and the reality about risks has been some 450 bans on fracking or the transporting of fracking wastes (Food & Water Watch) (see Chapters 12 and 13). The geographically broadest and most impactful is New York State's ban. It has a significant share of the Marcellus and Utica plays and, like Pennsylvania, is close to the nearby East Coast population centers where gas demand is high. Another, small, but possible bellwether ban is Denton, Texas, a town in the oil patch, demonstrating that when it comes to fracking, familiarity can lead to contempt. These bans, or attempted bans, have led to a backlash, with the legislatures of Texas (House Bill 40) and Oklahoma (Senate Bill 809) and (earlier) Pennsylvania[1] banning the practice (Act 13). Additionally, the legal viability of local bans remains uncertain due to challenges under the takings clause or improper home rule jurisdiction *vis-a-vis* the state.[2] Colorado, on the other hand, shows perhaps a better way, where a Governor-led task force brought stakeholders together to provide greater local input in exchange for avoiding the passage of bans. Such task forces have also been convened at the local level, with a prominent example being that of Santa Fe, New Mexico; however, even the Santa Fe effort has resulted in a lawsuit filed under the takings clause (*SWEPI LP v. Mora County, NM*). Such local pre-emption will likely continue to be an active area of action, with its ultimate effect yet to be seen.

If banning were to become widespread, industry impacts could be commensurably large. However, given robust production levels, we are nowhere near this point. What is happening is that state regulators – the primary body that addresses O&G externalities – are tightening their regulation of the industry (Groundwater Protection Council, 2014) and thereby raising production costs, although no studies are available to document by how much (see eg., Chapter 12).

In addition, in spite of all this regulatory tightening, the regulators are unlikely to stop there. Many groups that are not hostile to the O&G industry are calling for tighter or more efficient regulations, such as the IOGCC-convened STRONGER (State Review of Oil, 2014) review of state regulations and model best-in-class regulatory framework drafted by the Environmental Defense Fund (2014) and Southwestern Energy (see eg., Chapter 3).

Does the current round of tightening, and will future rounds, significantly choke back production through raising costs? The short answer is probably not. The reason, however, may have less to do with costs of regulation being a small fraction of total production costs and profits – no one knows how large these costs are! – than with the rapid pace of technological change that keeps reducing production costs and regulatory costs. Most prominent is the decision by many companies in some basins to recycle their fracking fluids and produced water, an approach that reduces freshwater withdrawals and material needs as well as associated traffic used to transport water to the site and remove wastewater (EPA, 2015) (see eg., Chapter 5). The overall productivity improvements can be seen in Figure 2.6, which documents increasing production in the face of declines in rigs (EIA, 2015c). As production of a new well declines quickly, these trends mean that each remaining rig is more productive than previous rigs, a combination of better geological

---

[1] Pennsylvania passed Act 13 in 2012, which set up an impact fee program to distribute money to local jurisdictions experiencing shale gas development in return for these jurisdictions giving up claims to stop the practice. In December 2013, the Pennsylvania Supreme Court ruled in *Robinson Township v. Commonwealth* that such a restriction on local authority was unconstitutional.

[2] The extent to which takings lawsuits are successful will likely continue to vary; most recently, we saw a lawsuit against a voter-approved ban dropped in San Benito County, California. For a bill making it easier to file for takings claims related to fracking bans, see H.R. 510 of the 114th Congress. For a broader discussion, see the following site: http://www.frackinginsider.com/litigation/a-new-weapon-in-the-fight-for-land-development-use-of-takings-lawsuits-to-challenge-local-fracking-bans/

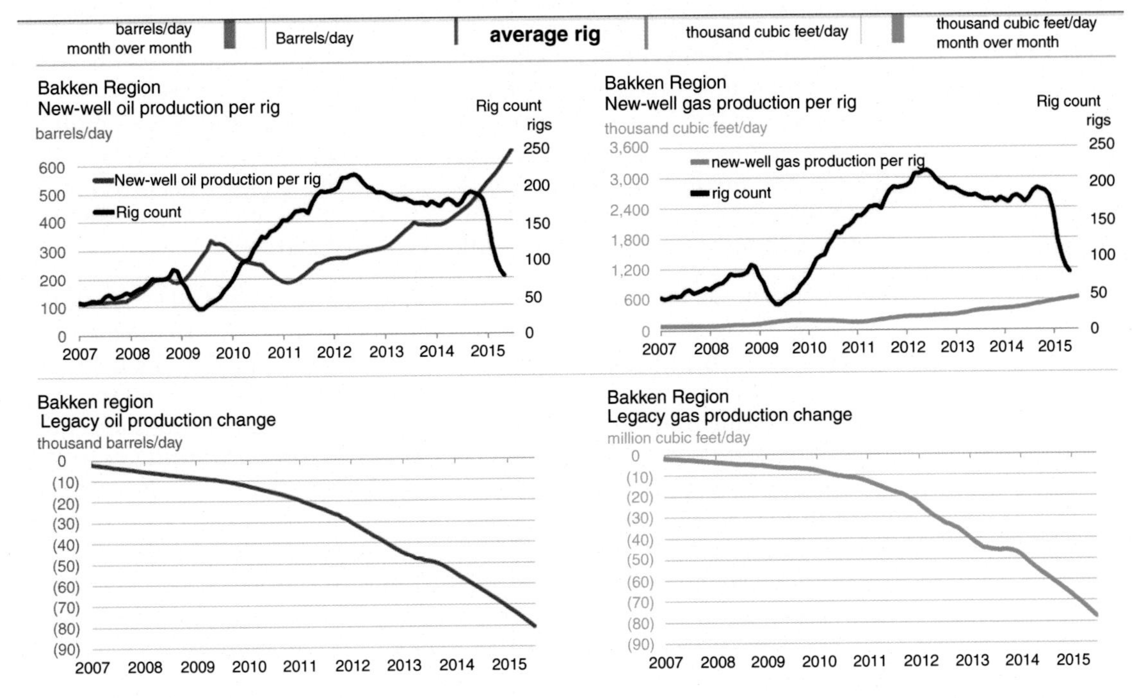

FIGURE 2.6
Increasing drilling productivity of new oil wells against the declining rig counts and legacy oil production in the Bakken. *Source: EIA (2015c).*

understanding and better techniques for fracking. Thus, even if regulatory compliance costs are rising, production costs are falling as an offset.

Climate Change

Compared with these community-level concerns, the threat of climate change and the regulatory risk surrounding it represents an existential threat to the industry. This threat is illuminated by research (McGlade and Ekins, 2015) showing that in order to meet global goals for temperature rise (2°C or less), 80% of global coal reserves, 50% of global gas reserves, and 30% of global oil reserves must not be burned by 2050. To get to that point from where we are currently may seem impossible, but the gradual disadvantaging of natural gas, coal, and oil through climate policy – particularly carbon pricing – could force major reductions in the size of the fossil fuel industry. Alternatively, breakthroughs in the cost of carbon capture and storage technology or other approaches to removing carbon from the waste stream could save the industry, although this is a contested point (McGlade and Ekins, 2015 research found that carbon capture and storage only allows for an increased fossil fuel reserve use of 6%).

In the shorter term, abundant, low-cost natural gas could act as a low-carbon energy "bridge" to the midterm future by lowering global greenhouse gas (GHG) emissions through fuel switching while renewable and other low-carbon energy technologies mature and become economically viable. Natural gas has also been suggested as a complement to renewables by filling in the load curve; this would come about from the ability of "fast-cycling" gas plants to backstop the intermittency of renewables. A tenet of the natural gas bridge case is that abundant gas and accompanying low gas prices will act through energy markets alone (i.e., in the absence of government imposed GHG mitigation policy) to displace fuels with higher carbon content and thereby reduce emissions (Brown and Krupnick, 2010; Burtraw et al., 2012; Paul et al., 2013).

Researchers at the Joint Global Change Research Institute (JGCRI) recently used their large-scale, global change assessment model (GCAM) to examine the natural gas bridge on a global and far longer timescale, finding that the hypothesis is questionable. The Joint Global Change Research Institute team ran the global change assessment model through mid-century (2050) under two gas scenarios (Flannery et al., 2013). In the first, natural gas is viewed from the year 2000, where large global gas resources exist but are too costly to exploit on a grand scale. In the second (circa 2010) the same gas resources exist, but they are extractable at considerably lower cost. The 2010 scenario is designed to mimic the current understanding of future gas availability and pricing. Not surprisingly, natural gas production and its use expands globally under the abundant gas scenario being 37% greater in 2050 than predicted under the more conservative supply and pricing case.

Consistent with both Resources for the Future studies, gas expands its share in all energy sectors with the greatest increase coming in electric power generation displacing coal and renewables (10% less coal, 10% less renewables). Emissions of CO_2 from coal decline 12%, but lower gas prices lead to lower electricity prices and a 3.7% increase in electricity consumption. Most importantly, there was no difference in CO_2 emissions between the conservative and abundant gas scenarios – that is, the widespread market penetration of gas did not lower CO_2 emissions in the absence of government mitigation policy. The result that abundant gas does not reduce emissions is due to offsetting factors. While gas does displace high-carbon coal, it also displaces zero-carbon nuclear and renewables. Importantly, lower priced gas leads to increased electricity generation from all fuel types.

All the previous studies examined only CO_2 emissions, where the edge is clearly in natural gas's favor over coal. The future for natural gas becomes cloudier still when leaks of natural gas itself are taken into account (see discussion in Chapter 3). In this case, unless there are low-cost options to reducing these leaks (which there may well be, see ICF International, 2014), ever tighter government GHG regulations could put the industry at risk.

CONCLUSIONS

The economics of the natural gas revolution involves sizing up the social benefits of cheaper natural gas and comparing them to the social costs. A few different takes on national benefits have been laid out above and disaggregated informally to sectoral and regional benefits, although one might add the improvement in energy security from not having to rely on imported natural gas, as was expected before the revolution. In addition, possible risks to the revolution, and therefore, reductions in expected future benefits, were catalogued. The costs of the revolution and primarily heightened environmental risks will be examined in the next chapters.

To summarize the findings, abundant and cheap natural gas will quite likely be a part of America's future unless climate change policy, in the absence of affordable carbon mitigation technology, forces it out. Rapid technological change continues to increase productivity and reduce development costs. And new sources of gas supply in Mexico, Canada, and elsewhere, made possible by fracking technologies, could come on if the price begins to creep up and infrastructure for its delivery can be developed.

All of this is good news for the US economy, as it is fueling a renaissance in some manufacturing sectors, creating local boom towns, holding down electricity price increases as cheap coal is being phased out and lowering energy bills for gas heating.

Nevertheless, there are economic risks. One is to LNG exporters, where expanded world supplies could lower Asian prices and strand export assets. Another, more fundamental, risk is in continued erosion of the industry's social license to operate. This license took a major hit from the Deepwater Horizon oil spill in the Gulf of Mexico (never mind that this was offshore oil rather than onshore natural gas, where the risks are far lower). And continued hits are coming from local communities attempting to ban the practice and subsequent attempts by state legislators to ban the bans. Continued uncertainties about the real risks posed by unconventional gas development dog the industry.

And, as noted, the existential risk comes from future climate change policy. In the shorter term, the industry needs to make clear their commitment to reducing methane enough to make natural gas a cleaner alternative to coal (see eg. Chapter 3). In the very long term, technologies to capture, transform, store, and use CO_2 will need to be in place for the industry to prosper.

References

Allcott, H., Keniston, D., 2014. Dutch Disease or Agglomeration? The Local Economic Effects of Natural Resource Booms in Modern America. NBER Working Paper No. 20508. National Bureau of Economic Research, Cambridge, MA.

American Chemistry Council (ACC), 2011. Shale Gas and New Petrochemicals Investment: Benefits for the Economy, Jobs, and US Manufacturing. ACC, Washington, DC.

American Chemistry Council (ACC), 2015. The Rising Competitive Advantage of U.S. Plastics. ACC, Washington, DC.

Auch, T., 2014. Crime and the Utica Shale. FracTracker Alliance. http://www.fractracker.org/2014/06/crime-utica-shale/

Barth, J.M., 2010. Unanswered Questions about the Economic Impact of Gas Drilling in the Marcellus Shale: Don't Jump to Conclusions. JM Barth & Associates, Inc, Croton on Hudson, NY.

Brown, C., 2013. State Revenues and the Natural Gas Boom: An Assessment of State Oil and Gas Production Taxes. National Conference of State Legislatures, Washington, DC.

Brown, S.P.A., Krupnick, A.J., 2010. Abundant Shale Gas Resources: Long-Term Implications for U.S. Natural Gas Markets. Resources for the Future, Washington, DC, RFF DP 10-41..

Brown, S. P.A., Yücel, Mine K., 2013. The Shale Gas and Tight Oil Boom: US States' Economic Gains and Vulnerabilities. Council on Foreign Relations, New York.

Burtraw, D., Palmer, K., Paul, A., Woerman, M., 2012. Secular trends, environmental regulations and electricity markets. Electr. J. 25 (6), 35–47.

Burtraw, D., Palmer, K., Pan, S., Paul, A., 2015. A Proximate Mirror: Greenhouse Gas Reductions and Strategic Behavior under the US Clean Air Act. Resources for the Future, Washington, DC, RFF DP 15-02..

Christopherson, S., Rightor N., 2011. How Should We Think About the Economic Consequences of Shale Gas Drilling? Working Paper Series. Cornell University, Ithaca, NY.

Considine, T.J., 2010. The Economic Impacts of the Marcellus Shale: Implications for New York, Pennsylvania, and West Virginia. American Petroleum Institute, Laramie, WY.

Considine, T.J., Watson, R., Blumsack, S., 2010. The Economic Impact of the Pennsylvania Marcellus Shale Natural Gas Play: An Update. Pennsylvania State University, University Park.

Considine, T.J., Watson, R.W., Considine, N.B., 2011. The Economic Opportunities of Shale Energy Development. The Manhattan Institute, New York.

Deloitte Center for Energy Solutions, 2013. The Rise of the Midstream: Shale Reinvigorates Midstream Growth. Deloitte Center for Energy Solutions, Washington, DC.

Ditzel, K., Plewes, J., Broxson, B., 2013. US Manufacturing and LNG Exports: Economic Contributions to the US Economy and Impacts on US Natural Gas Prices. Prepared by Charles River Associates for Dow Chemical Company. Charles River Associates, Washington, DC.

Ecology and Environment, 2011. Economic Assessment Report for the Supplemental Generic Environmental Impact Statement on New York State's Oil, Gas, and Solution Mining Regulatory Program. Prepared for New York State Department of Environmental Conservation. Ecology and Environment, Inc, Lancaster, NY.

U.S. Energy Information Administration, 2012. Annual Energy Outlook 2012 with Projections to 2035. DOE/EIA-0383(2012). DOE, Washington, DC.

U.S. Energy Information Administration, 2015a. Analysis of the Impacts of the Clean Power Plan. DOE, Washington, DC.

U.S. Energy Information Administration, 2015b. Annual Energy Outlook 2015 with projections to 2040. DOE/EIA-0383(2015). DOE, Washington, DC.

U.S. Energy Information Administration, 2015c. Drilling Productivity Report, June 2015. DOE, Washington, DC.

U.S. Energy Information Administration, 2015d. Monthly Energy Review (March 2015). Washington, DC: DOE. http://www.eia.gov/totalenergy/data/monthly/

EMF (Energy Modeling Forum), 2013. Changing the Game? Emissions Markets and Implications of New Natural Gas Supplies. Stanford University, Stanford, CA.

Environmental Defense Fund, 2014. Model Regulatory Framework for Hydraulically Fractured Hydrocarbon Production Wells.

U.S. Environmental Protection Agency, 2015. Assessment of the Potential Impacts of Hydraulic Fracturing for Oil and Gas on Drinking Water Resources (External Review Draft). U.S. Environmental Protection Agency, Washington, DC, EPA/600/R-15/047.

Flannery, B., Clarke, L., Edmonds, J., 2013. Perspectives From the Abundant Gas Workshop. Joint Global Change Research Institute, College Park, MD, http://www.globalchange.umd.edu/data/gtsp/topical_workshops/2013/spring/presentations/Edmonds_ImplicationsOfAbundantGas_2013-04-29.pdf.

Food & Water Watch. Local Actions Against Fracking. Food & Water Watch, http://www.foodandwaterwatch.org/water/fracking/anti-fracking-map/local-action-documents/ (accessed 24.06.2015).

Gopalakrishnan, S.H., Klaiber, A., 2014. Is the shale energy boom a bust for nearby residents? Evidence from housing values in Pennsylvania. American Journal of Agricultural Economics 96 (1), 43–66.

Groundwater Protection Council, 2014. State Oil and Gas Regulations Designed to Protect Groundwater Resources. State Oil and Gas Regulatory Exchange.

Haggerty, J., Gude, P.H., Delorey, M., Rasker, R., 2013. Oil and Gas Extraction as an Economic Development Strategy in the American West: A Longitudinal Performance Analysis, 1980–2011. Headwaters Economics, Bozeman, MT.

Houser, T., Mohan, S., 2014. Fueling Up: The Economic Implications of America's Oil and Gas Boom. Peterson Institute for International Economics, Washington, DC.

ICF International, 2014. Economic Analysis of Methane Emission Reduction Opportunities in the U.S. Onshore Oil and Natural Gas Industries. Prepared for Environmental Defense Fund. ICF International, Fairfax, VA.

IHS Global Insight (USA), 2011. The Economic and Employment Contributions of Shale Gas in the United States. IHS Global Insight (USA), Washington, DC.

Jacobsen, G., Parker, D., 2014. The economic aftermath of resource booms: evidence from boomtowns in the American West. Econ. J., DOI: 10.1111/ecoj.12173.

Kelsey, T.W., Shields, M., Ladlee, J.R., Ward, M., 2011. Economic Impacts of Marcellus Shale in Susquehanna County: Employment and Income in 2009. Marcellus Shale Education and Teaching Center, University Park, PA.

Kelsey, T.W., Riley, A., Milchak, S., 2012. Real Property Tax Base, Market Values, and Marcellus Shale: 2007 to 2009. Pennsylvania State University, University Park, PA.

Kowalski, L., Zajac, G., 2012. A Preliminary Examination of Marcellus Shale Drilling Activity and Crime Trends in Pennsylvania. Pennsylvania State University, University Park, PA.

Krupnick, A., Kopp, R., Hayes, K., Roeshot, S., 2014. The Natural Gas Revolution: Critical Questions for a Sustainable Energy Future. Resources for the Future Report (March).

Logan, J., Lopez, A., Mai, T., Davidson, C., Bazilian, M., Arent, D., 2013. Natural gas scenarios in the US power sector. Energy Econ. 40, 183–195.

Mason, C.F., Muehlenbachs, L.A., Olmstead, S.M., 2014. The economics of shale gas development. Annu. Rev. Resour. Econ. 7 (1), 100814-125023.

McGlade, C., Ekins, P., 2015. The geographical distribution of fossil fuels unused when limiting global warming to 2°C. Nature 517, 187–190.

Medlock III, K.B., 2012. Discussion of US LNG Exports in an International Context. Presentation at USAEE Houston Chapter Luncheon (September 12).

Muehlenbachs, L., Spiller E., Timmins C., 2015. The housing market impacts of shale gas development. Am. Econ. Rev. (forthcoming).

Ozpehriz, N., 2010. The State Taxation of Natural Gas Severance in the United States: A Comparative Analysis of Tax Base, Rate, and Fiscal Importance. Heinz School of Public Policy, Carnegie Mellon University, Pittsburgh, PA.

Paul, A., Beasley, B., Palmer, K.L., 2013. Taxing Electricity Sector Carbon Emissions at Social Cost. Resources for the Future, Washington, DC, Discussion Paper 13-23-REV..

Pifer, R.H., 2013. Marcellus Shale Development and Pennsylvania: Community Sustainability. Presented at the Widener University School of Law. Harrisburg, PA.

Plumer, B., 2013. Here's How the Shale Gas Boom Is Saving Americans Money. Wonkblog, Washington Post. www.washingtonpost.com/blogs/wonkblog/wp/2013/12/18/the-shale-gas-boom-is-saving-americans-money-but-how-much/

PricewaterhouseCoopers (PWC), 2012. Shale Gas: Reshaping the Chemicals Industry. www.pwc.com/en_US/us/industrialproducts/publications/assets/pwc-shale-gas-chemicals-industry-potential.pdf

Raimi, D., Newell, R., 2014. Shale Public Finance: Local Government Revenues and Costs Associated With Oil and Gas Development. Duke University Energy Initiative. Duke University, Durham, NC.

Richardson, N., Gottlieb, M., Krupnick, A.J., Wiseman, H., 2013. The State of State Shale Gas Regulation. Resources for the Future, Washington, DC.

Siikamäki, J., Krupnick, A.J., 2013. Attitudes and the willingness to pay for reducing shale gas risks. In: Alan, J., Krupnick (Eds.), Managing the Risks of Shale Gas: Key Findings and Further Research. Resources for the Future, Washington, DC, pp. 4–5.

State Review of Oil and Natural Gas Environmental Regulations, Inc. (STRONGER)., 2014. 2014 STRONGER Guidelines.

Tankersley, J., 2015. Job creation in U.S. plastics manufacturing rebounding. Washington Post, May 12. http://www.washingtonpost.com/business/economy/jobs-in-us-plastics-manufacturing-rebounding/2015/05/12/8eff27d0-f8ef-11e4-9ef4-1bb7ce3b3fb7_story.html

Weber, J.G., 2012. The effects of a natural gas boom on employment and income in Colorado, Texas, and Wyoming. Energy Econ. 34 (5), 1580–1588.

CHAPTER 3
Methane Emissions from the Natural Gas Supply Chain

David Richard Lyon
Environmental Defense Fund, Austin, TX; University of Arkansas, Environmental Dynamics Program, Fayetteville, AR, USA

Chapter Outline

INTRODUCTION

Natural gas production in the United States (US) increased 36% between 2005 and 2014, primarily due to the development of shale gas resources with horizontal drilling and hydraulic fracturing (EIA, 2015). Increased gas supply led to lower gas prices in the US, which contributed to a 21% decrease in coal-fired electricity generation over this time period (EIA, 2015) (see Chapters 1 and 2). The replacement of coal with natural gas for electricity generation has long-term climate benefits since natural gas combustion produces less carbon dioxide (CO_2) per unit of energy generated than coal. However, methane (CH_4), the primary constituent of natural gas, is a powerful greenhouse gas with 120 times the radiative forcing of CO_2 on a mass basis (IPCC, 2013). Methane has an effective atmospheric lifetime of approximately 12 years while a large portion of CO_2 emissions persists in the atmosphere for much longer. Therefore, the climate impact of methane is greater in the short term with a 20-year global warming potential (GWP) of 84 and a 100-year GWP of 25; with the inclusion of climate–carbon feedbacks, GWP values increase to 86 and 32, respectively (IPCC, 2013).

Alvarez et al. (2012) developed the concept of technology warming potential to compare the relative climate impacts over time of different technologies. The climate impacts of natural gas-fueled technologies are dependent on

Environmental and Health Issues in Unconventional Oil and Gas Development. http://dx.doi.org/10.1016/B978-0-12-804111-6.00003-0

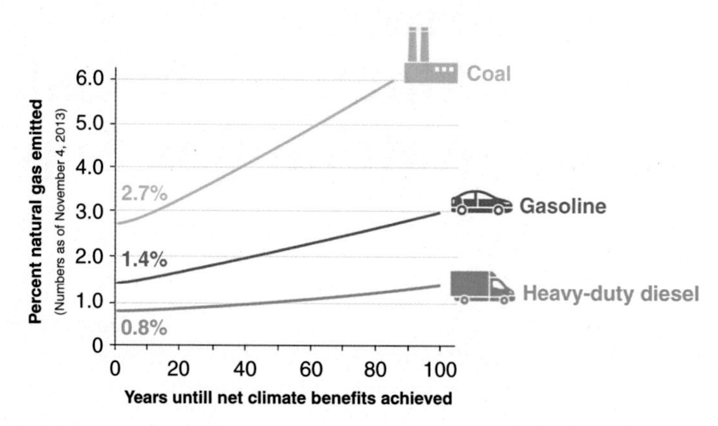

FIGURE 3.1
The short-term climate impacts of natural gas supply chain methane emissions can delay the climate benefits of switching from more carbon–intensive fossil fuels to natural gas. The years until net climate benefits are achieved for three technology conversions (coal-to-gas for electricity, gasoline-to-gas for light-duty vehicles, diesel-to-gas for heavy-duty vehicles) is shown as a function of the supply chain natural gas leak rate. For each technology, leak rates at or below the intercept ensure immediate climate benefits (e.g., 0.8% for heavy-duty natural gas vehicles).

methane emissions across the natural gas supply chain. As shown in Figure 3.1, replacement of a coal-fired power plant with a gas-fired plant will have immediate climate benefits if supply chain methane emissions are less than 2.7% of produced natural gas (Alvarez et al., 2012; updated by IPCC, 2013). At higher gas leak rates, switching from more carbon-intensive fossil fuels to gas will be worse for the climate in the near term due to the short-term climate impacts of methane. Although CO_2 emission reductions are necessary for long-term mitigation, reducing emissions of short-lived climate pollutants, such as methane, has the advantage of potentially delaying climate system tipping points and allowing more time for adaptation (Shoemaker et al., 2013). Consequently, it is critical to quantify and mitigate methane emissions from the natural gas supply chain to minimize its climate impacts and assure that gas displacement of other fossil fuels has immediate climate benefits.

EMISSION SOURCES

Methane emissions occur across the entire natural gas supply chain. The production sector includes the short-term development of new wells and the long-term operation of producing wells. Since many unconventional wells coproduce oil and natural gas, this chapter will consider Oil and Gas (O&G) production as a single sector, but focus on natural gas in downstream sectors. The gathering sector is a system of pipelines and compressor stations that transports gas from well pads to processing plants or transmission pipelines. The processing sector includes plants that treat gas to pipeline quality standards, which can include removal of water, CO_2, hydrogen sulfide, and natural gas liquids (heavier hydrocarbons, such as ethane). Some produced gas is close to pipeline quality standards and may bypass the processing sector and instead undergo minor treatment, such as dehydration, in the production or gathering sectors. The transmission and storage sector is a system of high-pressure pipelines and compressor stations that transports gas from gathering systems and processing plants to high-demand customers such as power plants and local distribution city gates. This sector also includes underground and liquefied natural gas (LNG) storage facilities that store gas for periods of higher demand. Finally, the local distribution sector is a system of pipelines and metering and regulating stations that delivers gas from city gates to customers, such as commercial and residential buildings.

Methane emissions from the natural gas supply chain can be grouped into three basic source classifications: vented emissions, fugitive emissions, and incomplete combustion emissions. Vented emissions are intentional releases related to normal operations or safety procedures. Fugitive emissions are unintentional releases from equipment leaks and malfunctioning equipment. Incomplete combustion emissions are fuel slippage in the exhaust of natural gas combustion sources.

Natural gas-powered pneumatic controllers are the largest source of vented emissions in the US. These devices use the energy of pressurized gas to power equipment controlling process variables such as liquid level and temperature (Simpson, 2014). Depending on their design, pneumatic controllers emit gas continuously or intermittently during normal operation. Similarly, pneumatic pumps use pressurized gas to power chemical injection or glycol pumps.

Intentional release of gas for safety or convenience is another large source of vented emissions. Blowdowns occur when equipment is depressurized prior to maintenance or to alleviate overpressurization. Associated gas venting is the release of gas produced by oil wells that lack a connection to gas-gathering systems. Some wells accumulate liquids in the wellbore that inhibit gas production; these wells may be temporarily vented to atmosphere to unload liquids, which also emits methane.

Well completions are a vented emission source specific to unconventional development. After a well is hydraulically fractured, excess fluid and proppant must be cleared from the wellbore. In the earlier years of unconventional development, completions were vented to the atmosphere, which released methane in flowback gas. As of January 2015, a US Environmental Protection Agency (US EPA) rule requires almost all hydraulically fractured gas wells, but not oil wells, to use reduced emission completions to capture completion flowback gas for sales instead of venting the gas to the atmosphere; a proposed rule would extend this requirement to oil wells (US GPO 40 CFR Part 60 Subpart OOOO) (see Chapter 12).

Hydrocarbon and produced water storage tanks have vented emissions referred to as flashing losses, working losses, and breathing losses. After fluid flows from a high-pressure separator to an atmospheric pressure tank, entrained gas is emitted as flashing losses. Working losses occur from displacement of vapors when a tank is filled. Breathing losses are caused by changes in ambient temperature or pressure. Similar to storage tank flashing, glycol dehydrators can emit methane entrained in wet glycol through the vent that releases water vapor. Tank emissions can be controlled with flares or vapor recovery units that combust or capture these losses.

Equipment leaks are fugitive emissions from poorly sealed or damaged components that are designed to have zero emissions, such as connectors and pipelines. Fugitive emissions can also occur from components that are expected to have some leakage (e.g., the vents of reciprocating compressor rod packing seals). Although these sources are designed to vent some gas during normal operation, their emission rates can increase with equipment wear. Malfunctions of venting sources or control equipment can cause fugitive emissions. For example, a malfunctioning pneumatic controller can emit beyond its designed rate or tank emissions can bypass a flare if the tank hatch is open.

Incomplete combustion emissions occur from engines, flares, or other equipment such as heaters that combust natural gas. Reciprocating compressor engines normally emit a greater portion of their fuel throughput than centrifugal compressors (EPA, 1995). Natural gas flares are often assumed to combust 98% of gas, but a recent study reported greater than 99.8% methane combustion efficiency for 11 flares in North Dakota and Pennsylvania (Caulton et al., 2014).

STATE OF KNOWLEDGE PRIOR TO UNCONVENTIONAL DEVELOPMENT

The US EPA annually publishes two sources with methane emission estimates from the US natural gas supply chain: the US Greenhouse Gas Inventory (GHGI; EPA, 2015a) and the Greenhouse Gas Reporting Program (GHGRP; EPA, 2015b) (also see Chapters 1 and 12). Both these sources rely heavily on data collected during a comprehensive study in the early 1990s by the US EPA and the Gas Research Institute that estimated 1992 US natural gas supply chain methane emissions were equivalent to 1.4 ± 0.5% of gross natural gas production (Harrison et al., 1996).

The GHGI is an annual report that includes estimates of US annual greenhouse gas emissions by source category from 1990 to 2 years prior of the publication year; it is published in fulfillment of the US' commitments under the United Nations Framework Convention on Climate Change. Emissions are reported at the national level for natural gas systems and petroleum systems except for the gas production sector, which is reported at the regional level. The GHGI estimates activity factors (equipment and facility counts) based on a combination of recent data, such as well counts, and assumptions, such as the number of pneumatic controllers per well, based on Harrison et al. (1996) and other studies. For most source categories, potential emissions are estimated by multiplying activity factors by emission factors (average emissions per equipment or site) based on Harrison et al. (1996) and other studies. Net emissions are calculated by subtracting estimated emission reductions, which are due to regulations and the voluntary US EPA Natural Gas STAR program, from potential emissions. Each year, US EPA updates the GHGI methodology and recalculates annual emissions back to 1990.

The GHGI has undergone several methodological changes in recent years that greatly changed estimates of natural gas system methane emissions. Most recent year emission estimates increased from 4.6 teragrams (Tg) methane in the 2010 GHGI to 10.5 Tg in the 2011 GHGI, primarily due to the addition of a source category for hydraulically fractured well completions and a methodological change resulting in increased emission estimates from liquid unloading. The 2013 GHGI emission estimate was reduced to 6.9 Tg due to another revision in liquid unloading methodology. A revised methodology for hydraulically fractured well completions reduced the 2014 GHGI estimate to 6.2 Tg. The 2015 GHGI estimates 2013 emissions from natural gas systems are 6.3 Tg with an additional 1.0 Tg from petroleum systems production. Assuming a supply chain average gas composition of 90% methane, 7.3 Tg methane is equivalent to 1.4% of gross natural gas production.

The GHGRP is a mandatory reporting program for US facilities with annual greenhouse gas emissions ≥25,000 metric tons carbon dioxide equivalents (CO_2e). Petroleum and natural gas facilities report under Subpart W, which includes onshore petroleum and gas production, offshore petroleum and gas production, gas processing, gas transmission, underground gas storage, LNG storage, LNG import/export, and gas local distribution (US GPO 40 CFR Part 98). These facilities also report incomplete combustion emissions under Subpart C. Gas gathering and boosting facilities report under Subpart C only, but the US EPA has amended the rule to also require reporting under Subpart W beginning in 2016. Facilities are defined at the site level except for onshore production, which are defined as a company's entire well pad assets in a basin, and local distribution, which are defined as a company's statewide distribution assets. Since 2011, reporters have been required to report annual emissions by source category and associated data at levels of detail varying by source and sector. Emissions must be estimated using methods prescribed by the rule for each source category, which include direct measurements, engineering equations, and emission factors. Similar to the GHGI, GHGRP emission factors are based primarily on Harrison et al. (1996). The GHGRP should not be viewed as a comprehensive inventory since it only includes facilities above the reporting threshold and excludes the gathering sector and some emission sources.

RECENT STUDIES

Research on natural gas supply chain methane emissions has resurged over the last few years due to concern regarding the climate impacts of increased natural gas development, particularly since a 2011 paper estimated that shale gas has a 3.6–7.9% leak rate (Howarth et al., 2011). US EPA GHGI and GHGRP emission estimates depend on data from Harrison et al. (1996) and other studies collected prior to the growth in unconventional O&G development. Numerous recent studies have investigated whether unconventional development or other changes in operational practices have increased or decreased methane emissions, including a series of 16 studies sponsored by the Environmental Defense Fund. Research can be divided into two categories based on their methodological approaches: top-down studies that use atmospheric measurements of well-mixed air to estimate emissions at a regional or larger scale, and bottom-up studies that measure emissions at the component or site level and sometimes extrapolate emissions to a regional or national scale using activity and emission factors. Bottom-up studies are summarized by sector in later sections, followed by top-down studies of total O&G emissions.

Production

Allen et al. (2013) used direct measurements to quantify emissions from equipment leaks, pneumatic controllers, and pneumatic pumps at 150 onshore production sites, 27 well completion flowbacks, 4 well workovers, and 9 liquid-unloading wells located across the US. Compared with GHGI estimates, emissions were higher for pneumatic controllers and equipment leaks, but lower for well completions. Two-thirds of well completions sent flowback gas to a sales line or flare, which controlled 99% of potential emissions. This demonstrates that well completion flowbacks, the main difference between conventional and unconventional production, can be effectively controlled with existing technologies.

Two follow-up studies to Allen et al. (2013) measured emissions from 107 liquid-unloading wells and 377 pneumatic controllers (Allen et al., 2015a,b). Liquid-unloading emissions were similar to the most recent GHGI estimates with the highest emissions from wells with more than 100 annual unloading events. Pneumatic controllers had 17% higher average emissions per device than the GHGI, and the average number of controllers per well was 2.7 compared with 1.0 in the GHGI, which indicates that the GHGI underestimates their national emissions by up to a factor of 3. A small number of controllers, many that were malfunctioning, had emission rates exceeding 6 standard cubic feet per hour (scfh). These devices comprised 19% of the population but accounted for 95% of emissions.

Allen et al. (2013) estimated emissions from the US natural gas production sector by scaling up measured sources using national activity factors and assuming GHGI estimates were accurate for unmeasured sources. National emission estimates were updated in the follow-up studies (Allen et al., 2015a,b) using 2012 activity factors and new pneumatic controller and liquid-unloading data. Production sector 2012 emissions were estimated to be 2185 gigagrams (Gg) methane, equivalent to 0.49% of gross gas production. The upper-bound emission estimate was 2815 Gg due to uncertainty in the pneumatic controller and liquid-unloading emissions. In comparison, GHGI and GHGRP 2012 production emissions were 1992 and 2200 Gg, respectively; the GHGI and Allen et al. estimates are for natural gas only, while the GHGRP includes O&G production.

Brantley et al. (2014) applied US EPA Other Test Method 33A, a mobile sampling method using point source Gaussian dispersion modeling, to estimate emissions at 210 production sites in the Barnett, Denver-Julesburg, Pinedale, and Eagle Ford basins. Emission rates were log-normally distributed and had a weak positive correlation ($R^2 = 0.083$) with gas production. Compared with Allen et al. (2013) and a study in the Barnett Shale (ERG, 2011), emission rate geometric means were about twice as high in Brantley et al. (2014), which may be due to the exclusion of tank-flashing emissions in these other studies or the bias of mobile sampling toward higher emitting sites with detectable downwind plumes.

Gathering and Processing

A recent study measured site-level emissions at 114 US gathering facilities and 16 processing plants using the dual-tracer correlation method (Roscioli et al., 2015; Mitchell et al., 2015). Gathering facility emissions were positively skewed with 12% of sites contributing 50% of emissions. Infrared camera surveys revealed that 20% of gathering sites had venting tanks and four times the average emissions of sites without substantial venting. Processing plants had higher average emission rates than gathering facilities (170 kg methane/h vs. 55 kg methane h^{-1}) but lower emissions proportional to gas throughput (0.075% vs. 0.2%). For both facility types, throughput was positively correlated with absolute emissions and negatively correlated with throughput-normalized emissions.

Marchese et al. (2015) estimated emissions from the US gathering and processing sectors using Monte Carlo simulations incorporating facility counts and emissions data from Mitchell et al. (2015). The number of US processing plants (606) is obtainable from national databases, but there is no comprehensive, national list of gathering facilities. The authors estimated the number of gathering facilities, 4549 (+921/−703), by comparing state permit databases to lists obtained from study industry participants. Gathering and processing (G&P) sector 2012 emissions were estimated to be 2421 (+245/−237) Gg methane compared with 1296 Gg in the 2014 GHGI and 180 Gg in the 2013 GHGRP. Processing sector emissions were estimated to be 546 Gg, lower than the GHGI (892 Gg), but higher than the GHGRP (179 Gg). Since the GHGI includes gathering within the production sector, the authors used industry participant facility equipment counts to allocate GHGI emissions between production and gathering. The study estimate of gathering sector emissions (1875 Gg) is almost five times the GHGI estimate (404 Gg) and several orders of magnitude higher than the GHGRP (0.5 Gg), which excluded vented and fugitive emissions from gathering facility reporting requirements. Gathering and Processing (G&P) sector emissions are equivalent to 0.47 ± 0.05% of gas production.

Transmission and Storage

Subramanian et al. (2015) quantified methane emissions at 45 transmission and storage (T&S) sector compressor stations using component-level direct measurements and the site-level dual-tracer correlation method. Study onsite estimates used infrared camera surveys to identify emitting sources followed by quantification with high-flow samplers; incomplete combustion emissions were estimated with US EPA AP-42 emission factors (EPA, 1995). Average tracer flux emission rate was 80 kg methane/h. Similar to the G&P sector, T&S sites had a skewed emission rate distribution with 50% of emissions from 10% of sites. The two highest emitting sites, which were defined as super-emitters, had much higher site-level emissions based on the tracer correlation method than aggregate measured component-level emissions. This discrepancy was caused by the presence of leaking compressor isolation valves that could not be accurately quantified with component-level measurements. For the 25 sites exceeding the GHGRP reporting threshold, study onsite emissions were 1.8 times higher than emissions estimated using prescribed GHGRP methods. This difference is due to GHGRP methodologies that exclude certain sources (e.g., reciprocating compressor rod-packing vents in pressurized, standby mode) and require use of inaccurate emission factors, including an incomplete combustion factor that is over 500 times lower than the AP-42 factor for reciprocating compressors.

Emissions from the US T&S sector were estimated in Zimmerle et al. (2015) with Monte Carlo simulations that integrated emissions data from Subramanian et al. (2015), detailed facility data from study industry participants, and GHGRP data. National emission estimates for 2012 were 1503 (+750/−283) Gg methane compared with GHGI and GHGRP estimates of 2071 and 200 Gg methane, respectively. In contrast to the GHGI, Zimmerle et al. (2015) estimated that there are fewer T&S stations in the US and the compressors at these sites have a greater proportion of lower emitting dry seal centrifugal compressors. T&S sector emissions are equivalent to 0.35% (+0.10/−0.07%) of T&S sector throughput.

Local Distribution

Lamb et al. (2015) measured methane emissions at 230 pipeline leaks and 229 metering and regulating (M&R) stations at 13 US local distribution systems. Sites were randomly selected from lists of pipeline leaks and M&R stations in representative areas provided by study industry participants. High-flow samplers were used to directly measure M&R components and in tandem with surface enclosures to measure underground pipeline leaks. There was an extremely skewed emission rate distribution for pipeline leaks with 1.3% of leaks contributing 50% of total emissions. Compared with Harrison et al. (1996), which is the basis of GHGI emission factors, both pipeline leaks and M&R stations had lower average emission rates. Nine M&R stations were measured in both Lamb et al. (2015) and Harrison et al. (1996), eight of which had lower emissions in the more recent study. Lamb et al. (2015) estimated 2011 US local distribution emissions using the new measurement data. Emission factors were developed for pipeline mains and services by pipeline material and M&R stations by operating pressure, and then multiplied by GHGI activity factors. GHGI emission estimates were used for customer meter and upset emissions. Local distribution 2011 emissions were estimated to be 393 Gg methane (95% upper confidence limit = 854 Gg), compared with 1329 and 640 Gg in the GHGI and GHGRP, respectively. In contrast to the GHGI, study estimates were about 3 times lower for pipeline leaks and 13 times lower for M&R stations. The substantially lower emissions were attributed to pipeline replacement, leak surveys, and station upgrades and maintenance since the 1990s.

Top-Down Studies

Many recent studies have used aircraft, tower, or satellite-based measurements to estimate total methane emissions in regions with O&G production and/or natural gas distribution (Pétron et al., 2012, 2014; Townsend-Small et al., 2012; Jeong et al., 2013; Karion et al., 2013; Miller et al., 2013; Kort et al., 2014; Wecht et al., 2014; Peischl et al., 2013). Some of these studies have estimated the fraction of emissions from O&G sources using source apportionment approaches including stable isotope and hydrocarbon ratios, or by subtracting bottom-up estimates of other sources (Townsend-Small et al., 2012; Pétron et al., 2014). There is large variability among basins in top-down estimates of methane emissions as a percentage of gas production. For example, aircraft mass balance studies have reported leak rates of 0.18–0.41% in the Marcellus, 1.0–2.1% in the Haynesville, 1.0–2.8% in the Fayetteville, 2.6–5.6% in the Denver-Julesburg, and 6.2–11.7% in the Uintah (Peischl et al. 2013; Pétron et al., 2014; Karion et al., 2013).

Top-down estimates of total and O&G methane emissions often have been higher than bottom-up estimates. Miller et al. (2013) analyzed methane observations with an atmospheric transport model and geostatistical inverse modeling to estimate US emissions. Anthropogenic emission estimates were 1.5 and 1.7 times higher than the GHGI and the Emissions Database for Global Atmospheric Research (EDGAR, JCR/PBL, 2011), respectively; O&G emissions in the south-central US were estimated to be 2.3–7.5 times higher than EDGAR. A review of research studies found that top-down estimates typically exceed bottom-up estimates; the authors estimate US methane emissions are 1.25–1.75 times higher than GHGI estimates (Brandt et al., 2014). The authors propose four hypotheses that may account for underestimation by O&G emission inventories: (1) bottom-up data are no longer representative of current technologies and practices, (2) bottom-up datasets have a small sample size and possible sampling bias, (3) skewed emission distributions result in sampled data with lower average emission rates than the population, and (4) inaccurate activity factors.

In October 2013, a coordinated research campaign in the Barnett Shale used simultaneous top-down and bottom-up approaches to quantify methane emissions from the O&G supply chain (Harriss et al., 2015). Bottom-up data included direct component measurements of T&S stations (Johnson et al. 2015) and local distribution M&R stations (Lamb et al., 2015). Aircraft- and vehicle-based approaches were used to quantify site-level emissions at O&G well pads, compressor stations, processing plants, and landfills by analyzing downwind plumes using the mass balance approach (Lavoie et al., 2015; Nathan et al., 2015), Gaussian dispersion modeling (Lan et al., 2015; Yacovitch et al., 2015), or the mobile flux plane method (Rella et al., 2015). Regional methane emissions were estimated using the top-down, aircraft-based mass balance approach during eight flights (Karion et al., 2015). Source apportionment data included carbon and hydrogen stable isotope and hydrocarbon ratios of canister air samples from different source types (Townsend-Small et al., 2015) and continuous ethane-to-methane ratios measured on mass balance aircraft (Smith et al., 2015). Top-down estimates of total methane emissions were 76 ± 13 Mg/h with 60 ± 11 Mg/h from fossil sources (Karion et al., 2015; Smith et al., 2015). Lyon et al. (2015) used data from the campaign and other sources to construct a spatially resolved methane emission inventory; bottom-up estimates of total and O&G methane emissions were 72.3 (+10.1/−8.9) Mg/h and 46.2 (+8.2/−6.2) Mg/h, respectively. This bottom-up estimate of O&G emissions was 1.5, 2.7, and 4.3 times higher than alternative estimates based on the GHGI, GHGRP, and EDGAR, respectively, primarily due to the inclusion of high-emission sites and more complete activity factors, particularly for gathering stations. Top-down and bottom-up estimates of the Barnett region natural gas supply chain leak rate (1.3–1.9 and 1.0–1.4%, respectively) are not significantly different. Zavala-Araiza et al. (2015) used well pad data from the campaign to introduce the concept of functional super-emitters defined by emissions proportional to gas production. Functional super-emitters comprised 15% of sites and contributed 77% of total production site emissions in the Barnett region.

Remaining Uncertainties

Compared with the GHGI, bottom-up studies have reported similar emissions for production and T&S (Allen et al., 2015a; Zimmerle et al., 2015), higher emissions for gathering (Marchese et al., 2015), and lower emissions for processing and local distribution (Marchese et al., 2015; Lamb et al., 2015). A forthcoming study will integrate recent data sources to estimate the magnitude and uncertainty of US natural gas supply chain emissions (Littlefield et al., Personal Communication). There are several sources of remaining uncertainty that may affect these emission estimates.

Positively skewed emission rate distributions are common to many types of O&G facilities and components. As shown in Figure 3.2, a relatively small number of high-emission sites sometimes referred to as super-emitters, contribute a large fraction of emissions. Recent studies have used statistical methods to quantify the uncertainty associated with skewed distributions; for example, the upper confidence limit of US local distribution emissions was over twice the central estimate mainly due to the majority of measured pipeline emissions coming from a very small number of leaks (Lamb et al., 2015). Additional research on

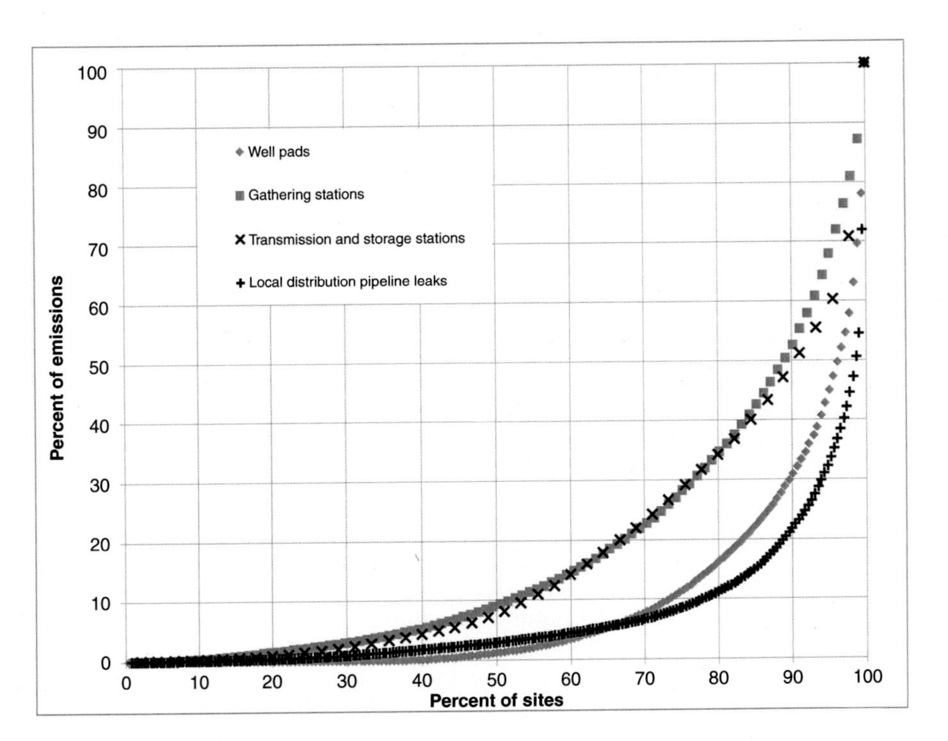

FIGURE 3.2
Facilities have highly skewed emission rate distributions with a small fraction of sites accounting for the majority of emissions.
Emission rate distributions from four source types are plotted as percent of sites in ascending order of emission rate versus percent of total emissions from sites at or below that rank. For example, the lowest emitting 50% of well pads contribute 1% of total emissions from measured sites, while the highest emitting 10% contribute 69% of total emissions. Data are from four recent studies. Sources: Rella et al. (2015); Subramanian et al. (2015); Mitchell et al. (2015); Lamb et al. (2015).

the prevalence, magnitude, and causes of super-emitters may reduce the uncertainty associated with sampling skewed distributions.

Activity factors for some equipment and facilities are poorly known. Allen et al. (2015a) reported that the average number of pneumatic controllers per well was 2.7 times higher than GHGI estimates. The number of gathering facilities is especially uncertain since these sites are subject to fewer reporting requirements. Marchese et al. (2015) improved estimates of US gathering facility counts but their lower and upper-bound estimates still vary by a factor of 1.4. More comprehensive reporting of activity factors is critical to reducing uncertainty in bottom-up emission estimates.

Several emission sources that were not measured by recent studies continue to have uncertainty associated with their previous emission estimates. Production sector storage tanks and compressors were not measured fully by Allen et al. (2013). Other studies have indicated that high emission well pads and gathering stations are often associated with tank venting (Brantley et al., 2014; Mitchell et al., 2015), which supports the need for additional data on tank emissions. Gathering pipelines are not known to have any published emission data. The GHGI uses data from Harrison et al. (1996) that was based on local distribution main leaks to estimate gathering pipeline emissions; Marchese et al. (2015) used the GHGI data in their updated estimate of gathering sector emissions. Gathering pipelines, which are less regulated than either transmission or local distribution pipelines, are potentially a large emission source that should be a focus of future research.

Emissions may also be associated with natural gas end use past the customer meter, such as leaks at power plants and incomplete combustion by residential gas furnaces. A top-down study estimated Boston region emissions are 2.7 ± 0.6% of gas throughput compared with a bottom-up estimate of 1.1% (McKain et al., 2015); the authors hypothesize that the gap may be partially due to end use emissions. A forthcoming study will report emission data from natural gas vehicles and fueling stations (Clark et al., personal communication).

Source apportionment methods such as stable isotope ratios cannot distinguish methane emissions from the natural gas supply chain and other fossil sources. Emissions from abandoned O&G wells and geologic seepage may be responsible for some of the gap between top-down and bottom-up estimates (Brandt et al., 2014). There are approximately 3 million abandoned and inactive wells in the US, but their emissions currently are not included in the GHGI. A recent study of 19 abandoned wells in Pennsylvania reported a highly skewed distribution with an average emission rate of 11 g methane/h (Kang et al., 2014). Due to the large number of abandoned wells, emission rates of this magnitude could result in substantial emissions, but there is high uncertainty because of the small sample size and limited geographic scope of the study. A forthcoming study will report emissions from over 100 wells in four states (Townsend-Small et al., Personal Communication). Geologic microseepage in hydrocarbon-prone basins is estimated to emit 10 Tg methane/year globally (Etiope and Klusman, 2002). Seepage likely contributes a portion of top-down estimates of fossil methane emissions, but high variability and lack of data from most regions precludes a reasonable estimate without further research.

CONCLUSIONS

Natural gas has climate benefits compared with more carbon-intensive fossil fuels, but these benefits may be reduced or delayed by methane emissions across the supply chain. The US EPA's most recent estimate of US natural gas and petroleum systems emissions is 7.3 Tg methane, equivalent to 1.4% of gross gas production. Much of the underlying data of the US EPA GHGI are based on a 1990s' study (Harrison et al., 1996) and therefore may not be representative of technological and operational changes associated with unconventional O&G development. Numerous recent studies have used top-down and bottom-up approaches to quantify O&G methane emissions. Bottom-up studies have estimated that supply chain emissions are roughly in line with GHGI estimates, but with higher or lower emissions from some equipment and sectors. Several top-down studies have reported higher emissions than GHGI estimates, which suggests bottom-up estimates may have remaining uncertainty, particularly from poorly characterized sources, such as storage tanks and gathering pipelines.

US natural gas supply chain methane emissions are likely of a magnitude such that coal-to-gas switching for electricity generation will result in immediate climate benefits, although regional variability in emission rates may cause gas from some basins to be worse for the climate in the short term. However, methane emissions may be high enough that there is short-term climate damage from other technology conversions, such as diesel to gas for heavy-duty vehicles, which require lower supply chain loss rates (Camuzeaux et al., 2015). Regardless of the current emission rate, the climate benefits of natural gas relative to more carbon–intensive fossil fuels can be increased by reducing methane emissions. Fortunately, emissions can be effectively controlled with technologies such as reduced emission completions (Allen et al., 2013). US oil and gas industry methane emissions can be reduced by 40% at a cost of less than $0.01 per thousand cubic feet of produced natural gas (ICF, 2014). Several studies have indicated that the majority of emissions come from a small number of sources, many of which are malfunctioning or have otherwise avoidable emissions, but the identity of these sources can be unpredictable. Therefore, comprehensive and frequent leak detection and repair programs to identify and mitigate these sources are critical to reducing emissions.

References

Allen, D., Torres, V., Thomas, J., Sullivan, D., Harrison, M., Hendler, A., Herndon, S., Kolb, C., Fraser, M., Hill, A., Lamb, B., Miskimins, J., Sawyer, R., Seinfeld, J., 2013. Measurements of methane emissions at natural gas production sites in the United States. Proc. Natl. Acad.Sci. 110 (44), 17768–17773.

Allen, D., Pacsi, A., Sullivan, D., Zavala-Araiza, D., Harrison, M., Keen, K., Fraser, M., Hill, D.A., Sawyer, R., Seinfeld, J., 2015a. Methane emissions from process equipment at natural gas production sites in the United States: pneumatic controllers. Environ. Sci. Technol. 49 (1), 633–640.

Allen, D., Sullivan, D., Zavala-Araiza, D., Pacsi, A., Harrison, M., Keen, K., Fraser, M., Daniel Hill, A., Lamb, B., Sawyer, R., Seinfeld, J., 2015b. Methane Emissions from Process Equipment at Natural Gas Production Sites in the United States: Liquid Unloadings. Environ. Sci. Technol. 49 (1), 641–648.

Alvarez, R., Pacala, S., Winebrake, J., Chameides, W., Hamburg, S., 2012. Greater focus needed on methane leakage from natural gas infrastructure. Proc. Natl. Acad.Sci. 109 (17), 6435–6440.

Brandt, A., Heath, G., Kort, E., O'Sullivan, F., Petron, G., Jordaan, S., Tans, P., Wilcox, J., Gopstein, A., Arent, D., Wofsy, S., Brown, N., Bradley, R., Stucky, G., Eardley, D., Harriss, R., 2014. Methane leaks from North American natural gas systems. Science 343 (6172), 733–735.

Brantley, H., Thoma, E., Squier, W., Guven, B., Lyon, D., 2014. Assessment of methane emissions from oil and gas production pads using mobile measurements. Environ. Sci. Technol. 48 (24), 14508–14515.

Camuzeaux, J., Alvarez, R., Brooks, S., Browne, J., Sterner, T., 2015. Influence of methane emissions and vehicle efficiency on the climate implications of heavy-duty natural gas trucks. Environ. Sci. Technol. 49 (11), 6402–6410.

Caulton, D., Shepson, P., Cambaliza, M., McCabe, D., Baum, E., Stirm, B., 2014. Methane destruction efficiency of natural gas flares associated with shale formation wells. Environ. Sci. Technol. 48 (16), 9548–9554.

Eastern Research Group,2011. City of Fort Worth Natural Gas Air Quality Study; City of Fort Worth: Fort Worth, TX. http://fortworthtexas.gov/gaswells/air-quality-study/final/

Etiope, G., Klusman, R., 2002. Geologic emissions of methane to the atmosphere. Chemosphere 49 (8), 777–789.

European Commission, Joint Research Centre (JRC)/Netherlands Environmental Assessment Agency (PBL), 2011. Emission Database for Global Atmospheric Research, release version 4.2. http://edgar.jrc.ec.europa.eu

Harrison, M.R., Shires, T., Wessels, J.K., Cowgill, R.M., 1996. Methane emissions from the natural gas industry; EPA: Washington, DC. http://www.epa.gov/methane/gasstar/documents/emissions_report/1_executiveummary.pdf

Harriss, R., Alvarez, R., Lyon, D., Zavala-Araiza, D., Nelson, D., Hamburg, S., 2015. Using multi-scale measurements to improve methane emission estimates from oil and gas operations in the Barnett Shale region Texas. Environ. Sci. Technol. 49 (13), 7524–7526.

Howarth, R., Santoro, R., Ingraffea, A., 2011. Methane and the greenhouse-gas footprint of natural gas from shale formations. Climatic Change 106 (4), 679–690.

ICF, 2014. Economic Analysis of Methane Emission Reduction Opportunities in the U.S. Onshore Oil and Natural Gas Industries. https://www.edf.org/sites/default/files/methane_cost_curve_report.pdf

IPCC, 2013. In: Stocker, T.F., Qin, D., Plattner, G.-K., Tignor, M., Allen, S.K., Boschung, J., Nauels, A., Xia, Y., Bex, V., Midgley, P.M. (Eds.), Climate Change 2013: The Physical Science Basis. Contribution of Working Group I to the Fifth Assessment Report of the Intergovernmental Panel on Climate Change. Cambridge University Press, Cambridge, United Kingdom and New York, NY, USA, p. 1535.

Jeong, S., Hsu, Y., Andrews, A., Bianco, L., Vaca, P., Wilczak, J., Fischer, M., 2013. A multitower measurement network estimate of California's methane emissions. J. Geophys. Res. 118 (19), pp.11,339–11,351.

Johnson, D., Covington, A., Clark, N., 2015. Methane emissions from leak and loss audits of natural gas compressor stations and storage facilities. Environ. Sci. Technol. 49 (13), 8132–8138.

Karion, A., Sweeney, C., Pétron, G., Frost, G., Hardesty, M.R., Kofler, J., Miller, B., Newberger, T., Wolter, S., Banta, R., Brewer, A., Dlugokencky, E., Lang, P., Montzka, S., Schnell, R., Tans, P., Trainer, M., Zamora, R., Conley, S., 2013. Methane emissions estimate from airborne measurements over a western United States natural gas field. Geophys. Res. Lett. 40 (16), 4393–4397.

Karion, A., Sweeney, C., Kort, E., Shepson, P., Brewer, A., Cambaliza, M., Conley, S., Davis, K., Deng, A., Hardesty, M., Herndon, S., Lauvaux, T., Lavoie, T., Lyon, D., Newberger, T., Pétron, G., Rella, C., Smith, M., Wolter, S., Yacovitch, T., Tans, P., 2015. Aircraft-based estimate of total methane emissions from the Barnett Shale region. Environ. Sci. Technol. 49 (13), 8124–8131.

Kang, M., Kanno, C., Reid, M., Zhang, X., Mauzerall, D., Celia, M., Chen, Y., Onstott, T., 2014. Direct measurements of methane emissions from abandoned oil and gas wells in Pennsylvania. Proc. Natl. Acad.Sci. 111 (51), 18173–18177.

Kort, E., Frankenberg, C., Costigan, K., Lindenmaier, R., Dubey, M., Wunch, D., 2014. Four corners: the largest US methane anomaly viewed from space. Geophys. Res. Lett. 41 (19), 6898–6903.

Lamb, B., Edburg, S., Ferrara, T., Howard, T., Harrison, M., Kolb, C., Townsend-Small, A., Dyck, W., Possolo, A., Whetstone, J., 2015. Direct measurements show decreasing methane emissions from natural gas local distribution systems in the United States. Environ. Sci. Technol. 49 (8), 5161–5169.

Lan, X., Talbot, R., Laine, P., Torres, A., 2015. Characterizing fugitive methane emissions in the Barnett Shale area using a mobile laboratory. Environ. Sci. Technol. 49 (13), 8139–8146.

Lavoie, T., Shepson, P., Cambaliza, M., Stirm, B., Karion, A., Sweeney, C., Yacovitch, T., Herndon, S., Lan, X., Lyon, D., 2015. Aircraft-based measurements of point source methane emissions in the Barnett Shale basin. Environ. Sci. Technol. 49 (13), 7904–7913.

Lyon, D., Zavala-Araiza, D., Alvarez, R., Harriss, R., Palacios, V., Lan, X., Talbot, R., Lavoie, T., Shepson, P., Yacovitch, T., Herndon, S., Marchese, A., Zimmerle, D., Robinson, A., Hamburg, S., 2015. Constructing a spatially resolved methane emission inventory for the Barnett Shale region. Environ. Sci. Technol. 49 (13), 8147–8157.

McKain, K., Down, A., Raciti, S., Budney, J., Hutyra, L., Floerchinger, C., Herndon, S., Nehrkorn, T., Zahniser, M., Jackson, R., Phillips, N., Wofsy, S., 2015. Methane emissions from natural gas infrastructure and use in the urban region of Boston Massachusetts. Proc. Natl. Acad.Sci. 112 (7), 1941–1946.

Marchese, A.J., Vaughn, T.L., Zimmerle, D., Martinez, D.M., Williams, L.L., Robinson, A.L., Mitchell, A.L., Subramanian, R., Tkacik, D.S., Roscioli, J.R., Herndon, S.C., 2015. Methane emissions from United States natural gas gathering and processing. Environ. Sci. Technol. 49 (17), 10718–10727.

Miller, S., Wofsy, S., Michalak, A., Kort, E., Andrews, A., Biraud, S., Dlugokencky, E., Eluszkiewicz, J., Fischer, M., Janssens-Maenhout, G., Miller, B., Miller, J., Montzka, S., Nehrkorn, T., Sweeney, C., 2013. Anthropogenic emissions of methane in the United States. Proc. Natl. Acad.Sci. 110 (50), 20018–20022.

Mitchell, A., Tkacik, D., Roscioli, J., Herndon, S., Yacovitch, T., Martinez, D., Vaughn, T., Williams, L., Sullivan, M., Floerchinger, C., Omara, M., Subramanian, R., Zimmerle, D., Marchese, A., Robinson, A., 2015. Measurements of methane emissions from natural gas gathering facilities and processing plants: measurement results. Environ. Sci. Technol. 49 (5), 3219–3227.

Nathan, B., Golston, L., O'Brien, A., Ross, K., Harrison, W., Tao, L., Lary, D., Johnson, D., Covington, A., Clark, N., Zondlo, M., 2015. Near-field characterization of methane emission variability from a compressor station using a model aircraft. Environ. Sci. Technol. 49 (13), 7896–7903.

Peischl, J., Ryerson, T., Brioude, J., Aikin, K., Andrews, A., Atlas, E., Blake, D., Daube, B., de Gouw, J., Dlugokencky, E., Frost, G., Gentner, D., Gilman, J., Goldstein, A., Harley, R., Holloway, J., Kofler, J., Kuster, W., Lang, P., Novelli, P., Santoni, G., Trainer, M., Wofsy, S., Parrish, D., 2013. Quantifying sources of methane using light alkanes in the Los Angeles basin, California. J. Geophys. Res. 118 (10), 4974–4990.

Pétron, G., Frost, G., Miller, B., Hirsch, A., Montzka, S., Karion, A., Trainer, M., Sweeney, C., Andrews, A., Miller, L., Kofler, J., Bar-Ilan, A., Dlugokencky, E., Patrick, L., Moore, C., Ryerson, T., Siso, C., Kolodzey, W., Lang, P., Conway, T., Novelli, P., Masarie, K., Hall, B., Guenther, D., Kitzis, D., Miller, J., Welsh, D., Wolfe, D., Neff, W., Tans, P., 2012. Hydrocarbon emissions characterization in the Colorado Front Range: a pilot study. J. Geophys. Res. 117 (D04304), 1–19.

Pétron, G., Karion, A., Sweeney, C., Miller, B., Montzka, S., Frost, G., Trainer, M., Tans, P., Andrews, A., Kofler, J., Helmig, D., Guenther, D., Dlugokencky, E., Lang, P., Newberger, T., Wolter, S., Hall, B., Novelli, P., Brewer, A., Conley, S., Hardesty, M., Banta, R., White, A., Noone, D., Wolfe, D., Schnell, R., 2014. A new look at methane and nonmethane hydrocarbon emissions from oil and natural gas operations in the Colorado Denver-Julesburg Basin. J. Geophys. Res. 119 (11), 6836–6852.

Rella, C., Tsai, T., Botkin, C., Crosson, E., Steele, D., 2015. Measuring emissions from oil and natural gas well pads using the mobile flux plane technique. Environ. Sci. Technol. 49 (7), 4742–4748.

Roscioli, J., Yacovitch, T., Floerchinger, C., Mitchell, A., Tkacik, D., Subramanian, R., Martinez, D., Vaughn, T., Williams, L., Zimmerle, D., Robinson, A., Herndon, S., Marchese, A., 2015. Measurements of methane emissions from natural gas gathering facilities and processing plants: measurement methods. Atmos. Meas. Tech. 8 (5), 2017–2035.

Shoemaker, J., Schrag, D., Molina, M., Ramanathan, V., 2013. What role for short-lived climate pollutants in mitigation policy? Science 342 (6164), 1323–1324.

Simpson, D., 2014. Pneumatic controllers in upstream oil and gas. Oil Gas Facil. 3 (05), 083–096.

Smith, M., Kort, E., Karion, A., Sweeney, C., Herndon, S., Yacovitch, T., 2015. Airborne ethane observations in the Barnett Shale: quantification of ethane flux and attribution of methane emissions. Environ. Sci. Technol. 49 (13), 8158–8166.

Subramanian, R., Williams, L., Vaughn, T., Zimmerle, D., Roscioli, J., Herndon, S., Yacovitch, T., Floerchinger, C., Tkacik, D., Mitchell, A., Sullivan, M., Dallmann, T., Robinson, A., 2015. Methane emissions from natural gas compressor stations in the transmission and storage sector: measurements and comparisons with the EPA greenhouse gas reporting program protocol. Environ. Sci. Technol. 49 (5), 3252–3261.

Townsend-Small, A., Tyler, S., Pataki, D., Xu, X., Christensen, L., 2012. Isotopic measurements of atmospheric methane in Los Angeles, California, USA: Influence of "fugitive" fossil fuel emissions. J. Geophys. Res. 117 (D07308), 1–11.

Townsend-Small, A., Marrero, J., Lyon, D., Simpson, I., Meinardi, S., Blake, D., 2015. Integrating source apportionment tracers into a bottom-up inventory of methane emissions in the Barnett Shale hydraulic fracturing region. Environ. Sci. Technol. 49 (13), 8175–8182.

United States Government Publishing Office, 2015. Electronic Code of Federal Regulations Title 40 Part 98 – Mandatory Greenhouse Gas Reporting. http://www.ecfr.gov/cgi-bin/text-idx?tpl=/ecfrbrowse/Title40/40cfr98_main_02.tpl

United States Government Publishing Office, 2015. Electronic Code of Federal Regulations Title 40 Part 60. Standards of Performance for New Stationary Sources: Subpart OOOO – Standards of Performance for Crude Oil and Natural Gas Production, Transmission and Distribution. http://www.ecfr.gov/cgi-bin/text-idx?SID=true&node=sp40.7.60.oooo

United States Energy Information Administration, 2015, Washington, DC. http://www.eia.gov/

United States Environmental Protection Agency, 2015. Inventory of U.S. Greenhouse Gas Emissions and Sinks: 1990-2013; EPA, Washington, DC. http://www.epa.gov/climatechange/ghgemissions/usinventoryreport.html

United States Environmental Protection Agency, 2015. Greenhouse Gas Reporting Program; EPA, Washington, DC. http://ghgdata.epa.gov/ghgp/main.do

United States Environmental Protection Agency, 1995. AP-42 Compilation of Air Pollutant Emission Factors; EPA, Washington, DC. http://www.epa.gov/ttnchie1/ap42/

Wecht, K., Jacob, D., Frankenberg, C., Jiang, Z., Blake, D., 2014. Mapping of North American methane emissions with high spatial resolution by inversion of SCIAMACHY satellite data. J. Geophys. Res. 119 (12), 7741–7756.

Yacovitch, T., Herndon, S., Pétron, G., Kofler, J., Lyon, D., Zahniser, M., Kolb, C., 2015. Mobile laboratory observations of methane emissions in the Barnett Shale region. Environ. Sci. Technol. 49 (13), 7889–7895.

Zavala-Araiza, D., Lyon, D., Alvarez, R., Palacios, V., Harriss, R., Lan, X., Talbot, R., Hamburg, S., 2015. Toward a functional definition of methane super-emitters: application to natural gas production sites. Environ. Sci. Technol. 49 (13), 8167–8174.

Zimmerle, D., Williams, L., Vaughn, T., Quinn, C., Subramanian, R., Duggan, G., Willson, B., Opsomer, J., Marchese, A., Martinez, D., Robinson, A., 2015. Methane emissions from the natural gas transmission and storage system in the United States. Environ. Sci. Technol., 9374–9383.

CHAPTER 4

A Review of Drinking Water Contamination Associated with Hydraulic Fracturing

Elyse Rester*, Scott D. Warner†

*Ramboll Environ USA, Senior Associate, Global Water Practice, USA;
†Ramboll Environ USA, Principal, Water Resources Management, Global Water Practice, USA

Chapter Outline

OVERVIEW

Background

Surface and groundwater contamination are important topics when discussing hydraulic fracturing (aka "fracking") for recovery of petroleum resources from tight shale formations or to enhance recovery from conventional subsurface reservoirs. Public and media attention on the issue was heightened in the late 2000s with the increased availability and use of newer technologies that allowed for more effective and efficient petroleum recovery (liquid and gas) from previously low permeability reserves (such as from shale) combined with the temporary surge in the price of oil in the market place (see

Environmental and Health Issues in Unconventional Oil and Gas Development. http://dx.doi.org/10.1016/B978-0-12-804111-6.00004-2

economics discussion in Chapter 2). Potential impacts to water resources, primarily groundwater, from various aspects of the fracturing process – including leakage of injectants and methane through well casing, migration of petroleum constituents through propagation of fractures, and unintentional (or intentional) release of wastewater from surface impoundments – created great public and regulatory attention to this quickly developed industry (see discussions in Chapters 5 and 6). This chapter serves to highlight facts and technical considerations relating to past and potential exposures from water associated with hydraulic fracturing.

The focus of this chapter is on surface and groundwater located within areas where hydraulic fracturing for petroleum recovery is used – water, which has historically, presently, or may in the future serve as a drinking water resource. This is consistent with United States (US) Environmental Protection Agency (EPA) concern that water bodies impacted by hydraulic fracturing are currently and will continue to be used in the future as sources of drinking water (EPA, 2015).

Proximity of Fracking Wells and Drinking Water Resources

Exposure pathways are more concentrated near the wells themselves, increasing the likelihood that were an exposure to occur it would do so within the vicinity of the well itself (see discussion of exposure pathways in Chapter 1). However, there are also exposure pathways that could lead to drinking water contamination outside the vicinity of a hydraulically fractured well (e.g., a truck carrying wastewater could spill) (EPA, 2015) (see discussions in Chapters 6 and 8). "Hydraulically fractured wells can be located near residences and drinking water resources. Between 2000 and 2013, approximately 9.4 million people lived [in the US] within one mile of a hydraulically, fractured well. Approximately 6,800 sources of drinking water for public water systems were located within one mile of at least one hydraulically fractured well during the same time period. These drinking water sources served more than 8.6 million people year-round in 2013" (EPA, 2015). In Pavillion, Wyoming (US), during an investigation into drinking water contamination as a result of fracking, "[EPA] found that dissolved methane concentrations in the domestic wells were higher near the gas production wells" (Cooley et al., 2012). Figure 4.1 shows the locations of public water systems (infiltration galleries, intakes, reservoirs, springs, and wells) in the US that are located within one mile of a hydraulically fractured well.

Recent legislation signed into law in California (US) in 2013 requires monitoring programs to insure protection of drinking water sources and characterize background concentrations. This legislation, Senate Bill 4, requires the State Water Resources Control Board to implement a groundwater-monitoring program and criteria in areas of oil and gas (O&G) operation (see further discussion on legislation in Chapter 12).

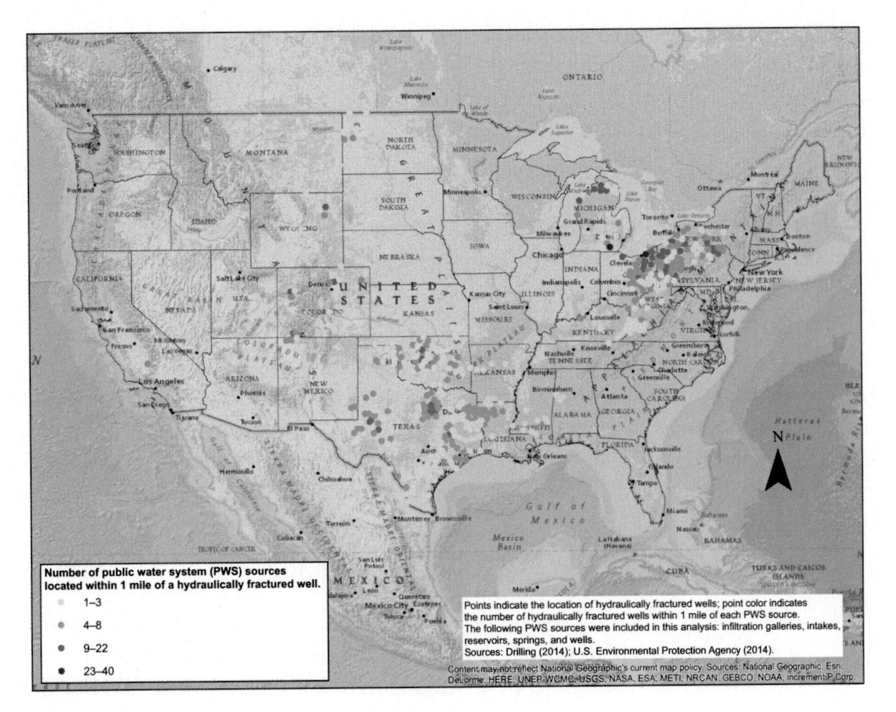

FIGURE 4.1
Public water systems (infiltration galleries, intakes, reservoirs, springs, and wells) in the United States located within 1 mile of a hydraulically fractured well.

EXPOSURE PATHWAYS

Potential surface and drinking water exposure pathways are outlined below, for each of the hydraulic fracturing water cycle components. The hydraulic fracturing water cycle is comprised of the following activities (EPA, 2015):

1. Water acquisition
2. Chemical mixing
3. Well injection
4. Flowback and produced water
5. Wastewater treatment and waste disposal

Water Acquisition

As droughts and competing interests for water increase, the need to protect this natural resource becomes even greater (see global water source discussion in Chapter 13). Chapter 5 discusses the water supply and options for recycling during hydrofracking and the balance between competing water demands, particularly in arid regions, a topic that has become a particularly hot button (see discussion of regulatory activity in Chapter 6). We focus here on the impacts on surface and groundwater quality as water is acquired for the hydraulic fracturing process.

Both groundwater and surface water can be used for hydraulic fracturing and the use of either if not done thoughtfully can cause a decrease in drinking source water quality. If groundwater in an aquifer is withdrawn in amounts exceeding the rates at which it is naturally recharged, negative pressure in the aquifer can result, which may cause infiltration of lower quality water or contaminants from adjacent formations. If surface water withdrawals are large enough to significantly reduce the flow in a stream, the stream's ability to mix or dilute future contaminants can be reduced (EPA, 2015).

Chemical Mixing

The fracking process utilizes a variety of chemicals, often called hydraulic fracturing fluids, to optimize the process of petroleum recovery. Volumes and chemical constituents vary by region, state, and site based on geography, availability, price, and preference. "An average of about 25 different chemicals are used in each hydraulic fracturing operation" (Long et al., 2015) and "no single chemical is used at all well sites across the country" (EPA, 2015).

The hydraulic fracturing fluid generally consists of three parts (EPA, 2015):

1. The base fluid, which is the largest constituent by volume and is typically water (usually 98% or more of the overall injected fluid volume);
2. The additives, which can be a single chemical or a mixture of chemicals; and
3. The proppant, solid material (usually sand) designed to keep hydraulic fractures open.

Determining the optimal chemical mixture and dose to achieve specific results (e.g., adjust pH, increase viscosity, limit bacterial growth), while maximizing the ability of the well to produce gas, requires knowledge, skill, and experience. These trade secrets are often cited during legislative debates involving hydraulic fracturing (see Chapter 12 on FracFocus program).

Potential health impacts from concentrated chemicals found in hydraulic fracturing fluids include cancer, immune system effects, changes in body weight, changes in blood chemistry, cardiotoxicity, neurotoxicity, liver and kidney toxicity, and reproductive and developmental toxicity (EPA, 2015). However, chemicals in fracking fluids are highly diluted. Studies would have to be conducted on a case-by-case basis to determine whether the dose and amount of exposure from a particular chemical mixture in a particular well at a particular time could cause any health impacts.

The routes by which hydraulic fracturing fluids may have the potential to contaminate a drinking water source include the following:

1. Spills from transport trucks (see further discussion on transportation in Chapter 9);
2. Spills from storage containers;
3. Well injection (discussed later in this chapter); and
4. Flowback and produced water (discussed later in this chapter).

Spills of fracking fluids have resulted from failure of equipment or container, human error, weather, and vandalism among others. "The most common cause [of hydraulic fluid spills] was equipment failure, specifically blowout preventer failure, corrosion, and failed valves" (EPA, 2015).

There is limited data available documenting the frequency of onsite spills of hydraulic fracturing fluids. In the US, national data does not currently exist; state-specific data is only provided by Colorado and Pennsylvania. "Frequency estimates from data and literature ranged from one spill every 100 wells in Colorado to between approximately 0.4 to 12.2 spills for every 100 wells in Pennsylvania. (Spill frequency estimates are for a given number of wells over a given period of time. These are not annual estimates nor are they for the lifetime of the well)" (EPA, 2015).

If a spill does occur, it must reach a surface or groundwater source to have the potential to directly affect drinking water quality. This can occur if a spill flows overland to a nearby surface water body, if a spill contaminates soil, which seeps into a surface water body, or if a spill permeates the soil to reach the groundwater below. "Of the 151 spills characterized by the EPA, fluids reached surface water in 13 (9% of 151) cases and soil in 97 (64%) cases. None of the spills of hydraulic fracturing fluid were reported to have reached ground water" (EPA, 2015) (see eg., Chapter 6). It should be noted that it would take a period of time for a spill to infiltrate soil and reach the groundwater below; and that given more time, it is unknown whether some of the spills characterized by the EPA would reach groundwater.

Well Injection

The proper construction, maintenance, and operation of wells is key to protecting groundwater from potential contamination. This is true for any type of well for any type of use, including hydraulic fracking. In hydraulic fracturing, cemented surface casings are used in the production wells to isolate groundwater from potential contamination from liquids and gases as they travel to the surface. "Impacts to drinking water resources from subsurface liquid and gas movement may occur if casing or cement are inadequately designed or constructed, or fail" (EPA, 2015). Figure 1.6 illustrates the components of a hydraulic fracturing well (see Chapter 1 for further detail). Well construction engineering requirements and designs are dictated by state regulations (see eg., Chapter 12). However, the rush to frack in the early days may have resulted in less oversight of well construction and thus inadequate casings may have been constructed. The American Petroleum Institute developed a set of guidelines to assure effective techniques (see also legislative oversight discussion in Chapter 12).

The vertical distance between the production zone and drinking water resources help protect against drinking water contamination. "Numerical modelling and microseismic studies based on a Marcellus Shale-like environment suggest that fractures created during hydraulic fracturing are unlikely to extend upward from these deep formations into shallow drinking water aquifers" (EPA, 2015).

"Hydraulic fractures produced in deep formations far beneath protected groundwater are very unlikely to propagate far enough upwards to intersect an aquifer. Studies performed for high-volume hydraulic fracturing elsewhere in the country have shown that hydraulic fractures have propagated no further than 2,000 ft. (610 m) vertically, so hydraulic fracturing conducted many thousands of feet below an aquifer is not expected to reach a protected aquifer far above" (Long et al., 2015).

Fracking for O&G located at depths similar to groundwater aquifers also occurs and requires added caution to protect against contamination. "The EPA's survey of O&G production wells hydraulically fractured by nine service companies in 2009 and 2010 estimated that 20% of 23,000 wells had less than 2,000 ft. (610 m) of measured distance between the point of shallowest hydraulic fracturing and the base of the protected ground water resources reported by well operators" (EPA, 2015).

When hydraulic fracturing occurs near existing production wells there is a potential for a fracture to intersect a nearby well. "These well communications, or 'frac hits', are more likely to occur if wells are close to each other or on the same well pad" (EPA, 2015). When a frac hit occurs and the offset well cannot withstand the forces from a nearby well fracturing event, offset well components may fail. This can cause the offset well to release fluids at the surface, which has the potential to cause contamination of surface or groundwater resources. "The EPA identified incidents in which surface spills of hydraulic fracturing-related fluids were attributed to well communication events" (EPA, 2015).

Older wells, often inactive, have a greater potential for failure if a frac hit occurs, since as discussed earlier, they may not have been constructed to the same standards we have today. "The Interstate O&G Compact Commission (www.iogcc.state.ok.us) estimates that over 1 million wells may have been drilled in the US prior to a formal regulatory system being in place and the status and location of many of these wells are unknown" (EPA, 2015).

Flowback and Produced Water

Flowback and produced water may contain some or all of the chemicals discussed in the previous chemical mix section (pH adjustment chemicals, biocides, etc.) and in addition "naturally occurring organic chemicals and radionuclides, metals, and other constituents of subsurface rock formations mobilized by the hydraulic fracturing process" (EPA, 2015).

The drinking water exposure pathways for flowback and produced water are the same as for the chemical mix and well injection (spills, well failure, etc.) (see Chapter 5).

No studies of spills during hydraulic fracturing have separated the numbers of chemical spills from flowback and produced water spills from wastewater spills. Studies have only been conducted on spills as a whole.

Wastewater Treatment and Waste Disposal

The pathways for wastewater treatment and waste disposal to impact drinking water resources are similar to those described earlier for spills and may range by mode of transport or form of storage. Drinking water resources can also be impacted if wastewater is not treated properly and discharged to surface water or reinjected, potentially contaminating groundwater. Wastewater treatment and waste disposal is discussed in detail in Chapter 5.

REGULATIONS OVERVIEW

Federal

While most of the regulation of hydraulic fracturing in the US is left to the individual states, several federal regulations protect water exposure as a result of fracking activities (Spence, 2010; Brady and Crannell, 2012) (see Chapter 12 for additional information):

- *Clean Water Act* – regulates disposal of wastewater into lakes, streams, or treatment plants.
- *Hazardous Materials Transportation Act* – regulates the transportation of hazardous chemicals.
- *Resource Conservation and Recovery Act* – regulates generation, transportation, treatment, storage, and disposal of hazardous waste.
- *Comprehensive Environmental Response, Compensation, and Liability Act* – establishes a federal "Superfund" to pay for the clean-up of accidents, spills, or other emergency releases of hazardous substances to the environment.
- *Safe Drinking Water Act* – protects the overall quality of public drinking water in the US.

The "Fracturing Responsibility and Awareness of Chemicals Act" also known as the "FRAC Act," which is comprised of companion bills S.1215 and H.R.2766, was introduced to the US Congress in 2009, though it has never received a vote. The FRAC Act would require disclosure of "the chemical constituents (but not the proprietary chemical formulas) used in the fracturing process" (Spence, 2010).

Federal regulations in the US require companies to keep material safety data sheets (MSDS) listing constituents ingredients and amounts, including toxic materials, at fracking sites (Spence, 2010). MSDS contain information about how to manage and respond to spills or chemical exposure from a specific chemical. The MSDS do not contain information on mixtures of chemicals nor the synergistic effects of chemical mixtures.

State

In the US, the states themselves are left to do much of the regulation of the hydraulic fracturing process.

The Energy Policy Act of 2005 added two exclusions to the definition of underground injection in the Safe Drinking Water Act: (1) The underground injection of natural gas for purposes of storage and (2) the underground injection of fluids or propping agents (other than diesel fuels) pursuant to the hydraulic fracturing operations related to oil, gas, or geothermal production activities. As a result, states are responsible for protecting drinking water aquifers from hydraulic fracturing contamination (Spence, 2010). "[No] federal statute regulates the disposal of fracking wastes on land or their injection into the ground: those methods of disposing of flowback water or other fracking wastes are subject only to state regulation" (Spence, 2010). In Chapter 12, the specific state and federal regulations are discussed in detail.

WATER EXPOSURE CASE STUDIES

Pavillion, Wyoming

Encana Oil and Gas (Encana, USA) bought the Pavillion field in Pavillion, Wyoming (US) in 2004; the field includes 169 gas-producing wells. In 2008, under Section 105(d) of the Comprehensive Environmental Response, Compensation, and Liability Act (CERCLA), citizens of Pavillion, Wyoming submitted a petition to the US EPA to conduct an investigation of possible drinking water contamination after noticing problematic smells and tastes of their well water.

The EPA, in conjunction with the Agency for Toxic Substances and Disease Registry (ATSDR), conducted a study over 3 years to identify and characterize contaminants in the Pavillion aquifer. The EPA conducted sampling at both shallow-monitoring wells (residential wells, stock wells, and municipal wells, ranging from 15 to 515 ft. deep) and two deep-monitoring wells (MW01 and MW02, at 765–785 ft. and 960–980 ft., respectively) (Folger et al., 2012).

The results of sampling for contaminants from the shallower portions of the aquifer reported benzene, xylenes, gasoline-range organics, and diesel-range organics. "Domestic water wells in the Pavillion area generally use groundwater from the shallower portions of the aquifer" (Folger et al., 2012). The EPA report indicated at least 33 surface disposal pits were likely sources for the contaminants detected in shallow groundwater: "detection [of these contaminants] in ground water samples from shallow monitoring wells near pits indicates that pits are a source of shallow ground water contamination in the area of investigation" (Folger et al., 2012).

Sampling results from deeper wells were the topic of subsequent debate. The following deep well–monitoring results were used to assess the source of drinking water contamination: high pH values, elevated potassium and chloride concentrations, detection of synthetic organic compounds, detection of petroleum hydrocarbons, and breakdown products of organic compounds. The following factors were also considered in the EPA's "lines of reasoning" when considering Pavillion drinking water contamination: well design and integrity of gas production wells, excursion of fractured fluids from sandstone units and along the wellbore, enhanced migration of natural gas, isotopic data, proximity of methane in domestic wells to production wells, methane concentrations highest near

MW01, shallow surface casing, lack of cement, sporadic bonding, and timing of citizen complaints (Folger et al., 2012).

While the EPA noted that factors other than hydraulic fracturing could cause specific contamination in drinking water (e.g., high potassium and chloride levels could be included in the range of natural variability in deeper portions of the aquifer) based on its "lines of reasoning" approach, the EPA believed the investigation supported the "explanation that inorganic and organic compounds associated with hydraulic fracturing have contaminated the aquifer at or below the depths used for domestic water supply" in the Pavillion area. The EPA also stated that its approach indicated that "gas production activities have likely enhanced the migration of natural gas in the aquifer and the migration of gas to domestic wells in the area" (Folger et al., 2012). "[EPA] found that dissolved methane concentrations in the domestic wells were higher near the gas production wells" (Cooley et al., 2012).

Stakeholder responses from industry groups raised issues with EPA sampling techniques and whether "chemicals used by EPA in drilling its monitoring wells may have affected the results of sampling the deep groundwater" (Folger et al., 2012). "The [Encana] press release stated that EPA's results from the investigation do not exceed state or federal drinking water quality standards for any constituent related to oil and gas development" (Folger et al., 2012). In its press release, Encana Oil & Gas (USA) Inc. called for an independent third-party review of the EPA data. A commentator from the Natural Resources Defense Council (NRDC), an environmental advocacy group, used the EPA results to call for "much stronger rules are needed and that is why NRDC supports federal regulation of fracking under the Safe Drinking Water Act" (Folger et al., 2012).

The Pavillion contamination investigation was one of the first studies on groundwater contamination from both "deep" and "shallow" wells. An article published by Pro Publica, an independent nonprofit news service, stated that findings from the EPA Draft Report "could be a turning point in the heated national debate about whether contamination from fracking is happening, and are likely to shape how the country regulates and develops natural gas resources in the Marcellus Shale and across the Appalachian states" (Folger et al., 2012).

The sampling methods and well construction performed by EPA have not held up under peer review, "EPA's data as well as results and conclusions appear to be compromised and unreliable" (Stephens, 2015). In June 2013, the EPA announced that the State of Wyoming would be conducting further sampling and investigation at the Pavillion site.

Dimock, Pennsylvania

In 2008 Cabot Oil and Gas Corporation (Cabot) began drilling for natural gas in Susquehanna County in Dimock, Pennsylvania (US). On January 1, 2009 an explosion in an outside drinking water well occurred due to methane buildup in the well. The Pennsylvania Department of Environmental Protection (PADEP) took the lead in the investigation of this explosion, and in February 2009 issued a notice of violation to Cabot, which stated that Cabot discharged natural gas into local waterways and failed to prevent natural gas from entering fresh groundwater (Cooley et al., 2012).

"In November 2009 (last amended December 2010), PADEP issued a consent agreement with Cabot for methane and metals removal systems for 18 private wells in the site area. The agreement calls for each well owner to enter into the agreement with Cabot. Until the treatment systems are installed, Cabot was to provide delivered water" (ATSDR, 2011). Only 6 of the 18 affected well owners signed the agreement and had water treatment systems installed. "Twelve well owners refused to sign agreements with Cabot and were part of a civil suit" (Cooley et al., 2012) (see Chapter 6).

"ATSDR Division of Regional Operations received the water sampling data for the 18 properties that are part of the consent order between Cabot and PADEP" (ATSDR, 2011). "Based on the maximum results for the approximately 18 wells sampled, levels of coliform bacteria, methane, ethylene glycol, bis(2-thylhexyl) phthalate (DEHP), 2-methoxyethanol, aluminium, arsenic, lithium, manganese, sodium, and iron were elevated about comparison values (CVs)" (ATSDR, 2011).

In December 2011, the EPA announced that Dimock water was safe to drink and allowed Cabot to discontinue providing drinking water to the 12 homes that did not receive treatment systems. However, several Dimock residents submitted sampling results to the EPA of their own testing and some indicated water was still polluted. As a result, the EPA agreed to retest the water at some of the homes in the area (Cooley et al., 2012). The Action Memo authorizing the EPA to begin residential well sampling at 60 homes in the area and provide alternative water supplies for 4 households with elevated contaminant levels was signed on January 19, 2012.

The New York State Department of Environmental Conservation (NYSDEC) met with PADEP in 2011 to understand what went wrong in several sites in Pennsylvania in order to help inform New York's 2011 Supplemental Generic Environmental Impact Statement process.

The problems identified at the Dimock site included excessive pressures and improperly or insufficiently cemented casings and equipment failures resulting from pressure surges when transferring water due to poor site design (Martens, 2011).

PREVENTION TECHNIQUES

As technologies emerge and best practices are optimized, potential water exposure can be more easily prevented. The more historical and baseline data are available for exposure studies, the better we can understand the sources of surface and groundwater contamination. Continued monitoring of aquifers, both pre- and postfracturing, is crucial to understanding the sources of groundwater contamination.

Site-specific actions can be taken to help alleviate risks. "Management of the rate and timing of surface water withdrawals has been shown to help mitigate potential impacts of hydraulic fracturing withdrawals on water quality" (EPA, 2015). Secondary containment for storage of fracking chemicals, flowback water, wastewater, etc. will help protect against contamination caused by equipment failure.

In the US, the States themselves are choosing to adopt similar prevention practices. "State programs exist to plug identified inactive wells, and work is ongoing to identify and address such wells" (EPA, 2015). A few specific state requirements targeted at preventing surface and groundwater contamination include (Martens, 2011):

- Cementing practices, testing, and use of intermediate casing;
- Stormwater permits required to include strict design measures to prevent failure of stormwater controls and potential fluid flow to offsite water sources;
- Environmental staff site visit prior to well pad construction to review well site layout to insure it is properly designed and to determine site-specific permit conditions;
- Pressure testing of fracturing equipment after installation and prior to hydraulic fracturing operations;
- Pressure testing of blowout prevention equipment;
- Using specialized equipment designed for entering the wellbore when pressure is anticipated; and
- A certified well control specialist to be present during postfracturing cleanout activities (see discussions for global practices in Chapter 13).

Just as companies are perfecting the chemistry of hydraulic fracking fluids as they gain experience, they are also improving skills in installation, inspection, maintenance, and operation of hydraulic fracturing equipment, which are valuable in reducing potential drinking water contamination.

CONCLUSIONS

Ground and surface water contamination has been associated with fracking activities, though the contamination often is related to operational issues and human error, rather than subsurface hydraulic fracturing itself. Contamination has occurred because of shallow fracturing (near levels of protected aquifers), poor cement casing in the wells themselves, or surface spills whether from equipment failure or human error. After conducting an assessment of the potential impacts of hydraulic fracturing for O&G on drinking water resources, the EPA states, "We did not find evidence of widespread, systematic impacts on drinking water resources in the United States" (EPA, 2015). Continued monitoring of groundwater, both pre- and posthydraulic fracturing as well as groundwater-sampling events covering longer durations, will be key to identifying and preventing drinking source water contamination in the future.

References

Agency for Toxic Substances and Disease Registry (ATSDR), 2011. Record of Activity/Technical Assist. UID# IBD7. Dimock Area, Dimock, Pennsylvania.

Brady, W.J., Crannell, J.P., 2012. Hydraulic fracturing regulation in the United States: the Laissez-Faire approach of the federal government and varying state regulations. Vt. J. Environ. Law 14, 39.

Cooley, H., Donnelly, K., Ross, N., Luu, P., 2012. Hydraulic fracturing and water resources: separating the frack from the fiction. Pacific Institute, Oakland, CA.

DrillingInfo, Inc. 2014. DI Desktop June 2014 download [Database]. Austin, TXA: DrillingInfo. http://info.drillinginfo.com/

Environmental Protection Agency (EPA), 2015. Assessment of the Potential Impacts of Hydraulic Fracturing for Oil and Gas on Drinking Water Resources. EPA/600/R-15/047a. External Review Draft.

Folger, P., Tiemann, M., Bearden, D.M., 2012. The EPA Draft Report of Groundwater Contamination Near Pavillion, Wyoming: Main Findings and Stakeholder Responses. Congressional Research Service, 7-5700.

Long, J., Birkholzer, J., Feinstein, L., 2015. An Independent Scientific Assessment of Well Stimulation in California Summary Report, An Examination of Hydraulic Fracturing and Acid Stimulations in the Oil and Gas Industry. California Council on Science and Technology.

Martens, J., 2011. New York State Department of Environmental Conservation, Fact Sheet: What We Learned from Pennsylvania. http://atlantic2.sierraclub.org/sites/newyork.sierraclub.org/files/documents/2013/02/pafactsheet072011.pdf

Spence, D., 2010. Fracking Regulations: Is Federal Hydraulic Fracturing Regulation Around the Corner? Energy Management Brief, Energy Management and Innovation Center, McCombs School of Business, University of Texas at Austin, p. 22.

Stephens, D.B., 2015. Analysis of the groundwater monitoring controversy at the Pavillion, Wyoming natural gas field. Groundwater 53 (1), 29–37.

CHAPTER 5

Water Use and Wastewater Management: Interrelated Issues with Unique Problems and Solutions

Daniel J. Price*, **Carl Adams, Jr****
*Ramboll Environ US, St. Louis, MO, USA; **Ramboll Environ Brentwood, TN, USA

Chapter Outline

As discussed in Chapter 1, hydraulic fracturing (fracking) is a process in which a fluid is injected at high pressure into low permeability rock containing oil and gas (O&G) deposits to fracture the rock and release liquid or gas. The fracking fluid typically is a mixture of water, proppants (sand or ceramic beads), and treatment chemicals to enhance water properties. The pressure creates fractures kept open by the proppants, which allow O&G to flow from the pore spaces to production wells. Hydraulic fracturing is not new, the first experimental use was in 1947 and it was used commercially shortly thereafter in vertical wells. Unconventional plays pair directional drilling with hydraulic fracturing, which allows for tapping the source rock and more effective production from thin reservoir units.

Water use and wastewater management are so closely related that they can be considered two sides of the same coin. This chapter will explore the relationship between the large amounts of water required for hydraulic fracturing in horizontal wells and the management of the wastewaters generated from the process due to the large amounts of water used. This demand for large volumes of water

Environmental and Health Issues in Unconventional Oil and Gas Development. http://dx.doi.org/10.1016/B978-0-12-804111-6.00005-4

and the requirement to manage the resulting wastewaters has created the following two major issues, which are the focus of this chapter:

- Water sourcing requirements particularly in areas experiencing water stress, and
- Proper management (capture, handling, treatment, and disposal) of the recovered wastewater.

WATER USE

The amount of water used for fracture stimulation in unconventional plays is determined in large part by the type of geological formation being fracked. The documented volumes of water required in any given O&G play vary somewhat depending on site-specific factors such as lithology; however, in general it takes approximately 100,000 to 1 million gallons of water to drill the initial horizontal well and approximately 2–4 million gallons to complete fracking activities at one horizontal well (Ground Water Protection Council, 2009). Multiple horizontal wells or legs are typically drilled off one vertical well and each horizontal leg includes multiple fracking events or stages; with the average being approximately 16 stages per horizontal well (Triepke, 2014). According to the United States Environmental Protection Agency (US EPA), there were between 70 billion and 140 billion gallons of water used in hydraulic fracturing activities in the United States (US) in 2010 (Rigzone, 2013). The recent 2014 US EPA "Assessment of the Potential Impacts of Hydraulic Fracturing for Oil and Gas on Drinking Water Resources," estimated hydraulic fracturing activities account for less than 1% of total water use and consumption in the US (US EPA, 2014, 2015). That said, the 2011 total annual water consumption for fracking in the Barnett shale, the largest play in Texas, is equal to about 9% of the annual water consumption for the city of Dallas (Nicot and Scanlon, 2012).

When hydraulic fracturing is complete and the pressure released, between 10% to 60% of the water injected into the well during the fracking process can discharge back out of the well (known as flowback wastewater) within the first few hours to weeks (Easton, 2014). Water will continue to be produced (produced water) for the life of the wellhead and can account for approximately 30–70% of the volume of the injected fracturing fluids (Ground Water Protection Council, 2009). A portion, sometimes a substantial portion, remains in the subsurface formation and is never recovered.

Water Sources

Water for hydraulic fracturing can be sourced from surface water, groundwater (fresh and saline/brackish), treated wastewater effluent, or water recycling facilities (CERES, 2014). Alternative sources of water for fracking include: treated mine water, waste cooling water from electrical power plants, and storm water form storm water control basins. There is some pending legislation in the US, most recently moving through the Pennsylvania Senate (Bill 875), to encourage the use of treated mine water in drilling operations (Law360, 2015). The legislation would provide immunity from liability to mine operators when they provide water that is used outside of the mine site for O&G development.

Within the US, the primary source of water used in fracking varies from shale play to shale play. Surface water is the preferred source in the Bakken of North Dakota, where groundwater sources can be limited (NDSWC, 2014), and in the Marcellus play where surface water sources are plentiful. In the Marcellus surface water is typically obtained from dedicated pumping stations on local creeks and is transported to the well pads via trucks. In arid, rural areas of the country, groundwater sources tend to make up larger portions of the water supply. Groundwater is the primary source of water to fulfill fracking needs in areas where the availability of surface water is limited. Groundwater sources account for 90% of the water supplied for fracking activities in the Eagle Ford shale play of southwest Texas (Arnett et al., 2014).

Water Stress

Since 2011, nearly half of the hydraulically fractured wells in the US were in regions with high or extremely high water stress. Extremely high water stress, using the World Resource Institute (WRI) definition, means over 80% of available surface and groundwater is already allocated for municipal, industrial, and/or agricultural uses. For example, hydraulic fracturing is a major source of water consumption in the arid Eagle Ford shale region of south Texas, and represents approximately one-quarter of all water consumption in Dimmit County, Texas, located in the heart of the Eagle Ford shale region. Water use for hydraulic fracturing in the Eagle Ford is highest in the US, with water use per well averaging approximately 4.4 million gallons, which is roughly equivalent to the amount of fresh water that New York City uses every 6 minutes (WRI, 2014). Groundwater depletion challenges already exist within the Eagle Ford, where 90% of the water demand for fracturing wells is being provided from groundwater sources. Local aquifer levels have declined from 100 ft. to 300 ft. over the past several decades before the demand from hydraulic fracturing activities (CERES, 2014). Development in the Eagle Ford is expected to continue to expand at a rapid rate with production potentially doubling over the next 5 years (CERES, 2014).

Similarly, groundwater stress has been observed in the Permian Basin of west Texas. Although average water use per well is much lower than in the Eagle Ford, more than 70% of the Permian Basin's wells are in extreme water stress areas (CERES, 2014). This area includes portions of the Permian Basin, an area that overlies parts of the depleted Ogallala Aquifer (High Plains aquifer), which historically has been tapped for agricultural irrigation purposes. Since 1940, the total volume of water in storage in the High Plains aquifer has declined by a volume of water roughly equivalent to two-thirds of the water in Lake Erie or approximately 266 million acre-ft. (Stanton et al., 2011) The rate at which the decline is occurring continues to increase, between 2000 and 2007 the average annual depletion rate was more than twice that recorded during the previous 50 years. The depletion is most severe in Texas, where the water table in the

High Plains aquifer has dropped from 100 ft. to 150 ft. (National Geographic's Freshwater Initiative, 2012). Water use associated with hydraulic fracturing in the Permian Basin has been forecast to double by 2020 (CERES, 2014).

Nearly all hydraulic-fracturing water use in California is located in regions of extremely high water stress. Many of the smaller shale plays throughout the US are also located in high- and extremely high-water stress regions, including the Piceance (western Colorado), Uinta (Utah), Green River (Wyoming), San Juan (New Mexico), Cleveland/Tonkawa (Oklahoma/Texas), and Anadarko Woodford (Oklahoma) basins (CERES, 2014).

One would think that recycling of recovered wastewater would be the normal practice in areas that have both a high volume of hydraulic fracturing activity and high-to-extreme water stress; however, that is most often not the case. The highest rates in the US for reuse of recovered wastewater are found in the wetter Marcellus shale region of Pennsylvania, Ohio, and New York, where water needs for an entire fracking operation are equivalent to only about 17 days of average local rainfall during the drier times of year (Jenkins, 2013). In contrast to the drier regions of the US, however, nearly 80% of recovered wastewater is recycled in the Marcellus. One reason that producers in Pennsylvania are doing more recycling is because the disposal options are more expensive. This is based on the lack of nearby access to deep injection wells for proper disposal of spent water. Therefore, it is more economical to recycle and reuse the water than ship the water out of state to locations in Ohio for disposal via deep-well injection (US EPA, 2014).

Water Transport

Issues related to transportation of shale gas resources, including source water and recovered wastewater, are discussed in detail in Chapter 9. One important issue calling for discussion relates to the large volumes of water and wastewater being transported largely by truck in some areas of the US. One of the public's major complaints regarding fracking is the amount of truck traffic, to and from the well pad and the resulting noise, engine emissions, and the possibility for increased motor vehicle accidents (see Chapters 8 and 9 for further discussion). During the fracking activities, trucks are typically delivering water on a continuous basis in order to keep up with demand. Truck use for transporting water for fracturing wells and then again for ultimate disposal of the returned wastewater far outweighs the trucking needs for site preparation, rig, equipment, and other materials and supplies. This increased demand on roadways is most often in areas that may not have been designed for heavy truck use.

WASTEWATER MANAGEMENT

Returned wastewater refers to all returned fluid, both flowback and produced water, from a hydraulically fractured well.

- Flowback water is a blend of formation water and fracking fluid. Flowback water can vary in composition depending on the original composition of the stimulation water used, the chemicals added to fracturing stimulation water, and the composition of the formation being hydraulically fractured. By the end of the 2–3 week flowback period, the composition of the flowback water is largely the same as the natural formation water (Ramboll Environ, unpublished results, 2014).

- Produced water (sometimes not differentiated from Flowback water) is comprised of the natural water from the geological formation where fracture stimulation is being performed (formation water). The major constituent of produced water is sodium chloride salt. Other significant constituents typically include barium, strontium, and magnesium. Depending on the formation, produced water can also contain significant quantities of radioactive radium, often referred to as naturally occurring radioactive material (NORM) (Ramboll Environ, unpublished results, 2014).

The volume of returned wastewater (flowback and produced) will decrease over time from each individual well and the rate of decrease will vary depending on characteristics specific to the well.

Treatment and Disposal Options for Returned Wastewater

Options for the treatment or disposal of recovered wastewater include: offsite treatment or disposal, disposal via reinjection onsite, evaporation, and treatment and reuse or recycling onsite. Recycling of at least some flowback water has become more commonplace throughout the industry. However, produced water is generally hauled offsite for disposal, typically via deep-well injection, due to the excessively high total dissolved solids (TDS) concentration (12–18%), primarily related to the salt content of the liquid. Each of these options is briefly discussed in the following paragraphs.

OFFSITE TREATMENT/DISPOSAL

Historically in the US, returned wastewaters have been hauled offsite to either publically owned treatment works (POTWs), commercial wastewater disposal facilities, or commercial injection wells. The vast majority of treatments at POTWs and commercial wastewater disposal facilities include:

- Primary treatment consisting of grit removal and primary clarifiers for suspended solids removal.
- Secondary treatment consisting of activated sludge biological treatment for removal of organics and ammonia.
- Disinfection, typically by filtration.

Some POTWs also include tertiary treatment consisting of sand filtration. Many commercial treatment systems use biological methods to treat the water. However, in a POTW or commercial treatment system the large amounts of inorganic salts present in flowback and especially in produced water are essentially just diluted rather than removed. As a result, this practice has been banned in several states in the US.

OFFSITE TREATMENT/REUSE

Centralized treatment of wastewater is emerging as a viable solution in managing water sourcing and wastewater treatment. Centralized treatment facilities handle both the flowback wastewater and produced wastewater from wells within a given region (typically a 40–50 mile radius). Pipelines connect wellheads directly to the central treatment plant. The targeted usage requirements for that wastewater are specified and the wastewater is then processed to meet that usage. Once processed, the wastewater is then piped directly to the targeted well site for reuse. The centralized wastewater management concept is beginning to gain momentum. In North America, over a dozen centralized wastewater treatment facilities servicing shale O&G drilling are now either up and producing, or in development (Easton, 2014).

REINJECTION

In many parts of the US, flowback and produced water generated from the hydraulic fracturing process are disposed off primarily by reinjection into the subsurface by deep-well injection using Class II disposal wells (see Chapter 12 for discussion of State regulations). The Underground Injection Control program defines Class II wells as those that inject fluids associated with oil and natural gas production, primarily salt water (brine). Some examples include:

- *Marcellus shale (Pennsylvania)*: The majority of the fracking related wastewater produced in Pennsylvania is disposed via Class II injection wells located in Ohio. Due to the expense of transport and disposal via injection wells, in the Marcellus Shale play approximately 80% of returned wastewater is recycled for reuse in the drilling and fracking process (Sommer, 2014).
- *Eagle Ford shale (Texas)*: Deep-well injection via Class II injection wells is the primary means of disposal of flowback and produced water (Lyons and Tintera, 2014). Flowback and produced wastewaters from this region are typically trucked to commercial disposal wells in the area.
- *Mississippi Lime play (Kansas)*: Due to the significant amounts of water produced in this play, reportedly 5–20 barrels of water for every barrel of oil produced (Ramboll Environ, personal communication with Kansas Department of Health and Environment, 2015), the Class II disposal wells are colocated with the drilled O&G production wells (more often centrally located to several drilling pads). This allows for the high volume of wastewater produced to be hard-piped to the disposal wells, which has the additional benefit of decreasing truck traffic.

Underground Injection Control regulations require that Class II disposal wells be completed into formations that are isolated from usable quality groundwater and sealed above and below by impermeable rock formations. Criteria used to identify suitable zones for underground injection of hydraulic fracturing derived wastewaters include (Kansas Geological Survey, 2005):

- The presence of thick confining units above the rock formations into which the wastewater will be injected.

- The presence of simple geologic structure, free of faulted and fractured zones that might allow the disposed fluids to migrate.
- Saltwater must be present in the injection zone (no freshwater aquifers).
- Sited in an area of low seismic risk.
- Negligible or low groundwater flow rates in the injection zone.
- A long groundwater flow path to areas of groundwater discharge.

A comprehensive siting study that includes an evaluation of potable aquifers in the area and an evaluation of potentially suitable injection zones is required prior to the installation of a disposal well (40 CFR Part 146 Subpart C). Typically, disposal via deep-well injection does not require treatment of the wastewater prior to injection.

The most significant disadvantage associated with deep-well injection is the possibility of induced seismic activity. Although, due to the energy used to fracture rocks for release of oil or natural gas during hydraulic fracturing activity, there is a possibility of producing microseismic events of a magnitude less than 2.0, this level of seismic activity tends not to be perceived. Felt earthquake activity (generally greater than a magnitude 3.0) resulting directly from the actual hydraulic fracturing process has been confirmed from only one location in the world, a site in England (National Research Council of the National Academies, 2012). However, induced seismic activity at a magnitude of 3.0 or greater has been documented associated with deep-well injection (National Research Council of the National Academies, 2012). Geologic units typically permitted and drilled specifically for the purpose of disposal of oil- and gas-associated wastewater are well below freshwater zones and below production zones. However, fluids injected under pressure in these deeper zones, should they be near a fault, can act as a friction-reducing agent, allowing a fault to move. The Texas Railroad Commission recently ordered four wastewater disposal well operators to conduct testing after a record 4.0 magnitude earthquake shook North Texas (Law360, 2015). The issue of induced seismic activity is discussed in more depth in Chapter 11.

Other disadvantages of using deep-well injection include:

- Transportation costs (discussed earlier in this chapter).
- The continuous need of freshwater for hydraulic fracturing operations (no reuse of returned wastewater).
- The lack of availability of Class II injection wells in some parts of the country. For example, most of the produced water disposed off from operations in the Pennsylvania portion of the Marcellus shale is disposed off in Ohio, requiring the transport of water over hundreds of miles (discussed earlier in this chapter).

EVAPORATION

Evaporation of flowback and produced water is widely practiced in arid areas, where evaporation rates exceed precipitation rates. The diversion of flowback and produced water to large settling basins/impoundments is primarily used in the western US and Australia. Misters, which spray flowback and produced wastewater, are commonly employed to speed up the evaporation process.

Although effective in eliminating the liquid phase of the returned wastewater, this option can result in significant drawbacks. Factors to be considered when using or considering evaporation as a suitable disposal method include:

- *Underlying soil and groundwater*: The permeability of the underlying soil needs to be considered as it relates to migration of potential contaminants into the subsurface. If the underlying soil consists of a highly permeable material (sand), the potential exists that shallow groundwater beneath the evaporation basin could be impacted during the evaporation process. Lining the impoundment with synthetic materials like polyethylene or treated fabric or compacted clay would eliminate the migration pathway but increase the cost of this disposal option.
- *Water collection strategy:* Depending on the spacing of the wells the strategy adopted for water collection may make a difference in the cost viability of using evaporation as a disposal method. If the wells are in close proximity, the use of single or multiple impoundments for each well may be more costly than other disposal options. Using a centralized set of collection impoundments for a cluster of wells in close proximity would be more cost effective and limit potential future liabilities by consolidating waste handling operations.
- *Land availability:* The use of evaporation basin requires significant amounts of land versus other disposal and reuse options that involve the use of tankage for water storage.
- *Sludge disposal:* Even with evaporation, some waste material remains. Evaporation operations can generate significant quantities of waste sludge that could contain coprecipitated radium in sufficient quantities to require special handling and disposal as a radioactive waste.
- *Emissions*: Evaporation basins emit air pollutants derived from both the volatile fraction of the original fracking chemicals and any of the chemical compounds naturally present in the gas (lighter hydrocarbons, hydrogen sulfide, etc.). Potential air emissions could be a significant issue depending on the locations of the basins in relation to populated areas.
- *Potential reuse:* If the evaporation basins are being used for both disposal and storage for potential reuse there are other issues that need to be considered. For example, stored wastewater could be negatively impacted from the addition of solids via windblown sand and dust or the addition of contaminants from surface water runoff.

Some additional siting studies will need to be conducted to evaluate these issues prior to implementing evaporation as a form of disposal.

REUSE

New treatment technologies have made it possible to recycle and reuse the water recovered from hydraulic fracking. Typically, most treatment technologies focus on flowback water treatment due to the high salt content of produced water, which makes recycling very difficult and costly. In most cases, simple filtration of flowback water using mesh layered filters (such as those produced by Pure Filter Solutions (http://www.pfilters.com/)) will generate a sufficient quality of water for reuse as makeup water for the fracturing fluid (Ramboll Environ, unpublished results, 2014). Other technologies that have been evaluated for the treatment of hydraulic fracturing wastewater include use of membranes (reverse osmosis [RO] and nanofiltration systems), electrocoagulation, and evaporation/crystallization. All of these water treatment options serve to separate out cleaner, reusable water from the waste that requires offsite disposal in compliance with all associated applicable regulations. Additional discussion of these technologies is provided in the following paragraphs.

- *Membranes (RO and nanofiltration systems)*: RO and nanofiltration systems use membranes to remove dissolved solids. In these processes, water is pushed across the membranes using high-pressure pumps. The membranes separate dissolved ions and large molecular weight molecules, leaving a permeate stream with a relatively low concentration of dissolved solids. An RO system prevents the passage of nearly all inorganic salts; however, nanofilters allow up to 70% of the monovalent salts (e.g., sodium chloride) to pass through the membrane with the permeate. To avoid clogging the filters, the rejected solids are continuously purged from the system in a concentrate stream known as a reject stream. The volume of this reject stream (the concentrated salts) can be quite high; from 15% to 25% of the influent flow are often seen from treatment of freshwater.
  Drawbacks of using membrane systems are that they can require a significant amount of pretreatment in order to prevent scaling, and fouling, and the large amounts of membrane reject may require additional treatment or special disposal. The largest drawback is that the salinities of produced water typically exceed the pressure capabilities of the membranes, making this option nonviable for treatment of most produced water. RO membranes cannot handle TDS levels of greater than 7%.
- *Electrocoagulation*: Electrocoagulation is conducted by applying an electrical charge to water that then allows suspended matter to form an agglomeration. It effectively removes suspended solids to submicrometre sizes, breaks emulsions such as oil and grease or latex, and oxidizes and eradicates heavy metals from water without the use of filters or the addition of separation chemicals. Within the O&G industry in general, and the hydraulic fracturing world specifically, Halliburton Company (Dallas, Texas) is the only supplier capable of providing this treatment in the field. The drawbacks to this technology are that it is very expensive and it generates significant quantities of waste sludge that could contain coprecipitated radium in sufficient quantities to require special handling as a radioactive waste.
- *Evaporators*: Evaporators treating flowback and produced water create a relatively clean, low-TDS condensate stream. They can be mechanically or thermally driven and also act as concentrators to produce either a syrup or dry salt cake. Drawbacks of using evaporators are largely cost-related.

While evaporators are normally designed to incorporate heat recovery to improve efficiency, they still require significant energy inputs. Evaporators are also typically a significant capital expense. In addition, the concentrate syrup or salt cake would likely require disposal as a radioactive waste due to the concentration of NORM (as discussed previously). When produced water is being treated, evaporators typically require the use of exotic materials due to the significant chloride content.

Chemical characteristics of flowback and produced water that can adversely affect fracking performance include TDS primarily inorganics consisting of sodium chloride salts, multivalent cations such as calcium, barium, iron, and sulfate. Additionally, while it is not an issue in slickwater fracking, boron can cause undesirable cross-linking during gel fracks. Some possible treatment methods to address these chemical characteristics that can cause adverse effects for recycling are discussed here:

- *Scaling agents*: High concentrations of scaling agents (calcium, barium, and strontium), are present in flowback and produced water. Removal of these scaling agents, via chemical addition and precipitation, results in the generation of a significant quantity of waste sludge likely to contain coprecipitated radium in sufficient quantities to require special handling as a radioactive waste. Therefore, most often antiscalant chemicals are added to fracturing fluids that keep these constituents in solution and prevent potential plugging in the well during subsequent fracking phases. An evaluation of scaling potential of recycled flowback and produced water found that, even without any antiscalant chemicals, scaling is not usually an issue. This is true even when using 100% flowback and produced water for fracking. Scaling becomes an issue only when sulfate from freshwater sources is introduced to the mixture. However, even if 100% of the returned wastewater were recycled, there would still be a need for significant quantities of makeup water, most often from freshwater sources. A benefit to using freshwater for makeup water is it decreases the overall content of TDS via dilution.
- *Iron*: Iron can adversely impact friction reducer performance and can lead to fouling in the well. Since iron is readily oxidized, the preferred treatment method is the addition of air or peroxide to oxidize the iron and subsequent removal by simple filtration.
- *Boron*: While it is not an issue in slickwater fracking, the presence of boron can cause excessive or undesired crosslinking during gel fracks. This issue can be remedied using a commercially available ion exchange resin produced by Dow Chemical Company (Midland, Michigan), which serves to selectively remove boron from the flowback and produced water.
- *TDS*: In slickwater fracking, friction reducers are added to the fracturing fluid in order to lower headloss during fracking. The lower headloss translates to a reduced pumping energy requirements and lower pressures in the well casing. Cationic and anionic friction reducers have similar performance at low TDS levels, but cationic friction reducers are effective and

produce acceptable results at higher TDS levels (up to 30%) (Ramboll, Environ unpublished results, 2014). Anionic charged friction reducers, which are normally used, lose their effectiveness as TDS levels increase. Cationic friction reducers are more expensive than anionic friction reducers; however, fracking with higher TDS waters could offset the increased chemical costs by reducing pumping pressures due to increased static head. It has been demonstrated that the use of cationic friction reducers produces acceptable results even with 100% recycled wastewaters (Ramboll Environ, unpublished results, 2014).

There are several options for the treatment of returned wastewater. The selection of feasible options for reuse of hydraulic fracturing wastewaters depends on multiple factors related to the geographic location, volume produced, chemical characteristics of the formation water, and chemical additives present. Potential reuses include: onsite as makeup water for the fracturing fluid, offsite agricultural irrigation, and offsite industrial.

Onsite Use

Flowback water typically requires very little treatment for onsite use as fracturing fluid makeup water as compared with produced water. This of course is dependent on the composition of the formation being fracked and the original composition of the stimulation fluid. Unfortunately, produced water is typically much more saline with corresponding higher TDS than flowback water and therefore requires greater treatment in order to reuse onsite. However, the treatment required for onsite reuse is likely considerably less than that will be required under other reuse scenarios.

As previously discussed, the quality of the water used in fracture stimulation can affect performance of the chemicals added to enhance the fracturing process. The industry has not set standards for initial water quality and the industry standards for acceptable recycle water quality standards continued to evolve. However, as part of a study conducted by West Virginia University on behalf of the United States Department of Energy's National Energy Technology Laboratory (NETL), an industry contact group (ICG) was identified and a detailed questionnaire was sent to the members to obtain information regarding wastewater quality criteria they believed were necessary for recycling (Ziemkiewicz et al., 2012).

The consensus from the ICG concerning acceptable water quality indicated total suspended solids should be below 20 μm to avoid sediment accumulating in the pore space and well bore. The chemical water quality parameter limits listed in Table 5.1 are levels identified by the group at which performance of the fracturing stimulation water is not impeded and scale accumulation is not a concern. A realty check as to how much treatment is required to meet these numbers can be obtained by a review of the average constituent concentrations in flowback water from the Marcellus region, which is also presented in Table 5.1. The average concentrations presented are based on the results from 225 samples collected from 36 separate wells in the Marcellus.

Table 5.1 ICG Chemical Criteria for Fracturing Water Make-up Versus Average Concentrations in Flowback Water from the Marcellus Region

Chemical Parameter	ICG Recommended Maximum Value	Average Concentration in Marcellus Flowback Water
Total hardness (as $CaCO_3$)	26,000	30,077
Total alkalinity (HCO_3)	300	428
Total dissolved solids (TDS)	50,000	109,156
Chlorides (Cl)	45,000	69,315
Sulfates (SO_4)	50	8.4
Calcium (Ca)	8,000	9,861
Magnesium (Mg)	1,200	1,330
Sodium (Na)	36,000	27,617
Potassium (K)	1,000	174
Total iron (Fe)	10	145
Barium (Ba)	10	6,506
Strontium (Sr)	10	4,477
Manganese (Mn)	10	Not analyzed

Note: All concentrations reported in parts per million (ppm).
Sources: Ziemkiewicz et al. (2012) and Ramboll Environ (unpublished results, 2014).

As illustrated in Table 5.1, the flowback water quality from the Marcellus appears to be of relatively high quality. For the vast majority of parameters the concentrations are in line or slightly above the ICG recommended maximum values. Notable exceptions are the scaling agent concentrations, specifically the cations barium and strontium, which are two orders of magnitude greater than the recommended values. Treatment would be required to either remove these scaling agents (precipitation) or to keep them in solution to avoid scaling issues.

The use of lower quality, nonpotable groundwater sources, water with greater than 10,000 ppm TDS, can alleviate many of the concerns related to water supply for other purposes (domestic and agricultural). However, this water is often found deeper than potable water aquifers, and as with recycled flowback water may have additional constituents, such as scaling agents, present at high concentrations that could require treatment prior to use.

Agricultural Use

In the US, over 80 million gallons a day of hydraulic fracturing wastewater is managed under US EPA's Clean Water Act's National Pollutant Discharge Elimination System (NPDES) for beneficial reuses such as agricultural irrigation.

Under the Clean Water Act's Subpart E of 40 CFR Part 435 (44 FR 22075, Apr. 13, 1979, as amended at 60 FR 33967, June 29, 1995) (Clean Water Act, 1995), the NPDES permit system allows for the specialized reuse of wastewater from O&G facilities located in the western half of the United States (west of the 98th meridian). To qualify for this exception, the wastewater must contain less than 35 mg/L of oil and grease and be used either for agriculture or livestock watering (Shariq, 2013) (see Chapter 12).

One of the largest impediments for reusing hydraulic fracturing wastewater in agricultural settings is the salt content, which is too costly to completely remove during treatment. Most agriculture does not tolerate high saline environments well. Therefore, the use of brackish water for irrigation is only recommended in areas that receive a minimum of 10–12 in. or more of rainfall per year. This amount of rainfall allows for the salt to be driven down past the root level, flushing the soil. An exception to this statement is reuse of wastewater in areas where plants thrive in brackish soil.

The ability for crops to incorporate the unique mixture of chemicals found in returned wastewater from hydraulic fracturing activities has not yet been studied extensively. However, arsenic (often a constituent in returned wastewater) has been shown in several studies to bioaccumulate in rice plants and organic hydrocarbons have also been identified in wheat plants grown in contaminated soil.

Industrial Use

The reuse of returned wastewater for industrial purposes depends on the presence of water-demanding industrial operations in the vicinity of the well. Depending on what is being manufactured, some industrial facilities have very specific water requirements and pretreat the water they receive from municipal potable water supplies. Again the primary contaminant of concern is the high salt levels. Therefore, it would be best if the nearby industry could use "dirty water" requiring less treatment. Paper mills are one of the industries that can use water with a high salt content. It is unlikely that industry would pay for the water, but the savings that would be incurred by eliminating the cost for disposal could make this a viable reuse alterative.

Zero Discharge Systems

Traditionally, a zero discharge system (ZDS) would include technologies such as evaporation or deep-well injection, where there is no liquid discharge to a surface body of water. These systems often have other residuals such as solids or concentrated wastes that require disposal. Current ZDS approaches consist of a recycle strategy wherein all of the flowback and produced water are utilized as alternative source water for future fracking activities. The concept behind current ZDSs is to keep all the contaminants in solution to the extent possible, including NORM, and thus minimize waste handling and disposal. The new ZDSs incorporate a closed-loop storage system that uses a series of tanks, including some large capacity storage tanks, and treatment trailers to process the wastewater for reuse.

The operation of current ZDS technology for recycling wastewater involves a minimum amount of treatment, proper storage, logistics, and planning. During well flowback, most current ZDSs treat the flowback water using oxidation and filtration to remove iron and suspended solids, after which it would be stored in a covered tank. Produced water from this and other completed wells, if any, would also be treated and stored in the same tank or set of tanks. A periodic maintenance dosage of a biocide is added to prevent biogrowth in the storage tanks and/or closed-loop system. When the company is prepared to perform the next frack, samples of the stored water along with the makeup water source would be blended and tested for friction reducer performance in order to determine the appropriate blend and chemical dosages to use. A ZDS generates no significant amount of sludge or residue that requires offsite disposal since only iron and boron concentrations in the returned wastewaters need to be controlled. Figure 5.1 shows a schematic of a typical ZDS.

A ZDS can be operated as a completely closed-loop system, eliminating both evaporation concerns and contamination concerns from wind-blown solids. This technology is particularly well suited for scenarios where multiple wells are being drilled and completed in close proximity to one another where a ZDS could be sized accordingly to function as a centralized treatment facility. A centralized system would eliminate the need for treatment onsite and result in substantial cost savings. Industry experts believe that an optimized ZDS approach could lead to improved well completions (i.e., more gas and oil) (Ramboll Environ, personal communication, 2014).

Alternatives to Water-Based Fracking Agents

There are several alternatives to water-based stimulation fluids for hydraulic fracturing that eliminate the water issues addressed in this chapter. The most common and commercially applied alternatives are the use of oil-based (liquefied petroleum gas) and foam-based fracking fluids (JRC, 2013). These alternatives are typically applied for water-sensitive formations. Many shale formations are water-sensitive due in part to the clay content and the ionic composition of the clay. Alternative fracking agents are also applied in environments where water is scarce. However, water scarcity needs to be quite severe to make the economics work. Water is typically a much cheaper alternative and it is better at bringing pressure against, and ultimately breaking up, rock due to the fact that water is "virtually incompressible."

OIL-BASED FLUIDS

Oil-based fracking fluids (diesel) were the first high-viscosity fluids used in hydraulic fracking operations. Liquefied petroleum gas (LPG) has been used as stimulation fluid for approximately 50 years. It was developed for conventional reservoirs before being adapted to unconventional reservoirs. LPG is an abundant byproduct of the natural gas industry and its properties of low surface tension, low viscosity, low density, and solubility within naturally occurring reservoir hydrocarbons have reportedly lead to more effective fracture lengths and the ability to more evenly distribute proppant, resulting in higher production

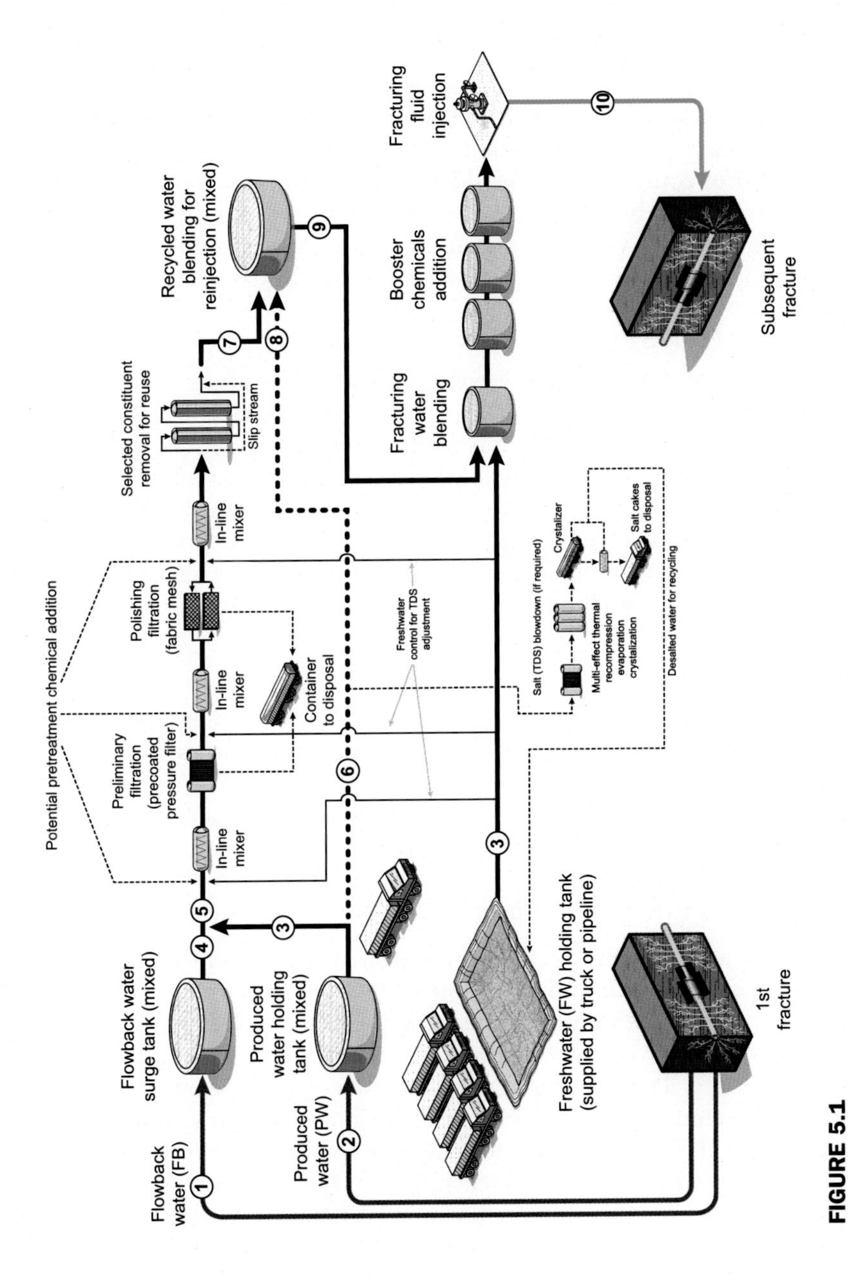

FIGURE 5.1
Zero discharge system. *Copyright 2013 ENVIRON Holdings, Inc.*

from a well. When fracking, the LPG remains liquid/gel, but after completing the process it returns to a gas phase and combines with the reservoir gas. This means that the fracking fluids are totally recovered within days of stimulation, creating economic and environmental advantages by reducing clean-up, waste disposal, and postjob truck traffic (GasFrac, 2013).

The main drawback of this technology is the risks/safety hazards associated with handling large amounts (several hundred tons) of flammable material. It is therefore a more suitable solution in environments with low population density, provided of course that the workers safety can be guaranteed. Investment costs are estimated to be higher because LPG initially costs more than water and is pumped into well at very high pressure requiring specialized equipment, and after each fracking it has to be liquefied again (Rogala et al., 2013).

FOAM-BASED FLUIDS

For water-sensitive formations and environments where water is scarce, foams have long been considered as one of the best fracking fluids (Neill et al., 1964; Komar et al., 1979; Gupta, 2009). Foams tend to be used commercially for fracking in shale gas reservoirs. They require lower (or no) water consumption, cause less damage in water sensitive formations, reduce the amount of chemical additives required, and result in less liquid to recover and handle after the fracking process. In addition to higher costs, a major potential disadvantage is the low proppant concentration in the injection fluid, which results in decreased fracture conductivity.

SUMMARY

Only a portion of the hydraulic fracturing stimulation fluid returns to the surface. A portion, sometimes substantial, remains in the subsurface formation and is never recovered resulting in the need to constantly replenish the water supply to continue fracking operations. This constant demand for water, approximately 2–7 million gallons per well, can create problems particularly in arid areas of high water stress where fresh groundwater resources have declined due to agricultural use prior to the onset of hydraulic fracturing activities in the area.

Although it would seem that recycling of recovered wastewater would be the normal practice in areas with a high volume of hydraulic fracturing activity where high to extreme water stress conditions exist; this is most often not the case. The highest rates for reuse of recovered wastewater are found in the wetter Marcellus shale region of Pennsylvania, Ohio, and New York. Operators, particularly those in arid regions of high water stress, are beginning to use nonpotable water (>10,000 ppm TDS) for drilling and hydraulic fracturing whenever practicable to minimize concerns related to water supply for other purposes (domestic and agricultural). However, this nonpotable water is often found deeper than potable water aquifers, and may have additional constituents, such as scaling agents, present at high concentrations that could require additional treatment prior to use.

Recently, recycling of at least some flowback water has become more commonplace throughout the US. However, produced water is generally still hauled offsite for disposal, typically via deep-well injection (Class II disposal wells), due to the excessively high TDS concentration primarily related to the salt content. Although use of deep-well injection is currently the most common disposal practice, issues related to this practice including induced seismic activity and transportation concerns, may drive more companies to seriously consider the reuse and recycling of returned wastewater.

New treatment technologies have made it possible to recycle and reuse the water recovered from hydraulic fracking. Typically, most treatment technologies focus on flowback water treatment due to the high salt content of produced water, which makes recycling it very difficult and costly. Technologies that have been evaluated for the treatment of hydraulic fracturing wastewater include use of membranes, electrocoagulation, and evaporation/crystallization. All of these water treatment options serve to separate out cleaner, reusable water from the waste that requires offsite disposal. Potential reuses include onsite as makeup water for the fracturing fluid, offsite agricultural irrigation, and offsite industrial; with onsite reuse by far the most common. Approximately 80% of the returned wastewater in Marcellus shale is recycled for reuse as hydraulic fracturing makeup water. The reused wastewater comprises approximately 18% of the injected volumes in the Marcellus as compared with approximately 5% in the Barnett shale in Texas. The driving factor for reuse in the Marcellus is likely due to the small number of disposal wells in Pennsylvania (US EPA, 2014) and the associated high costs for transport and disposal via deep-well injection in Ohio.

Water-free fracking still remains an early stage technology and is not currently widely used in the industry. Use of alternative methods is beginning to take off in the arid regions of Texas where competition for water is becoming more critical. As costs for water increase and regulators continue to focus on the large quantity of water used for fracking wells, the concept of water-free fracking will likely gain popularity.

The most promising technologies appear to be the ZDS approach wherein all of the flowback and produced water is utilized as an alternative source water for future fracking activities. The ZDS concept is to keep all the contaminants in solution to the extent possible, including NORM, and thus minimize waste handling and disposal. The systems incorporate a closed-loop storage system that uses a series of tanks, including some large-capacity storage tanks, and treatment trailers to process the wastewater for reuse. Although ZDS methods have higher initial costs than some alternative disposal methods, over time zero discharge recycling systems will provide operational cost savings based on the deferred cost of water acquisition and minimization of disposal costs.

References

Arnett, B., Healy, K., Zhongnan, J., LeClere, D., McLaughlin, L., Roberts, J., Steadman, M., 2014. Water Use in the Eagle Ford Shale: An Economic and Policy Analysis of Water Supply and Demand. A Report to Commissioner Christi Craddick, Texas Railroad Commission.

CERES, 2014. Hydraulic Fracturing and Water Stress: Water Demand by the Numbers. CERES Report, February 2014.

Clean Water Act. Subpart E, Agricultural and Wildlife Water Sue Subcategory, 40 CFR Part 435, 44 FR 22075 April 13, 1979, as amended at 60 FR 33967, June 29, 1995.

Easton, J., 2014. Environmental Science & Engineering Magazine, Centralized Treatment of Fracking Wastewaters Becoming a Viable Solution.

GasFrac. 2013. A completely closed system with automated remote operations. Available from: http://www.gasfrac.com/safer-energy-solutions.html

Ground Water Protection Council, ALL Consulting, 2009. Modern Shale Gas Development in the United States: A Primer.

Gupta, S., 2009. Unconventional Fracturing Fluids for Tight Gas Reservoirs. SPE Hydraulic Fracturing Technology Conference. Society of Petroleum Engineers, The Woodlands, Texas.

Jenkins, J., 2013. Energy Facts: How Much Water Does Fracking for Shale Gas Consume? http://www.theenergycollective.com/jessejenkins/205481/friday-energy-facts-how-much-water-does-fracking-shale-gas-consume

JRC, 2013. An overview of hydraulic fracturing and other formation stimulation technologies for shale gas production, JRC Technical Reports, Report EUR 26347 EN, 2013. https://ec.europa.eu/jrc/sites/default/files/an_overview_of_hydraulic_fracturing_and_other_stimulation_technologies_(2).pdf

KGS, 2005. Kansas Geological Survey. http://www.kgs.ku.edu/Publications/Bulletins/ED10/06_wells.html

Komar, C.A., Yost II, A.B., et al., 1979. Practical Aspects of Foam Fracturing in the Devonian Shale. SPE Annual Technical Conference and Exhibition. Las Vegas, Nevada, Not subject to copyright. This document was prepared by government employees or with government funding that places it in the public domain.

Law360, New York (June 22, 2015, 6:56 PM ET). Legislation moved through the Pennsylvania Senate Environmental Resources and Energy Committee Monday to encourage the use of treated mine water in natural gas drilling operations (Senate Bill 875).

Lyons, B., Tintera, J.J., 2014. Sustainable Water Management in the Texas Oil and Gas Industy. Atlantic Council Energy and Environmental Program.

National Geographic's Freshwater Initiative, 2012. Texas Water District Acts to Slow Depletion of the Ogallala Aquifer.

National Research Council of the National Academies, 2012. Induced Seismicity Potential in Energy Technologies.

NDSWC, 2014. Facts about North Dakota Fracking and Water Use, North Dakota State Water Commission, February 2014. http://www.swc.nd.gov/4dlink9/4dcgi/getcontentpdf/pb-2419/fact%20sheet.pdf

Neill, G.H., Dobbs, J.B., 1964. Field and Laboratory Results of Carbon Dioxide and Nitrogen in Well Stimulation.

Nicot, J.-P., Scanlon, B.R., 2012. Water use for shale-gas production in Texas, US Environ. Sci. Tchnol. 46, 3580–3586.

Rigzone, 2013. Fracking Goes Waterless: Gas Fracking Could Silence Critics, September 26, 2013. http://www.rigzone.com/news/oil_gas/a/129261/Fracking_Goes_Waterless_Gas_Fracking_Could_Silence_Critics

Rogala, A., Krzysiek, J., et al., 2013. Non-aqueous fracturing technologies for shale gas recovery. Physicochem. Prob. Mineral Process. 49 (1), 313–322.

Shariq, L., 2013. Uncertainties associated with the reuse of treated hydraulic fracturing wastewater for crop irrigation. Environ. Sci. Technol. 47 (6), 2435–2436.

Sommer, L., 2014. With Drought, New Scrutiny Over Fracking's Water Use. http://blogs.kqed.org/science/audio/with-drough-new-scrutiny-over-frackings-water-use/

Stanton, J.S., Qi, S.L., Ryter, D.W., Falk, S.E., Houston, N.A., Peterson, S.M., Westenbroek, S.M., Christenson, S.C., 2011. Selected Approaches to Estimate Water-Budget Components of the

High Plains, 1940 through 1949 and 2000 through 2009. USGS Scientific Investigations Report 2011–5183.

Triepke, J., 2014. Well Completion 101 Part 3 Well Simulation. http://info.drillinginfo.com/well-completion-well-stimulation/

USEPA, 2014. Assessment of the Potential Impacts of Hydraulic Fracturing for Oil and Gas on Drinking Water Resources.

USEPA Office of Research and Development, 2015. Assessment of the Potential Impacts of Hydraulic Fracturing for Oil and Gas on Drinking Water Resources. EPA/600/R-15/047c.

World Resources Institute (WRI), 2014. The Facts on Hydraulic Fracturing and Water Use.

Ziemkiewicz, P., Hause, J., Lovett, R., Locke, D., Johnson, H., Patchen, D., 2012. Zero Discharge Water Management for Horizontal Shale Gas Well Development, June 2012. West Virginia Water Research Institute, FilterSure, Inc., ShipShaper, LLP.

CHAPTER 6

A Primer on Litigation That Involves Alleged Water Well Contamination From Hydraulic Fracturing

Craig P. Wilson, Anthony R. Holtzman
K&L Gates LLP, Harrisburg, PA, USA

Chapter Outline

In the United States, there is a long history of litigation involving allegations that business operations caused contamination of residential water wells. Litigants have claimed that leachate from landfills entered their water wells, making the landfill operators liable.[1] They have claimed that hazardous substances escaped from underground storage tanks and migrated into their water wells, making the storage tank owners liable.[2] And, in cases that go back over a century, they have claimed that operations at oil and gas (O&G) well sites have caused brines to appear in their water wells, making the well operators liable.[3] Some of these litigants have been successful with their claims; others have not.

Building on this history, litigants today have renewed their focus on the O&G industry, with an emphasis, in particular, on its use of modern-day hydraulic fracturing techniques to extract natural gas from subsurface rock formations (see process discussion in Chapter 1). These techniques have enabled gas development projects to multiply in number and expand into new areas of the country, generating a variety of economic benefits. However, they have also led to an

[1] See, for example, *Jennett v South Macomb Disposal Auth.*, 2004 WL 2533649 (Mich. Ct. App. Nov. 9, 2004); *Artesian Water Co. v. Gov't of New Castle County*, 1983 WL 17986 (Del. Ch. Aug. 4, 1983).
[2] See, for example, *Felton Oil Co., L.L.C. v. Gee*, 182 S.W.3d 72 (Ark. 2004); *Exxon Corp. v. Yarema*, 516 A.2d 990 (Md. Ct. Spec. App. 1986).
[3] See, for example, *Collins v. Chartiers Val. Gas Co.*, 18 A. 1012 (Pa. 1890).

Environmental and Health Issues in Unconventional Oil and Gas Development. http://dx.doi.org/10.1016/B978-0-12-804111-6.00006-6

increase in litigation, with parties claiming that as a result of gas well drilling and hydraulic fracturing activities, their water wells have been contaminated with a variety of substances.

Although each of these cases is different from the others, they tend to involve common factual allegations, causes of action, requests for relief, and evidentiary issues. They also raise several distinctive legal and factual issues, some of which the courts have not yet settled in full. This chapter explores these matters.

FACTUAL ALLEGATIONS

In a typical pleading that is based on allegations of hydraulic fracturing-related water well contamination, the plaintiffs name multiple entities as defendants. Those entities typically include the owner of the gas well site and the well operator, and may include affiliates of the owner or operator, well site service providers (e.g., drilling and equipment rental companies), and even companies that manufacture or distribute chemicals that are used in the hydraulic fracturing process.

In making factual allegations against the defendants, the plaintiffs often lump them together, as a single unit, as opposed to being specific about which of them allegedly did which things at what times, or which allegations go with which of the legal claims. They also typically make conclusory allegations. The core allegation, for example, is often that the "defendants" (as a unit) drilled, constructed, hydraulically fractured, and operated gas wells in a way that "caused" groundwater contamination. The allegations typically contain little or no detail about the way that the defendants conducted these operations or how it allegedly caused the contamination.

Although these methods of pleading are arguably improper, courts have been hesitant to dismiss claims or cases (or require repleading) for that reason.[4]

In most cases, moreover, the plaintiffs assert that their allegedly contaminated groundwater supply is relatively close to the gas well operations. The distances range from less than 1000 ft. to several miles. They also contend that the contaminants in the groundwater include methane, ethane, minerals, and named and unnamed "hazardous substances" – which, in many cases, is a reference to chemicals found in hydraulic fracturing fluids (see eg., Chapter 4).

The plaintiffs also typically claim that as a result of the contamination of their groundwater supply, they have sustained a host of injuries, often including lost property value, "annoyance and discomfort," personal injuries, emotional distress, costs of remediating the water supply, costs of future monitoring of their health and physical well-being, or some combination of these concerns.

[4] See, for example, *Fiorentino v. Cabot Oil & Gas Corp.*, 750 F.Supp.2d 506, 513 (M.D. Pa. 2010) (turning aside argument that "Plaintiffs merely recite the elements of [their] claim in a conclusory fashion, rather than asserting facts that would satisfy the plausibility standard for a motion to dismiss").

CAUSES OF ACTION

As with their factual allegations, the plaintiffs' legal theories in these types of cases are often similar from one case to another.

The plaintiffs, for example, often assert statutory claims, alleging that by contaminating a groundwater supply, the defendants are liable under a statute that is designed to prevent and facilitate the remediation of releases of hazardous substances into the environment. An example of this type of statute is the Pennsylvania Hazardous Sites Cleanup Act (or HSCA), which is Pennsylvania's counterpart to the federal Comprehensive Environmental Response, Compensation, and Liability Act (or CERCLA). Sections 702 and 1101 of HSCA create a cause of action to recover response costs that were incurred in addressing releases of hazardous substances.[5] And Section 1115 of HSCA allows for "citizens' suits" to prevent or abate violations of the statute[6] (also see Chapter 12).

The plaintiffs also typically assert a claim for negligence. In order to prevail on that claim, they must establish that the defendants owed them a duty of care and breached that duty.[7] Generally speaking, in other words, they must show that in conducting their gas well operations, the defendants did not act in a reasonably prudent manner under the circumstances.[8] The plaintiffs also must establish that, as a result, they were injured (i.e., their groundwater supply was contaminated) and that the defendants' breach was both the actual cause of the injury (i.e., the injury would not have occurred but-for the breach) and the "legal," or proximate, cause of the injury (i.e., the breach was a "substantial factor" in bringing about the injury).[9] Under the rubric of negligence *per se*, moreover, the plaintiffs can establish that the defendants owed them and breached a duty of care by showing that the defendants violated a statutory or regulatory provision that was designed to protect them against injuries like the ones that they allegedly sustained.[10]

In addition, based on the notion that hydraulic fracturing activities are "*abnormally dangerous*," the plaintiffs often assert a claim for *strict liability* for abnormally dangerous activities. In order to prevail on that claim, they must generally show that, in light of the following six factors, the defendants' hydraulic fracturing activities were, in fact "abnormally dangerous":

1. the activities created a high degree of a risk of harm;
2. it was likely that the harm would be great;
3. it was not possible to eliminate the risk through the exercise of reasonable care;

---

[5] See 35 P.S. §§ 6020.702 & 6020.1101; see also *In re: Joshua Hill, Inc.*, 294 F.3d 482, 485 (3d Cir. 2002) (reciting elements for "prima facie case of liability under [Sections 702 and 1101] of HSCA").
[6] See 35 P.S. § 6020.1115.
[7] See Restatement (Second) Torts § 328A (1965).
[8] See Restatement (Second) Torts § 298 (1965).
[9] See Restatement (Second) Torts §§ 328A & 431 (1965).
[10] See Restatement (Second) Torts § 286 (1965).

4. the activities were not a "matter of common usage";
5. the activities were inappropriate for the area where they were conducted; and
6. the activities' value to the community was outweighed by their dangerous attributes.[11]

In addition, the plaintiffs must show that the hydraulic fracturing activities caused their alleged injuries and that those injuries are the kind that, because they were possible, made the activities abnormally dangerous in the first place.[12] If the plaintiffs satisfy all of these requirements, the defendants are liable *regardless* of how much care they exercised under the circumstances (i.e., regardless of whether they were actually negligent).[13]

Private nuisance claims are also typical. In order to prevail on a private nuisance claim, the plaintiffs must establish that as a result of the defendants' gas well operations, they suffered an invasion of the private use and enjoyment of their property (i.e., their groundwater supply) that was either intentional and unreasonable or unintentional but otherwise actionable under the rules of negligence or strict liability for abnormally dangerous activities.[14]

The plaintiffs, in addition, typically assert a claim for *trespass*. In order to prevail on that claim, they must establish that the defendants' gas well operations led to an unprivileged, intentional intrusion on their groundwater supply.[15]

Finally, the plaintiffs frequently assert a claim for medical monitoring. To succeed on that claim, they must show, in essence, that due to the defendants' operations, they were exposed to greater than normal levels of a hazardous substance, which caused them to experience a significantly increased risk of getting a serious latent disease, and that a medical monitoring regime exists that is reasonably necessary for potential early detection of the disease.[16]

REQUESTS FOR RELIEF

In a typical case of this type, the plaintiffs assert that if they prevail on one or more of their causes of action, the court should grant them various forms of relief.

The plaintiffs, for example, typically ask for compensation for injuries to their property that they allege resulted from the alleged water well contamination. In many states, if an injury to real property is "reparable," the measure of compensation is the cost of repairs, up to the diminution in value of the property.[17] In other states, the measure of compensation is "the diminution in the rental or

[11] See Restatement (Second) Torts § 520 & 520, cmt. (l) (1977).
[12] See Restatement (Second) Torts § 519 (1977).
[13] Id. at § 519(1).
[14] See Restatement (Second) Torts § 822 (1979).
[15] See Restatement (Second) Torts § 158 & 158, cmt. (c) & (e) (1965).
[16] See, for example, *Redland Soccer Club, Inc. v. Dep't of the Army*, 696 A.2d 137, 145–146 (Pa. 1997).
[17] See 22 Am. Jur. 2d Damages § 276; Restatement (Second) Torts § 929, cmt. (b) (1979).

usable value of the realty during the time span of the damage."[18] And, in some states, the plaintiffs can recover both types of compensation.[19]

If, on the other hand, the injury to real property is "permanent," the measure of compensation is typically the decrease in the fair market value of the property.[20]

In any event, if the court holds the defendants liable, the plaintiffs may generally recover damages for any annoyance and discomfort that they experienced as a result of the injury to their property.[21]

The plaintiffs also typically ask for compensation for personal injuries and emotional distress that they allege resulted from the contamination. The plaintiffs are generally entitled to recover compensatory damages for personal injuries, including medical expenses, if the court finds that the defendants are liable for those injuries.[22] In many states, however, the plaintiffs do not fit into one of the categories of people who may recover damages for "negligent infliction of emotional distress." In Pennsylvania, for example, the plaintiffs cannot recover these damages unless the defendants owed them a special contractual or fiduciary duty, they were physically impacted by the defendants' conduct, they were in a "zone of danger" and at a risk of immediate physical injury due to the defendants' conduct, or they witnessed a tortious injury to a close relative due to the defendants' conduct.[23] In Pennsylvania and most other states, moreover, the plaintiffs must prove that they exhibited physical manifestations of their alleged distress, for example, rashes or vomiting.[24] In a typical case, they cannot meet these requirements.

Another common form of requested relief is a *medical monitoring trust fund*. This remedy is available to the plaintiffs only if they prevail on a claim for medical monitoring. In that situation the court orders the defendants to finance a trust fund that the plaintiffs may use to cover the expenses of a monitoring procedure that is designed to determine whether, as a result of being exposed to elevated levels of a hazardous substance, they have developed a latent disease.[25]

The plaintiffs, in addition, often ask for *punitive damages*. In most states, they can recover those damages only if they prevail on one of their claims and prove that, in contaminating their water supply, the defendants acted in an intentionally malicious or recklessly indifferent manner.[26]

---

[18] See 22 Am. Jur. 2d Damages § 276.

[19] Id.

[20] See 22 Am. Jur. 2d Damages § 273.

[21] See Restatement (Second) Torts § 929(c) (1979).

[22] See generally Restatement (Second) Torts § 905 (1979).

[23] See, for example, *Doe v. Philadelphia Cmty. Health Alternatives AIDS Task Force*, 745 A.2d 25, 27 (Pa. Super. Ct. 2000).

[24] See Restatement (Second) Torts § 436A, cmt. (b) (1965).

[25] See, for example, *Redland Soccer Club*, 696 A.2d at 142 n.6 (noting that the trust fund "compensates the plaintiff for only the monitoring costs actually incurred").

[26] See Restatement (Second) Torts § 908, cmt. (b) (1979).

And the plaintiffs often seek compensation to cover the costs of their attorneys' fees for the litigation. Although, for most of their claims, this type of relief is not available, if they prevail on a statutory claim, the statute may enable them to obtain reimbursement from the defendants for the attorneys' fees that they incurred in prosecuting it.[27]

EVIDENTIARY ISSUES

In nearly every one of these cases, the key factual question is whether the defendants' gas well drilling or hydraulic fracturing activities *caused* (viz. *causation*) contaminants to enter the plaintiffs' water well. In order to prevail on *any* of their causes of action, in fact, the plaintiffs must demonstrate that there was a link between the gas well operations and the presence of contaminants in the well.[28] In attempting to make this showing, the plaintiffs must normally rely on a complex body of evidentiary materials and the inferences that can sometimes be drawn from them.

The evidence that is relevant to this question includes, for example, evidence of whether, based on regulatory requirements and industry standards, the gas well was designed properly, making it less likely to experience structural soundness or mechanical integrity problems (including leakage). This evidence may include, for example, well drawings, diagrams, and schematics, as well as testing regarding applicable standards and geologic conditions (see Chapters 4 and 12).

Also relevant to the causation question is evidence of whether the gas well was properly drilled, cased, cemented, and hydraulically fractured, lessening the chances that any substance escaped from it. This evidence may include well logs, drilling, and casing records and pressure test results for the well. It may likewise include documentation of any steps that were taken to correct any structural problems with the well that arose or were discovered after it was constructed.

Evidence of background conditions also bears upon the issue of causation. This evidence may include the results of any testing of the plaintiffs' water supply that occurred before the gas well was drilled. It may likewise include reports of known, predrill water well conditions in the general area where the plaintiffs' water supply is located (see eg., Chapter 4). The idea, of course, is that it is more difficult to prove a causal connection between the gas well operations and the

[27] Section 1115(b) of HSCA, e.g. provides that, in a "citizens' suit" under the statute, the court "may award litigation costs, including reasonable attorney and witness fees, to the prevailing or substantially prevailing party whenever the court determines such an award is appropriate." 35 P.S. § 6020.1115(b).

[28] See, for example, *Burnside v. Abbott Laboratories*, 505 A.2d 973, 978 (Pa. Super. Ct. 1985) ("An essential element of any cause of action in tort is that there must be some reasonable connection between the act or omission of the defendant and the injury suffered by the plaintiff."); see also *Acushnet Co. v. Coaters Inc.*, 937 F. Supp. 988, 1000 (D. Mass. 1996) ("forms of strict liability" related to abnormally dangerous activities "are still a part of the law of torts, and they do not dispense with the requirements of cause in fact and proximate cause – elements traditionally a part of every action in tort"); Restatement (Third) Torts: Physical & Emotional Harm § 28(a) & cmt. (a) (2010) (causation is element of any cause of action in tort).

presence of contaminants in a water supply if the contaminants were present in the water supply before the drilling began.

Isotopic and chemical analyses are also important. Using expert reports and testimony, the plaintiffs often try to demonstrate, for example, that their water well contained gas (methane, in particular) that had the same isotopic "fingerprint" as the gas that was produced from the gas well. Or they try to show that their water well contained chemicals that matched some of the ones that were used in the hydraulic fracturing process. These analyses are scientific and often highly complex (see eg., Chapter 5).

Apart from these evidentiary issues that relate to causation, there are typically evidentiary issues concerning alleged injury to property. Whether and to what extent any proven contamination of the plaintiffs' water well amounts to an injury to the plaintiffs' property, and whether the injury is reparable or permanent, often turns on the reports and expert testimony of qualified property appraisers.

And, if the plaintiffs are seeking compensation for personal injuries, that factor triggers questions about whether their alleged injuries can, and did, result from being exposed to the contaminants that they cite. From an evidentiary perspective, medical records, including records of the plaintiffs' predrill health status, and expert testimony from medical practitioners are important to addressing these questions.

DISTINCTIVE LEGAL AND FACTUAL ISSUES

Several distinctive, and to some extent unsettled, legal and factual issues have emerged from these types of cases.

Hydraulic Fracturing as an Abnormally Dangerous Activity

One such issue is whether hydraulic fracturing is an abnormally dangerous activity that can give rise to "strict" tort liability – that is, liability in the absence of negligence. In *Ely v. Cabot Oil & Gas Corporation*, the US District Court for the Middle District of Pennsylvania addressed this issue, concluding that hydraulic fracturing is not abnormally dangerous.[29]

In *Ely*, the plaintiffs alleged that a gas well operator and its affiliated service company engaged in operations at a gas well site that caused contamination of a number of groundwater supplies. Based on these allegations, they asserted claims against the operator and service company for HSCA relief, negligence, private nuisance, strict liability for abnormally dangerous activities, breach of contract, fraudulent misrepresentation, medical monitoring, and gross negligence. At the summary judgment stage of the case, the defendants asked the court to rule that hydraulic fracturing is not an abnormally dangerous activity. The court granted the request.

The court applied the six factors, mentioned earlier, for determining whether an activity is abnormally dangerous. It reasoned that "[t]he risks from a properly

[29] See 38 F.Supp.3d 518 (M.D. Pa. 2014).

drilled, cased and hydraulically fractured gas well are minimal."[30] It also explained that "[s]uch risks are substantially mitigated when due care is exercised."[31] It said that it was not willing to "credit the Plaintiffs' suggestion that natural gas drilling and hydraulic fracturing, is a 'novel' activity."[32] Nor, it said, could it embrace the plaintiffs' "assertion that wells drilled in accordance with valid leases and which were permitted by the Commonwealth's environmental regulatory body, and which otherwise complied with legal requirements with respect to setback limits, were nevertheless placed inappropriately."[33] And it catalogued a number of the economic benefits of hydraulic fracturing.

Although the US District Court for the Middle District of Pennsylvania concluded that hydraulic fracturing is not an abnormally dangerous activity, its decision is not controlling beyond the parties to that case. In the future, it and other courts throughout the country might reach the opposite conclusion.

Arbitration Versus Court

Another distinctive issue is whether the plaintiffs' claims should be decided in an arbitration proceeding, as opposed to a court.

In some cases, one or more of the plaintiffs have entered into an O&G lease with one or more of the defendants. The lease, moreover, typically calls for all disputes over damages that were allegedly caused by the "lessee's operations" to be resolved in arbitration. The plaintiffs, however, allege that the operations that caused their damages were not conducted on the leased premises, but instead on a nearby property, and therefore are not covered by the arbitration provision in the lease. In this situation, if the parties cannot agree on a forum, a court must determine whether the parties to the lease intended for "lessee's operations" to include activities that occurred away from the leased premises. In at least one case, the court, based on a detailed examination of the lease, concluded that they did not.[34]

A related issue is whether, if a court *does* send the claims to arbitration, any plaintiffs or defendants who are not parties to the O&G lease must nevertheless litigate the claims in the arbitration proceeding. Generally speaking, if the nonleasing parties have a sufficiently close relationship with the leasing parties (due to principles of agency law, corporate control, or the like) they must join the arbitration.[35] Otherwise, they need not.

---

[30] Id. at 529.

[31] Id. at 531.

[32] Id. at 532.

[33] Id.

[34] See *Stiles v. Chesapeake Appalachia, LLC*, No. 1346 MDA 2012 (Pa. Super. Ct. Slip Op. June 17, 2014).

[35] See, for example, *E.I. DuPont De Nemours and Co. v Rhone Poulenc Fiber and Resin Intermediates, S.A.S.*, 269 F.3d 187, 195–202 (3d Cir. 2001) (describing situations in which non-signatories to contract that contains arbitration clause may be bound by the clause); *Thomson-CSF, S.A. v. Am. Arbitration Ass'n*, 64 F.3d 773, 776 (2d Cir. 1995) ("we have recognized five theories for binding non-signatories to arbitration agreements: (1) incorporation by reference; (2) assumption; (3) agency; (4) veil-piercing/alter ego; and (5) estoppel").

The arbitration proceedings *versus* court proceedings issue is notable because each of those forums presents different potential advantages and drawbacks to litigants. In an arbitration proceeding, for example, one or more arbitrators preside and there is not a jury. The absence of a jury trial can be advantageous to the parties because in cases of this type, a jury could easily struggle with technical concepts (e.g., concepts of causation) and misunderstand the burden and standard of proof. In an arbitration proceeding, moreover, the rules of evidence are not strictly applied, which allows for evidence to be more thoroughly aired than in court. Arbitration proceedings, in addition, tend to be much faster than court proceedings, with litigants permitted to conduct only limited prehearing "discovery" of the evidence in one another's possession. And, once an arbitration proceeding is decided, the litigants may appeal the decision for only a handful of reasons (e.g., claimed fraud or bias in the arbitration process), which diminishes the likelihood of time-consuming appeals.

In a trial court, on the other hand, a judge presides and the parties have the right to a jury trial. While the proceedings are generally slower than arbitration proceedings, they involve much broader pretrial "discovery," which makes it easier for the litigants to ferret out evidence in support of their claims and defenses. During the trial, moreover, the rules of evidence are strictly applied, which (at least theoretically) prevents unreliable forms of evidence from infiltrating the decision-making process. And, once the trial court decides the case (through a jury verdict or otherwise) the decision, unlike an arbitration decision, may be challenged on appeal for essentially any facially legitimate reason – including that it is based on a legal or factual error. This broad right to appeal enhances fairness in the litigation process, but can lengthen the process.

Lone Pine Orders

In the trial court setting, cases of this type also raise the interesting question of whether and, if so, when the court should issue a "*Lone Pine order.*"

A Lone Pine order takes its name from a 1986 decision of the New Jersey Superior Court.[36] A trial court typically issues such an order during the discovery phase of a complex mass or toxic tort case in which the balance of the discovery process will be burdensome and there is a reason to believe that, during the trial, the plaintiffs will not be able to prove their claims. The order establishes that the case will not go forward unless the plaintiffs can make an initial evidentiary showing in support of their claims, including the element of causation.

A Colorado trial court issued an order of this type in *Strudley v. Antero Resources Corporation.*[37] In that case, the plaintiffs alleged that a gas well operator and some of its service providers undertook activities at a gas well site that caused contamination of nearby air, water, and soil. Based on these allegations, the plaintiffs asserted claims for negligence, negligence *per se*, nuisance, strict liability, and

---

[36] See *Lore v. Lone Pine Corp.*, 1986 WL 637507 (N.J. Super. Ct. Nov. 18, 1986).
[37] See 2012 WL 1932470 (Denver Co. Dist. Ct. May 9, 2012).

medical monitoring. Early in the discovery phase of the case, after the parties had made certain initial disclosures to one another, the trial court issued a *Lone Pine* order. In doing so, it noted the "significant discovery and cost burdens presented by a case of this nature" and that "ultimately [the plaintiffs] would need to come forward" with "expert opinions in order to establish their claims."[38] It also cited a Colorado regulatory agency's administrative determination that "plaintiffs' well water was not affected by O&G operations."[39]

In response to the Lone Pine order, the plaintiffs were unable to make an initial evidentiary showing in support of their claims, including causation. The trial court, therefore, dismissed the claims with prejudice.

On appeal, however, the Colorado Court of Appeals reversed the dismissal.[40] It determined that "[t]he circumstances surrounding the case were not shown to be so extraordinary as to require departure from the existing [procedural] rules" and entry into the use of a Lone Pine order.[41] "[T]he *Lone Pine* order," the court explained, "interfered with the full truth-seeking purpose of discovery" because "[a]lthough the initial disclosures provided the [plaintiffs] with some information related to their claims, the disclosed information was insufficient to enable them to respond fully to the *Lone Pine* order."[42] The court also noted that, unlike most cases in which courts have issued *Lone Pine* orders, "this was not a mass tort case," since it involved only four plaintiffs, four defendants, and one parcel of land.[43]

The Colorado Supreme Court, in turn, upheld the Court of Appeals' determination.[44] It reasoned that, unlike the Federal Rules of Civil Procedure, Colorado's rules of civil procedure do not "contain a grant of authority for complex cases or otherwise afford trial courts the authority to require a plaintiff to make a prima facie showing before the plaintiff fully exercises discovery rights under the Colorado Rules."[45] Allowing for trial courts to issue *Lone Pine* orders, it said, would: "interfere with the rights provided to litigants" and "produce consequences unintended by our rules by forcing dismissal before affording plaintiffs the opportunity to establish the merits of their cases."[46]

It is possible, going forward, that other trial courts will issue *Lone Pine* orders in cases of this type. To date, some trial courts have shown a reluctance to issue them. It remains to be seen how prevalent they will become and whether, and under what circumstances, they will be upheld on appeal.

---

[38] Id.

[39] Id.

[40] 350 P.3d 874 (Colo. Ct. App. 2013).

[41] Id. at 883.

[42] Id. at 881.

[43] Id. at 882.

[44] 347 P.3d 149 (Colo. 2015).

[45] Id. at 156.

[46] Id. at 159.

Causation

Perhaps the most prominent factual issue that these cases raise is the extent to which the plaintiffs will be able to prove a causal connection between the defendants' gas well drilling or hydraulic fracturing activities and groundwater contamination.

The plaintiffs' burden of proof is to meet the "preponderance of the evidence" standard, which is the equivalent of "more likely than not" inquiry.[47] And importantly, the plaintiffs can satisfy this burden by pointing to circumstantial (or second hand) as opposed to direct evidence of a causal connection.

The reliance on circumstantial evidence is illustrated by the Pennsylvania Superior Court's decision in *Hughes v Emerald Mines Corporation*.[48] In that case, the court upheld a jury's finding that a coal company's grouting operations had caused groundwater contamination, even though the finding was based solely on circumstantial evidence:

> Admittedly, the ground was not laid bare in an effort to trace the flow of grout at a depth of some ninety feet over an area of 600 feet of intervening ground.... There was plenty of evidence, however, from which a reasonable jury could deduce causation: the former twenty-five years of uninterrupted water supply; the physical proximity of the grouting operation to the ruined water wells; the timing of defendant's completion of its operations shortly before the resulting failure of the wells; the grout-like substance observed by witnesses in the muck removed from the well-bottoms; [and] the similar damages sustained by other well owners in the near vicinity at the same time.[49]

To date, many of the modern-day cases that are based on allegations of hydraulic fracturing-related water well contamination, if they haven't been settled by agreement, are still in the pretrial phase and no court, it appears, has found this type of causal connection. In several arbitration proceedings, however, the plaintiffs have assembled circumstantial evidence that led the arbitrator to conclude that problems with a gas well's mechanical integrity led to groundwater

[47] See, for example Restatement (Second) Torts § 433B, cmt. (b) (1965).

[48] 450 A.2d 1, 6 (Pa. Super. Ct. 1982).

[49] Id., see also *Reinhart v. Lancaster Area Refuse Auth.*, 193 A.2d 670, 672 (Pa. Super. Ct. 1963) ("Although there was no direct proof that any of the [fill] material actually dumped" by the defendants "found its way underground into plaintiffs' wells, the proximity of the operation to the wells, the depth of each, the nature of the fill as compared with the nature of the contamination, the time of fill in relation to the time the pollution was first noticed, and the results of a dye test to eliminate other causes were, we think, sufficient, circumstantially, to support the jury's finding that the landfill was the cause of plaintiffs' damage."); *Sunray Mid-Continent Oil Co. v. Tisdale*, 366 P.2d 614, 615 (Okla. 1961) (addressing whether oil well drilling caused groundwater contamination and stating: "We have held that negligence may be established by circumstantial evidence."); *Harper-Turner Oil Co. v. Bridge*, 311 P.2d 947, 951 (Okla. 1957) (noting that, while evidence of causal connection between gas well drilling and groundwater contamination was "to a great extent circumstantial, we regard it as sufficient to require the submission of the case to the jury").

contamination. That said, neither a court nor an arbitrator has connected the hydraulic fracturing process, in and of itself, to groundwater contamination and, while researchers have explored that issue in several recent studies, none of them has demonstrated such a connection.[50]

CONCLUSION

In the United States, the advent of modern-day hydraulic fracturing techniques has been accompanied by a number of lawsuits in which the plaintiffs allege that those techniques and gas well drilling, generally, have caused contamination of their residential water wells. These cases are the latest chapters in a long history of litigation involving allegations that business operations caused contamination of groundwater. Across the cases, the plaintiffs are advancing similar allegations, causes of action, and requests for relief and, therefore, triggering similar evidentiary issues. The cases, moreover, implicate a number of novel and complex legal and factual questions. Many of the cases are in their early stages and the theories and arguments continue to evolve. Only time will tell how the rest of the legal case history unfolds.

[50] See, for example, US Environmental Protection Agency, *Assessment of the Potential Impacts of Hydraulic Fracturing for Oil and Gas on Drinking Water Resources, External Review Draft* (June 2015) at ES-15: ("Numerical modeling and microseismic studies based on a Marcellus Shale-like environment suggest that fractures created during hydraulic fracturing are unlikely to extend upward from these deep formations into shallow drinking water aquifers."); Llewellyn, G.T., et al., *Evaluating a Groundwater Supply Contamination Incident Attributed to Marcellus Shale Gas Development, Proceedings of the National Academy of Sciences of the United States of America*, Vol. 112, No. 20 (April 2, 2015) (determining that groundwater contamination was likely caused by a casing problem in the gas well or surface release from a leaking wastewater pit).

CHAPTER 7

Occupational Health and Safety Aspects of Oil and Gas Extraction

Eric J. Esswein*, **, Kyla Retzer, Bradley King**, Margaret Cook-Shimanek†**
*University of the Witswatersrand, School of Public Health, Johannesburg, South Africa; **National Institute for Occupational Safety and Health (NIOSH), Western States Division, Denver, CO, USA; †University of Colorado at Denver and Health Sciences Center, Denver, CO, USA

Chapter Outline

WORKPLACE FATALITIES AMONG OIL AND GAS EXTRACTION WORKERS

Overview

The oil and gas (O&G) extraction workforce has an annual fatality rate seven times higher than all United States (US) workers (Mason et al., 2015). Manual labor, long work hours around heavy machinery, multiple contractors working on well sites simultaneously, and the constant movement of people and equipment (mainly on rural roads) contribute to elevated occupational fatality rates in this industry. Historically, the fatality rate for O&G extraction workers correlated with level of activity (e.g., number of active drilling rigs) (CDC, 2008). However, recent research published by the Centers for Disease Control and Prevention's National Institute for Occupational Safety and Health (CDC NIOSH) showed

Environmental and Health Issues in Unconventional Oil and Gas Development. http://dx.doi.org/10.1016/B978-0-12-804111-6.00007-8

that, despite a twofold increase in the number of workers and a 71% increase in active rigs during 2003–2013, there was a 36% decrease in the fatality rate (Mason et al., 2015).

Fatalities by Company Type and Establishment Size

O&G extraction fatality rates vary by type and size of company (Mason et al., 2015; Retzer et al., 2011). Using the North American Industrial Classification System (NAICS), worker deaths are classified by: (1) operators, who own or lease and operate the well, (2) well drilling contractors, and (3) servicing companies, who provide a variety of services completing and repairing the well. Drilling contractors have the highest fatality rate (44.6 per 100,000 workers), followed by servicing companies (27.9 per 100,000 workers). Operators have the lowest rate (11.6 per 100,000 workers) (Mason et al., 2015). Differences in fatality rates are likely explained by varying exposures to hazardous operations involved in drilling, completions, and servicing activities.

Consistent with findings in other industries, the smallest establishments (20 or fewer workers) have the highest fatality rates (Retzer et al., 2011). Small drilling contractors have fatality rates seven times greater than the industry as a whole. Small drilling contractors may employ older style rigs that have fewer engineered safety controls. Companies with 10 or fewer workers are also exempt from many federal occupational safety and health (OSH) regulations and may lack the resources to employ full-time safety and health professionals to provide safety and health training, implement policies and procedures, or cultivate a safety and health culture that is typically incorporated into larger companies.

The O&G extraction industry as a whole is exempt from several occupational safety and health regulations, including portions of the following Occupational Safety and Health Administration (OSHA) standards: process safety management of highly hazardous chemicals, benzene, hearing conservation, and lockout/tagout. The regulatory environment for this industry will be discussed in more detail in Chapter 12.

Transportation – The Leading Cause of Death to Workers

Transportation incidents are the leading cause of death among O&G extraction workers, accounting for nearly 40% of all work-related fatalities (Mason et al., 2015; Retzer et al., 2013). During 2003–2013, 479 workers died on the job in transportation incidents. The largest proportion of transportation incidents involved motor vehicle crashes, though some of the deaths are associated with aircraft and boats. During 2003–2013, there was a downward trend in the transportation fatality rate (Mason et al., 2015) (See Figure 7.1). However, it still remains the leading cause of death among O&G extraction workers.

A previous comparison of the motor vehicle fatality rates across industries found that O&G extraction had a rate only slightly lower than that for workers in the transportation and warehousing industry (7.6 vs. 9.3 deaths per 100,000 workers) (Retzer et al., 2013). Possible explanations for this elevated rate include frequent travel between well sites, driving on rural roads that lack

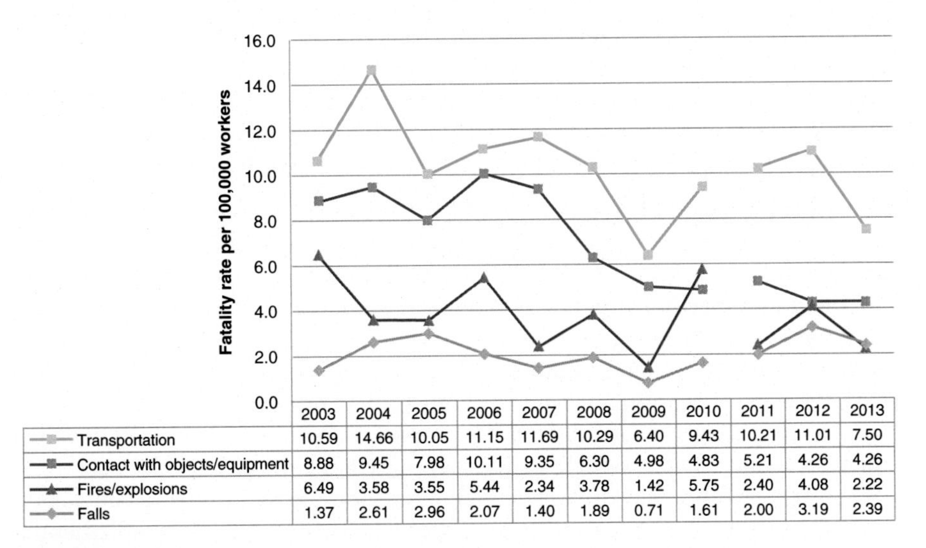

	2003	2004	2005	2006	2007	2008	2009	2010	2011	2012	2013
Transportation	10.59	14.66	10.05	11.15	11.69	10.29	6.40	9.43	10.21	11.01	7.50
Contact with objects/equipment	8.88	9.45	7.98	10.11	9.35	6.30	4.98	4.83	5.21	4.26	4.26
Fires/explosions	6.49	3.58	3.55	5.44	2.34	3.78	1.42	5.75	2.40	4.08	2.22
Falls	1.37	2.61	2.96	2.07	1.40	1.89	0.71	1.61	2.00	3.19	2.39

FIGURE 7.1
US O&G extraction worker fatality rates, leading causes of death, 2003–2013.
The break in 2011 represents revisions in the coding system. *Sources: National Institute for Occupational Safety and Health (NIOSH), using data from Bureau of Labor Statistics Census of Fatal Occupational Injuries and Quarterly Census of Employment and Wages.*

standard safety features, lack of safety belt use, extended and irregular work hours that contribute to driver fatigue, and truck traffic associated with the high volume of sand and water needed for hydraulic fracturing (Retzer et al., 2013; CDC, 2008; Mode and Conway, 2007). A majority of drivers who died in motor vehicle crashes were driving light duty vehicles (<10,000 lbs), mostly pick-up trucks, which are typically not regulated requiring the operator to have a commercial driver license (Retzer et al., 2013).

Besides work-related vehicle crashes, commuting to-and-from remote job sites has been identified by industry as a particular concern. There is no national surveillance system that tracks commuting related motor vehicle crashes. However, O&G extraction companies are beginning to incorporate off-duty motor vehicle safety as a part of their health, safety, and environment (HSE) programs because it is recognized that many motor vehicle fatalities occur off duty. As the geographical areas where large numbers of wells are being drilled shifts over time, often to more rural or remote areas, workers need to commute longer distances. Once their rotation (typically 2 weeks) is complete, workers often begin their commute home immediately, but are often fatigued from long shifts for multiple consecutive days (Rothe, 2008).

Research of Canadian oil field workers reported that fatigue and falling asleep at the wheel are common during long commutes to and from rural well sites (Rothe, 2008). Crash studies in the general population involving noncommercial and commercial vehicles have shown an increase in crash risks associated with decreased sleep duration (Hanowski et al., 2007; Connor et al., 2002; Cummings et al., 2001). In the US, a number of motor vehicle fatality reports for O&G extraction workers indicate drivers fell asleep while driving (Retzer et al., 2013). More research is needed on fatigue and its effect on worker safety and health. Factors likely affecting O&G extraction workers include inadequate sleep duration and poor sleep quality, long work days and commutes, and temporary lodging facilities with suboptimal sleeping environments.

One of the strategies being employed by O&G extraction companies to reduce risk to workers who drive are in-vehicle monitoring systems (IVMS). While the features of IVMS products vary widely, most systems monitor basic driving behavior, including speeding events, harsh braking events (an indicator of distraction), and harsh accelerations (an indicator of aggressive driving). Some monitors incorporate use of cameras, seatbelt monitors, and electronic logging of hours. Driver behavior reports are typically transmitted from an IVMS to designated persons for review on a regular basis. Companies report that use of IVMS helped them to reduce motor vehicle incident rates when combined with coaching of drivers who exhibit risky driving behavior. A review of O&G industry papers published on IVMS showed 50–90% reductions in crash rates after implementing IVMS (Retzer et al., 2013). Health and safety professionals who have used IVMS also report the ability to target their resources on drivers who need assistance to improve driving behavior, rather than costly company-wide initiatives. They also report that IVMS can help companies foster the concept of "driver social responsibility" on public roadways by endeavoring to avoid any negative impacts of O&G extraction in communities near well sites (International Association of Oil and Gas Producers, 2014a,b). However, an evaluation of the attitudes and opinions of truck drivers on the use of IVMS revealed that some drivers have concerns regarding privacy issues as well as the complexity, reliance, and reliability of the technology (Huang et al., 2005). The ultimate intent of IVMS is to safeguard employees when driving and not endanger other drivers on the road.

A second intervention proving useful for companies involves journey management (JM), a planned and systematic process of reducing transportation-related risks for a company's operations (Retzer et al., 2014). JM programs question the necessity of all trips and minimize the risks associated with necessary trips. Sophisticated JM software uses technology to anticipate, identify, and avoid hazardous road conditions. JM programs are widely implemented among operators but are just becoming incorporated into smaller US-based companies.

To reduce motor vehicle incidents in the O&G extraction industry worldwide, the IOGP released a Land Transport Safety Recommended Practice, #365. This guidance provides companies with the essential components of an effective land transport safety program. It is based on the best practices within industry and includes multiple tools for program implementation.

Other Leading Causes of Death

Contact/struck by/caught-in machinery injuries are the second leading cause of death to O&G extraction workers. Despite a lack of systematic, published epidemiological, or system safety analysis research attributing reasons for a recent reduction in the rate of contact injury deaths in drilling operations (Mason et al., 2015), engineering controls (e.g., new drilling rigs with automated pipe handling features) that remove workers from hazardous tasks are likely a factor. A 2014 NIOSH study compared injury rates on contemporary drilling rigs (e.g., automated pipe handling, alternating current motor drives) to older, manual rigs, and found a significant reduction in incidence of severe injuries for floormen and other workers whose tasks require their presence on the rig floor (Blackley et al., 2014). One of the most common types of fatal contact injury on drilling rigs is being struck by falling objects, primarily drill pipes or "tubulars" that weigh many hundreds of pounds each and are hoisted above the drilling floor and assembled together to form a drill string thousands of feet long.

Another leading cause of death is falling from height. Workers commonly work at height above the rig floor and also in the derrick. Adequate fall protection is required to ensure safety, but failures of anchors or guarding, or worker failure to anchor properly, can result in falls leading to fatalities.

NIOSH Fatalities in O&G Database

NIOSH and its partners (the National Occupational Research Agenda [NORA] Oil and Gas Sector Council) maintain a surveillance system called the "Fatalities in Oil and Gas Extraction (FOG) Database" to better understand factors involved in O&G extraction worker deaths (http://www.cdc.gov/niosh/topics/fog/). FOG helped identify nine deaths associated with tank gauging (discussed in next section) that could have been overlooked in traditional data sources. Topics of concern identified through FOG also include deaths from hydrogen sulfide exposure (also discussed in next section) and deaths due to dropped objects. Another trend identified through FOG is the increased number of fires or explosions associated with hot work or welding on produced water tanks. Produced water (i.e., brine or salt water) is not classified as hazardous material by the Department of Transportation and may not be placarded on tanks as hazardous, a fact which can leave workers unaware of the flammable/explosive and possibly toxic contents (hydrocarbons) that may be in these tanks (see discussion on transportation of O&G in Chapter 9). Lastly, FOG is able to track fatalities by specific operations and recently found that rig-up and rig-down tasks on drilling and workover rigs resulted in more deaths than any other operation in 2014.

Because accurate and comprehensive information is not available, FOG does not include injury or illness data. More work in the area of health surveillance is needed for this worker population.

RISKS FOR CHEMICAL EXPOSURES IN OIL AND GAS EXTRACTION WORKERS

Overview

There is a lack of systematic, peer-reviewed, industrial hygiene studies for land-based O&G extraction workers (NIOSH, 2010). This section describes the most common hazards and risks. It is also a call-to-action for industrial hygienists and safety professionals to understand that significant knowledge gaps exist. Workplace exposure assessment research is needed to understand the scope and magnitude of exposure risks and to prioritize controls to protect workers. This review addresses land-based O&G extraction. Offshore work typically involves stationary, fixed work areas unlike the dynamic and highly mobile nature of land-based operations. Risks for land-based operations differ because controls are typically designed and engineered into offshore platforms. Some exposure assessment studies exist for offshore worker; and implementation of administrative controls are easier for the "captive cohort" nature of offshore workers.

Hydrogen Sulfide

Hydrogen sulfide (H_2S or "sour gas"), is well known as a serious workplace exposure hazard in O&G extraction. Health effects associated with H_2S exposure range from acute and chronic respiratory and eye irritation to neurological effects and death from pulmonary edema or oxygen deficiency. Virtually all O&G extraction workers receive awareness-level training in recognition of H_2S hazards and the presence of olfactory warning properties (i.e., smell) can help with hazard identification at very low concentrations. Hydrogen sulfide occurs in crude oil and natural gas from the decay of organic matter (e.g., kerogen) in sedimentary formations. O&G wells vary greatly in H_2S content across basins and formations; consequently, exposure risks vary widely. Exposures to H_2S can occur during drilling, completions, and servicing operations. H_2S emission can also occur from water stored or brought on site contaminated with sulfite-reducing bacteria. Sulfur dioxide, produced from the combustion of H_2S, can also be an exposure hazard for workers in proximity to incomplete H_2S combustion. Precautions, such as flaring or use of H_2S scavenging equipment ("gas busters"), personal protective equipment, and H_2S personal monitors are used at wells with known H_2S contamination. Twenty-two fatalities in the O&G industry were identified from acute exposures to H_2S from 1984–1994 (Fuller and Suruda, 2000). During 2005–2014, there were nine H_2S fatalities identified from the NIOSH FOG database. A Canadian questionnaire study of 175 O&G extraction workers reported acute and chronic respiratory irritant symptoms. Some workers reported loss of consciousness from exposure to H_2S (Hessel et al., 1997).

Hydrocarbon Gases and Vapors

O&G extraction involves risks for inhalation and dermal exposures to hydrocarbons, including, but not limited to naphthalene, benzene, toluene, ethylbenzene, and xylenes (collectively known as N-BTEX), propane, pentane, butane, cyclohexane, methyl-cyclohexane, and n-heptane. Exposures to hydrocarbons can affect the eyes, lungs, and central nervous system. Hydrocarbon emissions from production tanks, if significant enough, can also lead to oxygen deficiency through simple displacement and generation of flammable atmospheres. A recently identified risk for fatalities in O&G extraction workers is from the combined acute effects of hydrocarbon exposures and oxygen deficient atmospheres during manual gauging and fluid collection at crude oil production tanks (NIOSH/OSHA, 2015). Nine worker deaths were identified from 2010–2014 involving manually gauging tanks ($n = 4$) or collecting fluid samples ($n = 5$) from open production tank hatches. Low molecular weight hydrocarbons and benzene (an occupational carcinogen) were identified in the blood of several decedents; exposure to H_2S was ruled out in these nine cases. One worker wore a multigas monitor that reached a nadir of 6.9% oxygen around the approximate time of death (normal concentration is 20.5%). This is consistent with the finding that very high concentrations of hydrocarbon gases and vapors can displace ambient oxygen and create a toxic environment in which one to two breaths leads to a limited duration of useful consciousness (Miller and Mazur, 1984). NIOSH issued science blogs about the hydrocarbon-related deaths and associated hazards (King et al., 2015; Snawder et al., 2014) and Hazard Alerts were developed by NIOSH and OSHA and the National Service, Transmission, Exploration & Production Safety (STEPS) Network (2015) in conjunction with its partners.

A Canadian study evaluated 1,547 full-shift and short-term area and personal breathing zone (PBZ) samples that were voluntarily submitted by five companies involved in O&G extraction, oil processing, and pipeline work in the province of Alberta (Verma et al., 2000). The highest risks for benzene exposures involved conventional gas extraction. Less than 1% of the exposures exceeded the province of Alberta occupational health criterion at the time of the study (1 part per million [ppm]) and 5% of samples were greater than the province of Alberta short-term exposure limit (STEL) of 5 ppm for benzene. A retrospective assessment (including interviews and modeling of monitoring and exposure assessment data) of benzene exposures in the Australian Petroleum Industry focused mainly on workers in midstream and downstream activities (transport and refinery), but did provide an estimate of exposure risk in upstream operations using 34 reference points. A mean benzene exposure of 0.05 ppm (range 0.001–0.06) was reported; risks for skin exposures were not evaluated (Glass et al., 2000).

Worker exposures to xylenes and other hydrocarbons can occur during O&G extraction, including work around mud pumps, shale shakers, and drill cuttings. Flowback operations can also present exposure risks when working around ponds, pits, and flowback tanks. Malaysian researchers reported that xylene is used as a solvent in well stimulation to dissolve accumulated organic plugs of paraffins from wellbores (Zoveidavianpoor et al., 2012). The researchers hypothesize that task-based activities may result in "higher exposures" but provide scant details relating to specific occupational exposure levels. A literature review of potential risks for exposures to hydrocarbons in drilling fluids reported that derrickmen, mud engineers, roughnecks, motormen, and laboratory supervisors are at risk from hydrocarbon exposure and, depending on exposure levels, potential health effects (Broni-Bediako and Amorin, 2010). Areas of exposure included the drilling floor, shale shakers, mud pits, sack room (mud mixing house), laundry facilities, and deck operational areas. In addition to inhalation, risks for dermal exposures, dermatitis, and skin irritation were also described. The presence of sensitizers (e.g., polyamine emulsifiers) and corrosive chemicals (e.g., zinc bromide) in drilling fluids is mentioned. No quantitative exposure assessment data were provided.

A Norwegian study at a drilling fluids test center investigated oil mist and vapor generation at shale shakers (Steinsvåg et al. 2011). Mist, vapour, and total volatile hydrocarbons were measured under different temperature parameters and base oils of differing viscosities (i.e., n-paraffin alkanes with boiling points of 210–260 °C and petroleum distillate base oil with boiling points of 250–325 °C). Results indicate statistically different mist and vapor concentrations between the two different drilling fluids. Oil mists generated using higher boiling point petroleum-based fluids were approximately two times higher than oil mists generated when using the lower boiling point alkane-paraffin fluid. Oil mists and vapor at shale shakers exceeded the Norwegian exposure limits of 1 mg/m^3 of air for oil mist, and 50 mg/m^3 for oil vapor, when fluid temperatures were greater than 50 °C. The researchers report that oil mist and vapors from shale shakers are difficult to control, but suggest that exposures could be reduced by cooling drilling fluids to less than 50 °C before they enter the shale shaker units, enclosing shale shakers and related equipment, and carefully considering which fluid system to use.

NIOSH researchers reported 17 workers gauging flowback and production tanks had higher risks for benzene exposures (full-shift TWA = 0.25 ± 0.16 ppm), compared with 18 workers at the same sites who did not gauge tanks (full-shift time-weighted average (TWA) = 0.04 ± 0.03 ppm [Figure 7.2]). The results showed that 2 of the 17 samples met or exceeded the American Conference of Governmental Industrial Hygienists (ACGIH) unadjusted benzene threshold limit value (TLV®) of 0.5 ppm and 6 of 17 samples exceeded an adjusted TLV of 0.25 ppm for the 12-h shift. Some task-based samples for benzene also exceeded the NIOSH short term exposure limit (STEL) for benzene (1 ppm as a 15-min

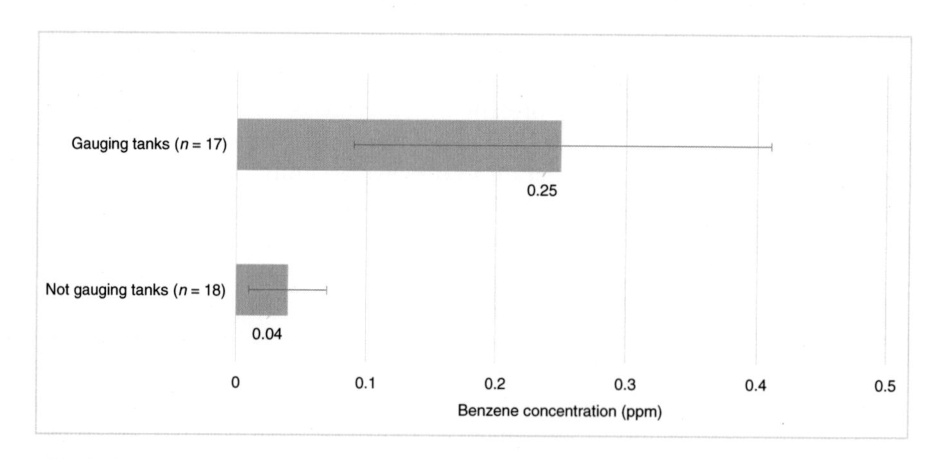

FIGURE 7.2
Time-weighted average, personal breathing zone benzene concentrations for workers gauging or not gauging tanks. *Source: Esswein et al. (2014).*

TWA). Direct-reading instruments detected peak benzene concentrations greater than 200 ppm at tank hatches. None of 35 full-shift PBZ samples exceeded the OSHA permissible exposure limit (PEL) for benzene of 1 ppm for general industry or 10 ppm for the O&G drilling, production, and servicing operations who are exempt from the OSHA benzene standard. Exposures to other measured hydrocarbons (e.g., toluene, ethyl benzene, and xylenes) did not exceed any established occupational exposure limits (Esswein et al., 2014).

Diesel Particulate Matter

Diesel engines are common on drilling, completions, and well-servicing sites. Diesel particulate matter (DPM) is emitted from diesel engines as a complex aerosol containing vapor and solid phase hydrocarbons, sulfates, oxides of nitrogen, and elemental and organic carbon particles. DPM is respirable and, therefore, capable of entering the gas exchange region of the lungs. Depending on the duration and magnitude of exposure, DPM can irritate the eyes and upper respiratory system, causing cough and phlegm production, and can exacerbate preexisting asthma (Pronk et al., 2009). Certain polyaromatic compounds in DPM have been determined by NIOSH to be occupational carcinogens (IARC, 2013). Neither OSHA, NIOSH, nor ACGIH have an occupational exposure limit for DPM, but the state of California, Department of Health Services, has an occupational exposure limit for DPM of 20 $\mu g/m^3$ of air as a TWA, referenced as elemental carbon, which is a surrogate for DPM exposures (CDHS, 2002). No published DPM exposure assessment research for O&G extraction workers was identified in the literature. NIOSH researchers conducted air sampling/exposure assessments for DPM from 2008–2012 at drilling, hydraulic-fracturing, well-servicing, and rig move operations. In this study, 104 air samples (48 PBZ and 56 area) were collected and analyzed for elemental carbon as a surrogate for DPM. Area air samples ranged from below the limit of detection (<LOD) to 68 $\mu g/m^3$ as a TWA; PBZ results were in a range of <LOD to 52 $\mu g/m^3$ as a TWA (authors' unpublished results). Exposure risk factors for DPM are believed to include: number, type, horsepower, and duration of operating diesel engines; controls, if any (e.g., low-sulfur fuels, engine filtration, the presence of diesel organic exhaust catalyst); temporal and spatial aspects of engine location relative to workers and workstations; wind direction and velocity; and weather conditions, such as inversions.

Respirable Crystalline Silica

Occupational exposures to respirable crystalline silica are associated with the development of silicosis, lung cancer, pulmonary tuberculosis, and airways diseases. Silicosis is a debilitating and often fatal lung disease that is preventable. Respirable crystalline silica exposures may also be related to the development of autoimmune disorders, chronic renal disease, and other adverse health effects (NIOSH, 2002). NIOSH researchers collected 111 full-shift PBZ samples for respirable crystalline silica at 11 hydraulic fracturing sites in 5 states over a 15-month period. Workers in 15 different job titles participated in the exposure assessment study (Esswein et al., 2013). Silica-containing dusts were visibly present at all sites

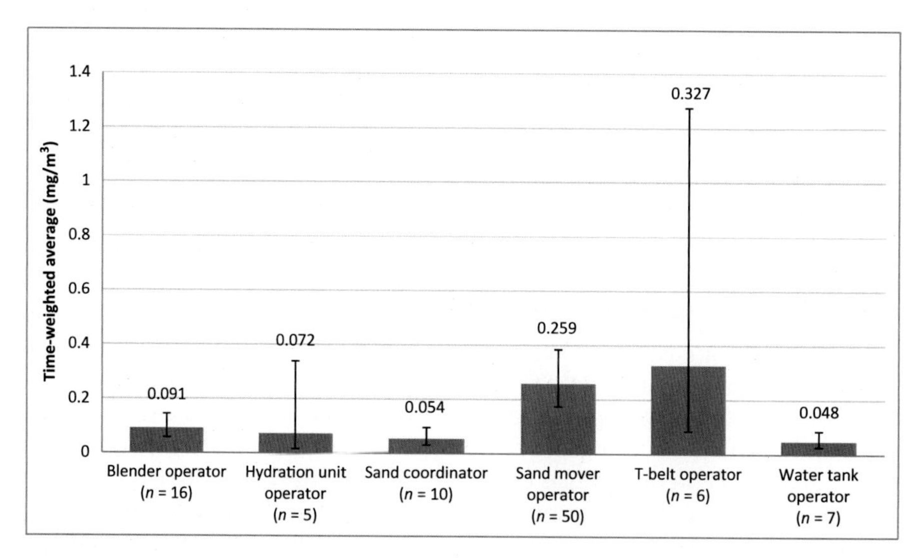

FIGURE 7.3
Respirable crystalline silica, time-weighted average geometric means (mg/m³) and 95% confidence intervals for 6 job titles of workers conducting hydraulic fracturing operations.
Source: Esswein et al. (2013).

during pneumatic transfer of sand. Sand mover and transfer belt operators had the greatest risk of exposure, exceeding the calculated OSHA PEL and the NIOSH recommended exposure limit (REL). The study reported: (1) 57 of 111 samples (51.4%) exceeded a calculated OSHA PEL, which varies based on the percentage silica in the sample, (2) 76 of 111 (68.5%) samples exceeded the NIOSH REL of 50 μg/m³ as a TWA, and (3) 93 of 111 (83.8%) exceeded the TLV of 25 μg/m³ established by the ACGIH. Mean exposure severities (exposure divided by occupational exposure limit), as compared with the NIOSH REL, ranged from less than 1 for 7 job titles (e.g., data van, pump truck, roving and wireline operators, and quality assurance/quality control (QA/QC) techs), to 10.44 for sand mover operators and 14.55 for transfer belt operators. A severity of "1" is equal to the respective occupational exposure criterion. Mean severities of exposures, compared with OSHA calculated PELs, ranged from < 1 for the same 7 job titles and 5.66 and 7.62 for sand mover operators and transfer belt operators, respectively. Workers with these same two job titles were determined to have the highest exposure risks for silica exposures based on calculated geometric means (Figure 7.3). Seven point sources of silica dust generation at hydraulic fracturing sites were identified and a variety of controls were recommended to limit crystalline silica emissions from the sources identified.

Metals

Risks for exposures to metals can occur from welding, cutting, grinding, repair and fabrication of machines and parts, and hardbanding or "hard facing" of tubulars to deter abrasion from drill pipe handling machinery. Hazards from fume and particulates can include parent or filler metals (e.g., chromium, cobalt, copper, iron, manganese, molybdenum, nickel, tungsten, vanadium, zirconium). No studies describing risks for workplace exposures to metals in O&G extraction during hardbanding, welding, fabrication, and repair operations were identified in the literature.

Six cases of elevated blood lead levels (BLLs) in workers and children from occupational and take-home exposures to lead-containing "pipe dope" were reported by the Oklahoma Childhood Lead Poisoning Prevention Program (Khan, 2011). The lead was reportedly from lead-containing pipe dope used as an assembly lubricant during well string makeup during drilling operations. Four children of three workers involved in drilling and well servicing were reported to have BLLs ranging from 18–22 μg/dL. Two of the workers associated with the children with elevated BLLs had BLLs of 29 and 39 μg/dL; the third worker did not get a BLL test. The CDC defines elevated BLLs as greater than 5 μg/dL in children and ≥10 μg/dL in adults (CDC, 2013). Environmental evaluations in the homes of the workers identified "elevated" lead surface concentrations in a laundry room, atop washing machines and furniture, and on work clothing and footwear.

Government and Industry Safety and Health Initiatives

Multiple government and industry efforts to address OSH in O&G extraction have been implemented in recent years. The National Service, Transmission, Exploration & Production Safety Network (STEPS) (www.nationalstepsnetwork.org) was established in South Texas in 2003 to focus on safety and health issues in the industry and to reduce injuries and fatalities in that region. STEPS has grown to 22 independent regional networks serving 20 states and has become the most effective tool for disseminating key safety and health information throughout the industry. The OSHA Oil and Gas Safety and Health Conference, the first national conference dedicated to the safety and health of workers in the oilfield, was created in 2008 and is held every two years. In addition, a standardized worker safety orientation called SafeLandUSA was developed. The SafeLandUSA orientation was incorporated into requirements for contract workers to access and work at well sites owned mostly by O&G operators. A new course in hazard recognition and regulatory health and safety standards for onshore O&G exploration and production operations called OSHA 5810 has been developed and is offered through OSHA Training Institutes and Education Centers throughout the US. There were also Federal Government efforts to address OSH concerns in this industry, including establishment of the CDC NIOSH National Occupational Research Agenda (NORA) Oil and Gas Sector Council (see Chapter 12). This program emphasizes collaborative efforts that led to partnerships with industry, government, and other stakeholders to conduct research in reducing the rate of injury and disease among workers in the O&G extraction industry by determining hazards and quantifying risks and ultimately implementing effective interventions. This program analyzes surveillance data and information gained from industry, workers, and safety organizations, conducts exposure assessment research on key safety and health issues, and develops and implements practical workplace solutions. OSHA has also initiated local, regional, and national emphasis programs specifically targeting aspects of the O&G extraction industry.

Disclaimer: The findings and conclusions in this report are those of the author(s) and do not necessarily represent the views of NIOSH.

References

Blackley, D.J., Retzer, K.D., Hubler, W.G., Hill, R.D., Laney, A.S., 2014. Injury rates on new and old technology oil and gas rigs operated by the largest United States onshore drilling contractor. Am. J. Ind. Med. 57, 1188–1192.

Broni-Bediako, E., Amorin, R., 2010. Effects of drilling fluid exposure to oil and gas workers presented with major areas of exposure and exposure indicators. Res. J. Appl. Sci. Eng. Tech. 2 (8), 710–719.

California Department of Health Services (CDHS), 2002. Health Hazard Advisory: Diesel Engine Exhaust. Hazard Evaluation System and Information Service, Occupational Health Branch, Oakland, CA. Available from: http://www.cdph.ca.gov/programs/hesis/Documents/diesel.pdf

Centers for Disease Control and Prevention (CDC), 2008. Fatalities among oil and gas extraction workers – United States, 2003–2006. MMWR 57 (16), 429–431.

Connor, J., Norton, R., Ameratunga, S., Robinson, E., Civil, I., Dunn, R., Bailey, J., Jackson, R., 2002. Driver sleepiness and the risk of serious injury to car occupants: population based case control study. Br. Med. J. 324 (7346), 1125–1129.

Cummings, P., Koepsell, T., Moffat, J., Rivara, F., 2001. Drowsiness, counter-measures to drowsiness, and the risk of motor vehicle crash. Inj. Prev. 7 (3), 194–199.

Esswein, E.J., Breitenstein, M., Snawder, J., Kiefer, M., Sieber, W.K., 2013. Occupational exposures to respirable crystalline silica during hydraulic fracturing. J. Occup. Environ. Hyg. 10 (7), 347–356.

Esswein, E.J., Snawder, J., King, B., Breitenstein, M., Alexander-Scott, M., Kiefer, M., 2014. Evaluation of some potential chemical exposure risks during flowback operations in unconventional oil and gas extraction: preliminary results. J. Occup. Environ. Hyg. 11 (10), D174–D184.

Fuller, D.C., Suruda, A.J., 2000. Occupationally related hydrogen sulfide deaths in the United States from 1984 to 1994. J. Occup. Environ. Med. 42 (9), 939–942.

Glass, D.C., Adams, G.G., Manuell, R.W., Bisby, J.A., 2000. Retrospective exposure assessment for benzene in the Australian petroleum industry. Ann. Occup. Hyg. 44 (4), 301–320.

Hanowski, R.J., Hickman, J., Furnero, M.C., Olson, R.L., Dingus, T.A., 2007. The sleep of commercial vehicle drivers under the 2003 hours-of-service regulations. Accid. Anal Prev. 39 (6), 1140–1145.

Hessel, P.A., Herbert, F.A., Melenka, L.S., Yoshida, K., Nakaza, M., 1997. Lung health in relation to hydrogen sulfide exposure in oil and gas workers in Alberta, Canada. Am. J. Ind. Med. 31 (5), 554–557.

Huang, Y., Roetting, M., McDevitt, J., Melton, D., Smith, G., 2005. Feedback by technology: attitudes and opinions of truck drivers. Transport. Res. 8, 277–297.

International Agency for the Research of Cancer, 2013. IARC Monographs on the Evaluation of Carcinogenic Risks to Humans: Diesel and Engine Exhausts and Some Nitroarenesvol. 105International Agency for Research on Cancer (IARC), Lyon, France.

International Association of Oil and Gas Producers (IOGP), 2014. Land Transportation Safety Recommended Practice: OGP Report No. 365 (Issue 2) London, UK. Available from: http://www.ogp.org.uk/pubs/365.pdf

International Association of Oil and Gas Producers (IOGP), 2014. Land Transportation Safety Recommended Practice, Guidance note 12. Implementing an in-vehicle monitoring program – a guide for the oil and gas extraction industry. London, UK. Available from: http://www.ogp.org.uk/pubs/365-12.pdf

Khan, F., 2011. Take home lead exposure in children of oil field workers. J. Okla. Med. Assoc. 104 (6), 252–253.

King, B., Esswein, E., Retzer, K., Snawder, J., Ridl, S., Breitenstein, M., Alexander-Scott, M., Hill, R., 2015. UPDATE: Reports of Worker Fatalities during Manual Tank Gauging and Sampling in the Oil and Gas Extraction Industry. NIOSH Science Blog. Available from: http://blogs.cdc.gov/niosh-science-blog/2015/04/10/flowback-3/

Mason, K., Retzer, K., Hill, R., Lincoln, J., 2015. Trends in occupational fatalities in oil and gas extraction. MMWR 64 (20), 551–554.

Miller, T.M., Mazur, P.O., 1984. Oxygen deficiency hazards associated with liquefied gas systems: derivation of a program of controls. Am. Ind. Hyg. Assoc. J. 45 (5), 293–298.

Mode, N., Conway, G.A., 2007. Working hard to work hard safely. Paper presented at Society of Petroleum Engineers, Exploration & Production, Environmental and Safety Conference. Galveston, TX.

National Institute for Occupational Safety and Health (US), 2010. NIOSH Fact Sheet: Field Effort to Assess Chemical Exposure Risks in Oil and Gas Workers. Denver, CO. Available from: http://www.cdc.gov/niosh/docs/2010-130/pdfs/2010-130.pdf

National Institute for Occupational Safety and Health (US), 2002. Hazard Review: Health Effects of Occupational Exposure to Respirable Crystalline Silica. NIOSH (US), Cincinnati, OH. Available from: http://www.cdc.gov/niosh/docs/2002-129/pdfs/2002-129.pdf

National STEPS Network, 2015. Tank Hazard Alert: gauging, thieving, fluid handling; how to recognize and avoid hazards. Available from: http://www.nationalstepsnetwork.org/docs_tank_gauging/TankHazardInfographicFinal04_22_15.pdf

Occupational Safety and Health Administration (US), 2015. Hazard Alert: Diesel Exhaust/Diesel Particulate Matter. OSHA (US), Washington, DC. Available from: https://www.osha.gov/dts/hazardalerts/diesel_exhaust_hazard_alert.html

Pronk, A., Coble, J., Stewart, P., 2009. Occupational exposure to diesel engine exhaust: a literature review. J. Expo. Sci. Environ. Epidemiol. 19 (5), 443–457.

Retzer, K., Hill, R., Conway, G., 2011. Mortality statistics for the US upstream industry: an analysis of circumstances, trends, and recommendations. Paper presented at Society of Petroleum Engineers Americas, Exploration & Production, Health, Safety, Security, and Environmental Conference. Houston, TX.

Retzer, K.D., Hill, R.D., Pratt, S.G., 2013. Motor vehicle fatalities in the oil & gas extraction industry. Accid. Anal. Prev. 51, 168–174.

Retzer, K., Tate, D., Hill, R., 2014. Journey management: a strategic approach to reducing your workers' greatest risk. Paper presented at Society of Petroleum Engineers Americas, Exploration & Production, Health, Safety, Security, and Environmental Conference. Long Beach, CA.

Rothe, J.P., 2008. Oil workers and seat belt wearing behavior: the Northern Alberta context. Int. J. Circumpolar. Health. 67 (2–3), 226–234.

Snawder, J., Esswein, E., King, B., Breitenstein, M., Alexander-Scott, M., Retzer, K., Kiefer, M., Hill, R., 2014. Reports of Worker Fatalities during Flowback Operations. NIOSH Science Blog. Available from: http://blogs.cdc.gov/niosh-science-blog/2014/05/19/flowback/

Steinsvåg, K., Galea, K.S., Krüger, K., Peikli, V., Sánchez-Jiménez, A., Sætvedt, E., Searl, A., Cherrie, J.W., Van Tongeren, M., 2011. Effect of drilling fluid systems and temperature on oil mist and vapour levels generated from shale shaker. Ann. Occup. Hyg. 55 (4), 347–356.

Verma, D.K., Johnson, D.J., McLean, J.D., 2000. Benzene and total hydrocarbon exposures in the upstream petroleum oil and gas industry. Am. Ind. Hyg. Assoc. J. 61 (2), 255–263.

Zoveidavianpoor, M., Samsuri, A., Shadizadeh, S.R., 2012. Health, safety and environmental challenges of xylene in the upstream petroleum industry. Energy Environ. 23 (8), 1339–1352.

CHAPTER 8

Public Health, Risk Perception, and Risk Communication: Unconventional Shale Gas in the United States and the European Union

Bernard D. Goldstein*, **Ortwin Renn****, **Aleksander S. Jovanovic**[†]
*Graduate School of Public Health, Pittsburgh, PA, USA, University of Cologne, Cologne, Germany; **SOWI V, Universität Stuttgart, Stuggart, Germany;
[†]Steinbeis Advanced Risk Technologies, Stuttgart, Germany

Chapter Outline

INTRODUCTION

The chapter reviews aspects of public health, public perception of risk, and communication in the United States (US) and Europe related to obtaining natural gas tightly bound to deep underground shale layers through technology known as unconventional gas development (UGD) (IRGC, 2013). UGD has developed primarily in the US through decades of progressive innovation. It represents the ability to tap tightly bound shale gas led to an unprecedented 35% increase in estimates of potential total US natural gas reserves from 2006 to 2008 with a continuing significant rise (Potential Gas Committee, 2015) (see Chapter 2 for further details). In Pennsylvania only 8 Marcellus shale wells were drilled in 2006 but by 2014 the total had grown to 8802 (Pennsylvania Department of Environmental Protection, 2015) (see Chapter 1, fig. 1.4).

Europe also has significant untapped reserves of tightly bound shale gas, estimated at about half that of the US (EIA, 2013; European Commission, 2014a,b)

Environmental and Health Issues in Unconventional Oil and Gas Development. http://dx.doi.org/10.1016/B978-0-12-804111-6.00008-X

(see Chapter 13 for additional discussion). The uncertainty of these assessments, as well as their possible adverse impacts, are key unknowns for shale gas development in Europe. The continuing evolution of public perception also is of major importance to the development of shale gas resources. Central to the acceptability of UGD is whether, and to what extent, UGD is a new technology, whether all the risks are known, and how are the risks perceived.

Among the reasons for the European Union (EU) to consider moving forward with UGD are an increasing dependency upon imports for natural gas (European Commission, 2014a), including heightened concerns about dependence upon Russia due to events in the Ukraine. Also of concern is the current large margin in the price of natural gas giving US industries a competitive advantage and contributing to the strengthening of the dollar. A recent review by the European Academies Science Advisory Committee lists three issues of importance to the eventual acceptability of shale gas in Europe: (a) implications of population density and water usage, (b) specific greenhouse gas emissions, and (c) public acceptance of shale gas development (EASAC, 2014).

Opposition to UGD exists in both the US and EU but has been more effective in Europe where many countries or parts of countries, including France, Germany, Bulgaria, Wales, and Scotland have imposed at least partial moratoriums on shale gas drilling (see eg., fig 1.1 in Chapter 1, Chapter 13). Growing opposition in the US recently led the State of New York to reverse its previous approval of UGD and install a ban (see Chapter 12). The reasons for these concerns as related to public health, risk perception, and risk communication are reviewed here, including an overview of the relatively limited literature on potential public health impacts of UGD and the role of public perception in response to shale gas drilling (e.g., Jovanovic et al., 2012). In addition to previously described considerations such as the precautionary principle, the role of litigation, and the response to innovation, we find that UGD adds another dimension to the differences between the EU and the US in public perception of risk – the ownership by private citizens of subsurface mineral rights in the United States, but not the EU.

Evidence of the remarkable increase in interest in the topic in both social media and in the scientific literature is shown in Figures 8.1 and 8.2, respectively.

Controversy about perception of risks related to UGD falls under two main headings. First, broader issues include disputes about the effect of exploitation of unconventional gas on global climate change, carbon footprints, and other sustainability issues. Methane releases during UGD affect the extent to which shale gas can effectively replace coal (Howarth et al., 2011; Jenner and Lamadrid, 2013) (see Chapter 3). It is also argued whether shale gas represents a bridging fuel to a carbon-free future or, on the contrary, decreases pressure on the needed development of a carbon-free economy (IRGC, 2013). Second, and a focus of this chapter, is that UGD has been connected to other possible risks, such as adverse health effects, seismicity, loss of water supplies, and decrease in air and water quality (European Commission, 2014a,b; Council of Canadian Academies, 2014; Health Effects Institute, 2015) (see discussions in Chapters 3,

FIGURE 8.1
Presence of the issue of fracking in social media: example of the Tweet-based RiskAtlas developed in the iNTeg-Risk project *iNTeg/Risk (2014)*

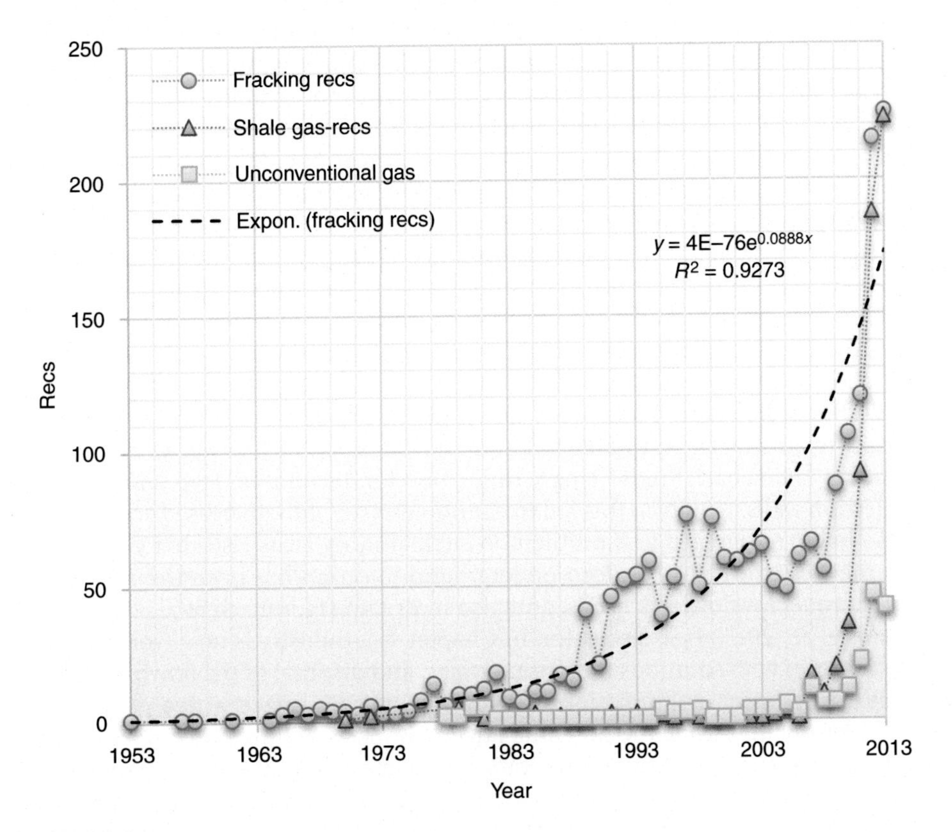

FIGURE 8.2
Annual number of fracking related papers published from 1953 to 2013 *Li et al. (2014)*

4, 5, and 11). Interest in UGD and its impacts is shared by many stakeholders' groups, often with different views: while public interest is manifest in diverse ways like websites, newspapers, and social media, with government interest being expressed through numerous reports, laws, and standards.

UNCONVENTIONAL GAS DRILLING

We use the term UGD to focus on current technology that has raised concerns about impacts on public health and the environment. UGD includes high-volume hydraulic fracturing (HVHF) and the turning of the vertical drill pipe within the relatively horizontal shale layer (see eg., fig. 1.2, Chapter 1). Holes are then blown in the well pipe through which water is forced under very high pressure. This hydraulic fracturing fluid contains about 0.5–1.0% of a chemical mixture, and 10–15% of physical proppant added to help remove the gas from the rock and to keep the fractures open (see Chapters 1 and 5).

The shale gas industry has repetitively emphasized to the public and to decision makers that the technology is safe because they have more than 50 years of experience in hydraulic fracturing (e.g., American Petroleum Institute, 2015). However, the current approaches to UGD are significantly different from those used in the past. Industry now can drill much deeper; use more than a 100-fold greater fluid with evolving chemical components under much higher pressure; and will drill and hydrofracture sequentially from the same drill pad in as many as a dozen different directions rather than in just one, thereby causing much longer periods of noise, light, and truck traffic. More important are the qualitative differences including the increasing problems of disposing of toxic flowback fluids (Vidic et al., 2013). Perhaps millions of liters of fluid flows back during the completion of the hydrofracturing process (see discussion in Chapter 5). Although in much smaller volumes per day, brine-laden water is produced during the lifetime of the well, which cumulatively adds a substantial volume to disposal challenges (Lutz et al., 2013). By 2011, in Pennsylvania, the very large volumes of flowback fluids could no longer be handled by the usual industrial waste treatment facilities (Wilson and Van Briesen, 2012; Ferrar et al., 2013a) leading to flowback fluids being sent to Ohio for disposal in deep underground injection wells. However, this led to earthquakes in Ohio. Increased seismicity is also being recognized as a problem in shale-drilling areas elsewhere, including in the western US where deep underground injection has been the major disposal route for flowback fluids, and also in Britain (Hornbach et al., 2015; Williams et al., 2015) (see discussion in Chapter 11). Industry is now working hard to come up with solutions for reuse, storage, and disposal of the flowback fluids, but this requires manipulation of the fluids on the surface where people live, thus greatly increasing the potential adverse impacts of unplanned releases into the environment. Two other qualitative differences from the past are the evolving use of a very different suite of hydrofracturing chemicals and, particularly pertinent to Europe and to the eastern US and Canada, UGD activities in areas with higher population densities unaccustomed to oil and gas (O&G) drilling.

Table 8.1 Potential Health and Safety Issues Related to UGD Activities

Issue	Comment
Safety and Health issues	• Both for the work force at the operating/drilling sites and the population in the surrounding adjacent areas
Air Pollution	• Worker and community exposure to hydraulic fracturing chemicals, silica, diesel exhaust, and drilling compounds • Community exposure to air toxics, including benzene, nitrogen oxides, diesel exhaust, ozone
Water Pollution	• Hydraulic fracturing chemicals, flowback and produced waters onsite or offsite, including transportation and storage, reactants and mixtures
Light and Noise Pollution	• Noise on the operating/drilling sites sometimes in populated areas
Psychosocial Effects	• Exacerbated by lack of transparency and trust issues, and by extent of familiarity with drilling
Community "Boomtown" Issues	• Influx of people without roots in the community; violence, drunken driving, drug abuse, venereal diseases

Health Risks

The potential health and safety issues presented by UGD activities are shown in Table 8.1. Workers are particularly at risk in this, as they are in other O&G drilling activities (see Chapter 7). Although US fatality rates are decreasing, they are still substantially higher than in the construction industry (Mason et al., 2015). Boomtown effects in local rural communities are associated with the influx of a relatively large number of well-paid young males without community roots; this may lead to violence, drunken driving, drug abuse, and venereal diseases (Jacquet, 2013).

Toxicological issues are related to three mixtures of chemical and physical agents: (a) the hydrofracturing agents, (b) the volatile shale hydrocarbons, and (c) the naturally occurring shale bed components that are dissolved by the hydrofracturing fluids and brought to the surface (Table 8.2). While attention has been focused mainly on the intentionally added hydrofracturing chemicals, these are less toxic than the other two mixtures that must be captured or disposed of safely. Of particular concern, and almost totally unstudied, is both the toxicology of combinations of these mixtures, and the potential formation of new reactants facilitated by the high temperatures in the deep underground shale layers (Goldstein et al., 2014).

Table 8.2 Agents of Toxicological Concern Associated With UGD Activities
Agents
1. Hydrofracturing agents
2. Hydrocarbons and gases present in shale: methane, ethane, propane, BTEX*, hydrogen sulfide
3. Natural constituents: brine components, barium, bromide, calcium, chloride, iron, magnesium, strontium, arsenic, radionuclides
4. Mixtures of any or all of above
Note: chemical reaction rates are increased at high temperatures (~250 °C in shale layers and perhaps also affected by high pressure and salinity)

*BTEX: benzene, toluene, ethylbenzene, and xylenes

Radionuclides are a particular concern and at times exceed the allowable level for usual disposal (Litvak, 2013). An association between household radon levels and UGD has been reported (Casey et al., 2015). Possible worker and/or community exposures may occur due to:

- Technically enhanced concentrations of radionuclides during the processing of flowback fluids and other drilling-related exposures.
- The selective absorption of radon daughters during the UGD process resulting in misleading radon levels (Nelson et al., 2015).
- The possibility, based upon the relatively short half-life of radon and its presence in natural gas, that higher levels of radon within homes with gas appliances will be due to the greater proximity of the natural gas sources to the highly populated northeastern US and the shorter time in the pipeline decreasing the extent of radioactive decay (Resnikoff, 2012). This conceivably might occur in the EU if local natural gas displaces natural gas obtained via pipelines from more distant areas.

The health risks and health benefits of UGD occur on different geographical scales. Benefits arise from replacement of coal with natural gas for energy development, thereby decreasing airborne particulates and other regional air pollutants and potentially decreasing greenhouse gas emissions that have a global health impact (see discussion in Chapter 3). The promised increased prosperity for shale regions also presumably would bring health benefits. In contrast, health risks primarily fall on workers and on community members who live in close proximity to drilling and wastewater disposal sites, although ozone formation from hydrocarbons and nitrogen oxides released during UGD can be a regional problem (Kemball-Cook et al., 2010).

The most consistent finding in virtually all reviews of the potential adverse health effects of UGD is the lack of sufficient information on which to draw conclusions (Korfmacher et al., 2013; German Advisory Council, 2013; Adgate et al., 2014; Shonkoff et al., 2014; Kovates et al., 2014; Cleary, 2012; Goldstein et al., 2013; Council of Canadian Academies, 2014; New York State Department of Health, 2014; Hays et al., 2015; Royal Society and Royal Academy of Engineering, 2012; Werner et al., 2015; Small et al., 2014). For example, the report of European Commission (2014b) states: "Assessment of health impacts is only starting, due to the novelty of the practice at the current scale." Even Public Health England, which supported moving ahead with shale gas drilling, noted "limited information" and "more studies are required." Their rationale for support of UGD included that in contrast to the United States, public health authorities in England would be closely involved (Harrison and Cosford, 2014; Goldstein, 2012). These many reviews often cite surveys of those living in proximity to shale gas drilling sites that have reported findings affecting many organ systems. These are often postal surveys or other approaches, which might bias the responders toward those who believe their health has been affected. Stress and stress-related symptoms are common, particularly in those who believe their health has been affected (Subra, 2010; Steinzor et al., 2013: Ferrar et al., 2013b). Among other findings, a household survey reported higher levels of upper respiratory and dermatological symptoms in proximity to gas wells (Rabinowitz et al., 2015). Higher levels of hospital admissions for cardiovascular disease in areas with higher UGD activity have been reported (Jemielita et al., 2015).

Several studies published or in abstract, have suggested differences in birth outcomes among mothers living in proximity to UGD well sites. The initial published study in Colorado found an association between maternal residential proximity to hydraulic fracturing sites and birth outcomes, including an increase in birth weight, an increase in reported congenital heart defects, and, to a lesser extent, neural tube defects (McKenzie et al., 2014). In contrast, Stacy et al. (2015) found a statistically significant decrease in birth weight and an increase in the percentage of small for gestational age babies in Southwestern Pennsylvania.

Two other studies not yet published in the peer-reviewed literature also report lower birth weights in association with shale gas drilling. Hill (2013) also noted a decrease in APGAR scores, a clinical measure of neonatal wellbeing. In a multistate sample, Currie et al. (2014) found low birth weight associated with proximity to well sites. These studies, while not definitive, strongly suggest the need for a large well-defined study of the potential adverse effects on birth outcomes in proximity to shale drilling sites.

A few studies have looked at health risks based upon an exceeding of allowable pollutant standards or utilizing risk assessment based upon measured or assumed exposures. Esswein et al. (2013, 2014) reported exceedance of the workplace silica exposure standards for workers involved with using silica as a proppant, and the benzene standard for wastewater workers (see Chapter 7 for additional discussion of workplace issues). McKenzie et al. (2012), in measurements in a community downwind of hydrofracturing sites in Colorado found higher exposures in those living closer to the site of various volatile chemicals. They concluded that subchronic exposures during well completion presented the highest risks, with a subchronic hazard index of 5 being driven primarily by exposure to trimethylbenzenes, xylenes, and aliphatic hydrocarbons; a slight increase in cumulative cancer risk, primarily due to benzene, was also noted. Recently, Paulik et al. (2015) reported an increase in polycyclic aromatic hydrocarbons in air samples near UGD sites with a resultant cancer risk that exceeded the United States Environmental Protection Agency (US EPA) acceptable risk standard. The risk of water contamination has also been explored (Meiners et al., 2012; Rozell and Reaven, 2012; EPA, 2015; Krupnick, 2012) (see discussion, Chapter 4).

Exposure Assessment

A particular barrier to the study of health risks has been the lack of adequate exposure assessment of humans or ecosystems. This lack of exposure data partially reflects complacency based on the assertion that the gas industry has been doing hydrofracturing safely for more than 65 years (American Petroleum Institute, 2015). In the US, information gaps in part reflect legal exceptions to usual federal oversight requirements that have been granted to the O&G industry and an unwillingness of congressional and state supporters of UGD to provide funding for such research (Gamper-Rabindran, 2014; Goldstein, 2014).

However, the lack of exposure data also reflects a challenge pertinent to any industrial source that is individually relatively moderate in size but is widely and multiply distributed in local geographical areas, has differing effects on a local, regional, and global scales, causes release to air, water, and soil, has major temporal variations in emissions, and, most importantly, is known to vary greatly from site to site thereby complicating generalization from one site to the next. The various reasons for site-to-site differences are listed in Table 8.3). GIS measurements of distance from the well site, or inverse distances from multiple well sites, have been used as indirect measures of exposure (Hill, 2013; McKenzie et al., 2014; Rabinowitz et al, 2015; Stacy et al., 2015). While providing useful information, this approach is limited as environmental releases may occur at offsite facilities, such as those handling liquid waste. GIS also works best when all sites are reasonably equivalent in environmental releases rather than in the current situation, which shows extreme variability. Evidence of marked site-to-site variation can be seen in the studies of Allen et al. (2013) of methane emissions from cooperating industries during two periods of relatively intense release: completion of the hydrofracturing process and intermittent "unloading." They found that methane emissions over an entire completion flowback event ranged from less than 0.01 Mg to more than 17 Mg. Even greater variation

Table 8.3 Selected Reasons for Variability in Environmental Releases Among UGD Sites

- Different safety culture.
- Different geology and other local site-specific issues.
- Different drilling technology.
- Different hydraulic fracturing chemicals.
- Different shale gas collection and distribution techniques.
- Different flowback disposal techniques.

was noted for unloading. Average methane release is a suitable measure for a monitoring goal related to climate change as atmospheric mixing of methane occurs globally over many years (see further discussions in Chapter 3). However, to protect individuals living nearby, for whom the methane release would also be accompanied by more toxic and higher molecular weight shale gas components such as benzene, the number of sites to be sampled would be a function of the expected variability among sites, which depends on the factors listed in Table 8.3. Further evidence of the challenges to classic exposure assessment imposed by UGD comes from an expert elicitation study by Resources for the Future in which the very high total of 264 routine "risk pathways" and 14 accident pathways were posited (Krupnick, 2012). As a result of these many challenges there is a need to develop and use biological markers of exposure in humans or ecosystems believed to be at risk. This might include pets or agricultural animals (Bamberger and Oswald, 2012).

These impediments may be less of a problem in the EU where government ownership of subsurface O&G rights will likely lead to intensive UGD in just a few local areas, as is now under way in England, thus facilitating standard ambient monitoring (see discussion in Chapter 13).

Risk Perception and its Role in EU/US Differences

The public's perception of UGD has been a key issue in its acceptance. Noteworthy have been campaigns by environmental groups, with the cooperation of industries requiring clean water, leading to reversals of apparent earlier acceptance of shale gas drilling in Europe and in New York State. As a generalization, the United States has moved ahead more rapidly than the EU on using UGD to obtain tightly bound shale gas. However, the differences within the EU among various countries, and the differences within the United States among various states, exceed the overall EU/US differences (see discussion in Chapter 13). In both the EU and the United States, some countries or states have banned UGD while at least one other is attempting to move ahead. In the United States, those states that have long-term active O&G industries, like Colorado, Texas, and Oklahoma, are using UGD to exploit tightly bound shale gas deposits. In contrast, in the northeastern US states, where UGD is new, there has been a varied response. Pennsylvania has been very active in pursuing UGD while New York, after much indecision, has decided against UGD for now. Similarly, West Virginia has active UGD while Maryland has yet to make a decision about it.

Scientific understanding of the factors involved in the public's understanding and acceptance of risk has been a subject of significant study with key advances in recent decades (Lofstedt, 2015; Fischhoff, 2013; Wachinger et al., 2013; Renn and Benighaus, 2013; Rosa et al., 2014). Among the major factors involved is the extent of trust in informants and the public's degree of familiarity with the risk (Wachinger et al., 2013; Lofsedt, 2005). Social amplification of risk (Kasperson et al., 1988) is apparent in the US, both in the greater acceptability of UGD in western states where O&G drilling is more routine (Deloitte, 2012) and in the lack of trust and transparency (Ferrar et al., 2013b; Goldstein, 2015). Specific issues include the use by shale gas advocates of strategies such as obscuring adverse outcomes in technical language – notably the repeated insistence that hydrofacturing has never led to groundwater contamination. This may be correct as long as hydrofracturing is narrowly defined as the successful release of hydrofracturing fluids deep underground. However, to the general public, hydrofracturing is "fracking," and fracking is anything that is part of the UGD process. The public's question is whether their water will be contaminated with hydrofracturing agents, and it is immaterial whether it occurs from an overturned truck or a well casing failure – neither of which industry defines as hydrofracturing despite knowing the public's understanding of the term (Goldstein, 2015; Everley, 2013). Further, repeatedly being told that hydrofracturing does not cause water contamination seems incompatible with media accounts about communities whose water has been contaminated with hydrofracturing chemicals (see discussion in Chapters 4 and 9). Based on interviews with industry informants, Heikkila et al. (2014) reported that some in the industry recognized that a lack of transparency was hurting. George P. Mitchell, the industry founder of hydrofracturing, coauthored an op-ed stating that the industry had "attempted to gloss over" legitimate public concerns (Bloomberg and Mitchell, 2012).

Reflecting public perception, the presumed spatial impact of UGD activities on health is partially mirrored by housing prices. A study in Pennsylvania found a strong negative impact associated with the early stages of shale exploration that was dependent both on the proximity and intensity of shale activity (Gopalakrishnan and Klaiber, 2013). Negative effects on property values occurred in proximity to roads, presumably reflecting high truck traffic; and for properties that used private water wells, presumably due to water quality concerns. A negative impact on the price of houses with water wells was also reported by Muehlenbachs et al. (2014) (also see discussion in Chapter 2).

The literature comparing the EU and the US on environmental issues focuses on such issues as whether there is a European preference for precaution versus a US dependence on a stronger posthoc litigation culture to punish the wrongdoer (Wiener, 2003; Hammitt et al., 2005; Jasanoff, 2005; Renn and Elliott, 2011). Also, frequently noted is a greater US public acceptability of new technology, such as genetically modified organisms. Some of these factors are likely to be at least partially responsible for the differences between the US and the EU on shale gas. Other well-known factors affecting risk perception can readily be seen as part of the varied US response to UGD. In the western US, most UGD occurs at a distance from population centers. However, substantial UGD drilling does occur in the more populated Dallas/Fort Worth area of Texas. However, this population has had long familiarity with living in proximity to O&G drilling and is more accepting than the population in the northeastern US, which has population densities and a general lack of familiarity with O&G drilling that is closer to much of the EU that has shale gas deposits (Deloitte, 2012). Thus, both mistrust and lack of familiarity appear to be at play in leading to opposition to UGD in the US.

Evidence of mistrust of the shale gas industry has been noted in a number of studies. Ferrar et al., 2013b, in unstructured interviews of individuals in

Pennsylvania who themselves believed their health had been affected by UGD, noted that issues related to trust and transparency were far more likely to be stated than were physical stressors. Among the latter, most frequently cited was noise (45%) while 79% stated that they were denied information or given false information, and over 50% reported corruption, that their concerns or complaints were ignored, or that they were being taken advantage of.

A survey was made allowing comparison of the situation in the US and Europe, in the framework of the German national project ENERGY-TRANS (Jovanovic et al., 2014) The survey and the respective ENERGY-TRANS project discussion paper looked at the issue of public perception of new energy technologies, using the example of hydraulic fracturing and UGD as an example of a controversial energy technology. In Germany and in many other European countries, there are doubts whether UGD should be accepted in the same way as in the US and what might be the possible consequences of such acceptance. The main goal of the ENERGY-TRANS survey has been to explore acceptance of hydrofracturing by professionals and stakeholders active in the area, and to compare it with the acceptance in other countries, particularly the survey performed by Groat and Grimshaw (2012) in Texas. In addition, the study tried to explore the existence of possible regional clusters of acceptance or rejection of hydraulic fracturing (see eg., Chapter 13).

The respondents, mainly from Germany, with strong participation from other EU countries and the US, answered questions related to the technology itself, to the environmental and other risks, and to regulations already in place and/or needed. The main results of the survey show that hydrofracturing technology is not rejected as such, but that the related environmental and health risks are causing concerns that require comprehensive, transparent, and understandable participation-based governance and regulation. Only with such regulations in place, can a broad public and social acceptance be attained.

The results of the direct comparison show (see Table 8.4 for details) that the respondents tend to be fairly well informed and look for balanced opinions, believing that an issue like hydrofracturing will benefit from better regulation and enforcement policies and instruments. Specific issues, such as water scarcity tend, however, to be more important in Texas, which is relatively arid.

In addition to the previously mentioned survey, direct online website polling was performed on University of Stuttgart students in 2013–2014, asking two questions:

1. Fracking for Germany? Would you consider fracking (a method of gas exploitation also known as hydraulic fracturing) as a SOCIALLY ACCEPTABLE alternative in the German national energy policy?
2. Fracking in your backyard? Would you accept a drilling installation needed for fracking in the vicinity of your home or working place (within, say, 1 km radius) if offered an adequate compensation (e.g., reimbursement, free gas supply, or the like)?

Both questions received about 20% of votes, indicating a readiness to consider hydrofracking as an alternative, even if it would imply “fracking in one’s own

Table 8.4 Comparison Between the United States and Europe Based on Surveys by Groat and Grimshaw (2012) and Jovanovic et al. (2014)

Element of the Survey	United States	Europe/Germany
Number of participants (rounded)	1500	500
Participants with a university degree	64%	90%
Main political orientation of respondents	Right 41 %	Left 35%
Agreement with the statement "*Some scientists and environmental groups claim hydraulic fracturing is dangerous and should be banned, but the industry insists it is a safe technology*"	72%	53%
Knowledge about the basics of the technology – "drilling at depths from 1000 to 3000"	38%	45%
Knowledge about regulation/legislation – "*Do you believe that in Germany/USA, an oil or gas operator is required to perform an environmental study before drilling?*"	Yes 77%	Yes 76%
Priorities – example "Water shortages and clean drinking water" (see also Table 8.5)	86%	28%
Do you think that state and/or national officials are doing enough to require disclosure of the chemicals used in natural gas drilling? Would you say they are?	Not doing as much as they should 47%	Not doing as much as they should 32%

Table 8.5 Priorities in Germany/Europe

Option: Which of the Following Statements Best Expresses Your View About Where Germany Should Focus its Energy Production in the Future? (Multiple Options)	%
Water shortage and clean drinking water are real concerns	28
Germany should put the emphasis on first developing new energy sources that require the least water and have minimal water pollution	59
Energy supply needs should override concerns about water shortages and water pollution	32
Germany should proceed first with developing energy sources even if they may pollute water or create shortages	9

backyard." This mix of perceptions was also reflected in comments such as on the negative side: "Refer to the US experience and learn the lesson!" or "No, I think it is important to feel free to drink the water out of the tube" and "risks seem bigger than potential benefits to me." More balanced was the response "We will always find a way to burn more gas so it is senseless to look for more. Rather, we should change our lifestyle and consume less." On the pragmatic side was: "Yes; however, the adequate compensation should be really adequate, which means a contribution which covers at least the loss in value of my house plus the loss of value of my surrounding area."

While factors related to precaution, innovation, and litigation, described earlier, can be expected to play a role in EU/US differences that cut across broad areas of environmental concern, there is an additional factor that may be most important in explaining the relative differences in the response to UGD. The US is almost unique, and certainly different from EU countries, in that the surface property owner, rather than the state, also owns subsurface rights to O&G unless these subsurface rights were originally sold to another nongovernmental owner. The private owner of the rights thus knows in advance of potential substantial economic rewards if UGD is permitted by the government. The impact of financial benefit was clearly seen in a random digit dialing survey, comparing over 500 residents of adjacent Pennsylvania counties with marked difference in UGD activity, which included 7 questions related to shale gas out of 128 in total. In the high UGD county, 29.9% of the survey population themselves or a family member had leased land for drilling. Among this group 36.7% reported themselves strongly in favor of shale drilling, while among those in the same county who were not family leaseholders, only 18.7% were strongly in favor. The latter was similar to the low UGD county in which 16.3% of nonleaseholders reported themselves strongly in favor, suggesting that the general economic benefit to the high UGD county, which was heavily advertised by industry, had relatively little effect on those not directly benefitting (Kriesky et al., 2013).

THE PATH FORWARD

Our major conclusions are that there has been inadequate study of the potential adverse health consequences of UGD; that to maximize the potential benefits of obtaining tightly bound shale gas much more attention is needed in order to understand the public's perception and to fairly and transparently involve the public in the decision process; and that a major reason for the current greater acceptance of shale gas drilling in the US as compared with the EU is that private ownership of subsurface drilling rights in the US provides more of a financial opportunity to private landowners.

Additional health and environmental research may be forthcoming. Industry and other UGD supporters appear to recognize that lack of information has led to actions that were adverse to the industry. The Governor of New York State changed the decision to allow UGD in parts of the state after a review by the state Department of Health made the precautionary conclusion that "Until the science provides sufficient information to determine the level of risk to public health from HVHF to all New Yorkers and whether the risks can be adequately managed, DOH recommends that HVHF should not proceed in NYS." (Kaplan, 2014; Zucker, 2014). A 2013 decision by the Pennsylvania State Supreme Court to strike down parts of a heavily proindustry state law included language by the Chief Justice "... the absence of data also suggests that the Commonwealth has failed to discharge its trustee duty of gathering and making available to the beneficiaries complete and accurate information" (Robinson Township, 2013). Perhaps coincidental, the American Petroleum Institute launched its first Request for research Applications (RFA) for shale gas health studies in 2014 after the Pennsylvania decision, although that is now being held in abeyance.

A common issue in responding to the environmental and health issues of UGD, both in Europe and the US, is the requirement to adapt public health and risk communication strategies to the needs of different policy actors. In Europe, "energy policy" is considered at both the national and EU levels. That UGD could be an important part of the EU and national strategies to create backup solutions for possible difficulties related to gas supplies from Russia is far from fully explored. The EU has no real supranational policy on UGD (despite activities such as the European Science and Technology Network on Unconventional Hydrocarbon Extraction; European Commission, 2014b). National energy policies differ among EU countries and are not necessarily aligned. In the US, UGD has moved ahead more rapidly, primarily due to the relative abundance of accessible tightly bound shale gas resources and to industry initiatives coupled with a relatively laissez-faire government policy. While the US Department of Energy has provided some research support, there has been little evidence of national policy guidance as a major force, and, until recently, little interest at the national or state government level in investigating potential adverse health consequences (Goldstein et al., 2013). Correspondingly, public perception of energy related benefits leads to a better acceptance on the national level in the US than in the EU, but this can probably abruptly change in the case of an energy supply crisis. Policy action is also hampered in the US by the fragmentation and lack of a well-defined peer structure within the UGD industry (Nash, 2013).

The "risk/safety policy" level defining and implementing shale gas policies is primarily the regional level and national states in the EU or the individual states within the US. Hence, the battle takes place on the level of states. Public concern about secrecy concerning hydrofracturing agents has been evident. More than 20 US states have adopted rules requiring companies to disclose hydrofracturing chemicals (see Chapters 1, 4 and 12). The EU countries are similar, but just as in the US, work must still be done to develop regulations that can get around the industry's claims of trade secrecy, which O&G companies invoke to resist public disclosure of the chemicals. Despite the recent increase in US reporting requirements approximately 70% of hydrofracturing mixtures contain at least one component that is not subject to disclosure (EPA, 2015). Further, with less fanfare, industry has been successful in achieving secrecy for flowback agents despite their relatively higher toxicity (Goldstein, 2015). The industry is in a tougher position in the EU, due to the pressure of the "precautionary principle."

Ensuring public acceptance of large projects such as UGD requires creating a positive stance in the majority of the population concerned. Nevertheless, even when a lot of effort is invested in creating a positive image, large projects are generally not well accepted. This is especially true if the emphasis in the process of ensuring public acceptance is focused on the general benefits and objectives of the respective project rather than on benefits to the individual or to those for whom one cares. The main elements needed to achieve a durable acceptance or at least tolerance of the new projects are those related to self-motivation (IRGC, 2013; Renn and Schweizer, 2014; Fiske, 2010), namely understanding, controlling, self-enhancing, and perceiving one's own benefit or benefit to those for whom one cares. Accordingly, if one understands the need and value of UGD, receives the information about its pros and cons, understands the risks for the environment, health, and the society, he/she will be more likely to accept it. The acceptance will be increased if the transparency of the decision-making and participation of the people concerned is ensured (International Energy Agency, 2012; IRGC, 2013; Jovanovic, 2015; North et al., 2014).

The same will be the case if one has the feeling that her/his own actions can influence the project (be in a way "in control" of it): he or she will be more likely to actively accept it. In this context, a paradox can take place: the more options to act one is given, the more one will believe that his actions can "count" and the project promoters must take into account that this can also lead to the lack of acceptance. However, even this is better than the alternative leading to tolerance based on a fatalistic feel that "nothing can be done," or tolerance based on fear in front of authorities. If, in addition, one manages to clearly identify and possibly document the benefits to the people directly concerned or to those for whom they care or personally know, then acceptance will increase further. Finally, acceptance will be boosted if the people concerned can identify themselves with the project primarily through acceptance that the new project fits into the social and cultural surroundings well. This requires community involvement.

A comprehensive acceptance improvement strategy should target all four aspects mentioned earlier. Communication, information, and dialogs must address them explicitly, focusing on the main message that the possibly disputed technology could act as enrichment for the local environment. Can this be claimed for hydrofracturing? Community understanding of the technology is often a problem, burdened by the lack of trust (the main providers of the information and explanations are often the exploitation companies). People do not "feel in control" when hydrofracturing is concerned. Local benefits are often limited (people at the fracking sites do not profit from the installation on the long-term and in a stable way – even the possible short-term benefits often cause problems, such as the "community 'boomtown' issues"). Finally, people can better identify with a project, which would bring known technology, lasting investment, and employment for say 20 years (e.g., a conventional gas exploitation site), than with a project, which will bring unfamiliar technology, poorly known risks, and disappear from the community within a couple of years (e.g., a hydrofracturing site). Good governance and regulation can help overcome some of the previously mentioned difficulties, but this is generally not the case for UGD as it appears to be less regulated and more susceptible to a lack of governmental control. In fact, the UGD is definitely a multidimensional energy problem, which, accordingly, should be analyzed as such: along several dimensions. For example, one could use the features (factors, "dimensions") proposed in the Porter, Gee and Pope (2015) report. This report clearly shows that the US "strengths" in factors like "property rights," "innovation," "hiring and firing," "capital markets," and others, can well explain the background of the US fracking boom (and some of the differences in perception between the US and the EU). In addition, the same factors may also be able to sustain operation even in economically rough times. Despite all of the considerations discussed earlier related to the willingness of the public and of governments to accept UGD, a major factor will be industry initiative, which is highly conditioned by international factors governing the price and the market for gas and oil (see eg., Chapter 2).

References

Adgate, J.L., Goldstein, B.D., McKenzie, L.M., 2014. Critical review: potential public health hazards, exposures and health effects from unconventional natural gas development. Environ. Sci. Technol. 48 (15), 8307–8320.

Allen, D.T., Torres, V.M., Thomas, J., Sullivan, D.W., Harrison, M., Hendler, A., Herndon, S.C., Kolb, C.E., Fraser, M.P., Hill, D.A., et al.,2013. Measurements of methane emissions at natural gas production sites in the United States. Proc. Natl. Acad. Sci. 110 (44), 17768–17773.

American Petroleum Institute, 2015. API launches new ads promoting America's shale revolution. http://www.api.org/news-and-media/news/newsitems/2015/jan-2015/api-launches-new-ads-promoting-americas-shale-revolution

Bamberger, M., Oswald, R.E., 2012. Impacts of gas drilling on human and animal health. New Solut. 22, 51–77.

Bloomberg, M.R., Mitchell, G.P., 2012. Op Ed: Fracking is Too Important to Foul Up. Washington Post.

Casey, J.A., Ogburn, E.L., Rasmussen, S.G., Irving, J.K., Pollak, J., Locke, P.A., Schwartz, B.S., 2015. Predictors of indoor radon concentrations in Pennsylvania, 1989–2013. Environ. Health Perspect., Advance Publication April 9, 2015. http://dx.doi.org/10.1289/ehp.1409014.

Cleary E., 2012. Chief Medical Officer of Health's Recommendations Concerning Shale Gas Development in New Brunswick. Office of the Chief Medical Offer of Health, New Brunswick Department of Health. http://www2.gnb.ca/content/dam/gnb/Departments/h-s/pdf/en/HealthyEnvironments/Recommendations_ShaleGasDevelopment.pdf

Council of Canadian Academies, 2014. Environmental Impacts of Shale Gas Extraction in Canada; Report of the Expert Panel on Harnessing Science and Technology to Understand the Environmental Impacts of Shale Gas Extraction, May, 2014, Ottawa, Canada. http://www.scienceadvice.ca/en/assessments/completed/shale-gas.aspx

Currie, J., Deutch, J., Greenstone, M., Meckel, K.H., 2014. The Impact of the Fracking Boom on Infant Health: Evidence from Detailed Location Data on Wells and Infants [abstract]. American Economic Association Annual Meeting.

Deloitte, 2012, Deloitte Survey – Public Opinions on Shale Gas Development: Positive Perceptions Meet Understandable Wariness. http://www.deloitte.com/assets/Dcom-UnitedStates/Local%20Assets/Documents/Energy_us_er/us_er_ShaleSurveypaper_0412.PDF Or Deloitte, 2012 http://www.ogfj.com/articles/2011/12/deloitte-survey.html

EASAC-European Academies Science Advisory Committee. Shale Gas Extraction: Issues of Particular Relevance to the European Union, 2014. https://www.google.com/?gws_rd=ssl#q=EASAC+-European+Academies+Science+Advisory+Committee.+Shale+Gas+Extraction:+Issues+of+Particular+Relevance+to++the+European+Union+October+2014

EIA, 2013. Technically Recoverable Shale Oil and Shale Gas Resources: An Assessment of 137 Shale Formations in 41 Countries Outside the United States. US Energy Information Agency http://www.eia.gov/analysis/studies/worldshalegas/

Environmental Protection Agency, 2015. EPA's Study of Hydraulic Fracturing for Oil and Gas and Its Potential Impact on Drinking Water Resources. http://www2.epa.gov/hfstudy

Esswein, E.J., Breitenstein, M., Snawder, J., Kiefer, M., Sieber, W.K., 2013. Occupational exposures to respirable crystalline silica during hydraulic fracturing. J. Occup. Environ. Hyg. 10 (7), 347–356.

Esswein, E.J., Snawder, J., King, B., Breitenstein, M., Alexander-Scott, M., Kiefer, M., 2014. Evaluation of some potential chemical exposure risks during flowback operations in unconventional oil and gas extraction: preliminary results. J. Occup. Environ. Hyg. 11 (10), D174–D184.

European Commission, 2014a. Communication from the Commission to the European Parliament, the Council, the European Economic and Social Committee and the Committee of the Regions on the exploration and production of hydrocarbons (such as shale gas) using high volume hydraulic fracturing in the EU. Available from: http://eur-lex.europa.eu/legal-content/EN/TXT/PDF/?uri=CELEX:52014DC0023R(01)&from=EN

European Commission, 2014b. European Science and Technology Network on Unconventional Hydrocarbon Extraction. https://ec.europa.eu/jrc/uh-network

Everley, S., 2013. How anti-fracking activists deny science: water contamination. Energy In Depth, Available from: http://energyindepth.org/national/how-anti-fracking-activists-deny-science-water-contamination/.

Ferrar, K.J., Michanowicz, D.R., Christen, C.L., Mulcahy, N., Malone, S.L., Sharma, R.K., 2013a. Assessment of effluent contaminants from three facilities discharging Marcellus Shale wastewater to surface waters in Pennsylvania. Environ. Sci. Technol. 47, 3472–3481.

Ferrar, K.J., Kriesky, J., Christen, C.L., Marshall, L.P., Malone, S.L., Sharma, R.K., Michanowicz, D.R., Goldstein, B.D., 2013b. Assessment and longitudinal analysis of health impacts and stressors perceived to result from unconventional shale gas development in the Marcellus Shale region. Int. J. Occup. Environ. Health 19 (2), 104–112.

Fiske, Susan, 2010. Social Beings. Core Motives in Social Psychology, 2. Edition John Willey, New York, 89 ff.

Fischhoff, B., 2013. The sciences of science communication. Proc. Natl. Acad. Sci. USA 110 (3), 14033–14039.

Gamper-Rabindran, S., 2014. Information collection, access and dissemination to support evidence-based shale gas policies. Energy Technol. 2, 977–987.

German Advisory Council, 2013. Sachverständigenrat für Umweltfragen (SRU): Fracking zur Schiefergasgewinnung – Ein Beitrag zur energie- und umweltpolitischen Bewertung. SRU, Berlin.

Goldstein, B.D., 2014. The importance of public health agency independence: marcellus shale gas drilling in Pennsylvania. Am. J. Public Health 104, e13–e15.

Goldstein, B.D., 2015. Relevance of transparency to sustainability and to Pennsylvania's Marcellus Shale Act 13. In: Dernbach, J.C., May, J.R. (Eds.), Shale Gas and the Future of Energy: Law and Policy for Sustainability. Edward Elgar Publishing Limited, Cheltenham.

Goldstein, B.D., Kriesky, J., Pavliakova, B., 2012. Missing from the table: role of the environmental public health community in governmental advisory commissions related to Marcellus Shale drilling. Environ. Health Perspect. 120, 483–486, 2012.

Goldstein, B.D., Bjerke, E.F., Kriesky, J., 2013. The challenges of unconventional shale gas development (UGD): So what's the rush? Symp. Green Technol. Infrastruct. 27, 149–186.

Goldstein, B.D., Brooks, B.W., Cohen, S.D., Gates, A.E., Honeycutt, M.E., Morris, J.B., et al.,2014. The role of toxicological science in meeting the challenges and opportunities of hydraulic fracturing. Toxicol. Sci. 139 (2), 271–283.

Gopalakrishnan, S., Klaiber, H.A., 2013. Is the shale boom a bust for nearby residents? Evidence from housing values in Pennsylvania. Am. J. Agric. Econ. 96 (1), 43–66.

Groat, C.G., Grimshaw, T.W., 2012. Fact-Based Regulation for Environmental Protection in Shale Gas Development. The Energy Institute, University of Texas, Austin, Texas, US, Available from: http://barnettprogress.com/media/ei_shale_gas_regulation120215.pdf.

Hammitt, J.K., Wiener, J.B., Swedlow, B., Kall, D., Zhou, Z., 2005. Precautionary regulation in Europe and the United States: a quantitative comparison. Risk Anal. 25, 1215–1228.

Harrison, J., Cosford, P., 2014. Public Health England's reply to editorial on its draft report on shale gas extraction. Brit. Med. J. 348, g3280.

Hays, J., Finkel, M.L., Depledge, M., Law, A., Shonkoff, S.B., 2015. Considerations for the development of shale gas in the United Kingdom. Sci. Total Environ. 512-512, 36–42.

Health Effects Institute, 2015. Strategic Research Agenda on the Potential Impacts of 21st Century Oil and Gas Development in the Appalachian Region and Beyond (draft). http://www.healtheffects.org/UOGD/UOGD.htm

Heikkila, T., Pierce, J., Gallaher, S., Kagan, J., Crow, D.A., Weible, C.M., 2014. Understanding a period of policy change: the case of hydraulic fracturing disclosure policy in Colorado. Rev. Policy Res. 2 (31), 65–85.

Hill, E.L., 2013. Unconventional Natural Gas Development and Infant Health: Evidence From Pennsylvania. Charles Dyson School Applied Economics and Management, Cornell University, Ithaca, NY, Available from: http://dyson.cornell.edu/research/researchpdf/wp/2012/Cornell-Dyson-wp1212.pdf.

Hornbach, M.J., DeShon, H.R., Ellsworth, W.L., Stump, B.W., Hayward, C., Frohlich, C., Oldham, H.R., Olson, J.E., Magnani, M.B., Brokaw, C., et al.,2015. Causal factors for seismicity near Azle, Texas. Nature Commun. 6, 6728.

Howarth, R.W., Ingraffea, A., Engelder, T., 2011. Should fracking stop? Nature 477, 271–275.

iNTeg/Risk, 2014. Early Recognition, Monitoring and Integrated Management of Emerging, New Technology related Risks. http://www.integrisk.eu-vri.eu/

International Energy Agency, 2012. Golden Rules for a Golden Age of Gas: World Energy Outlook Special Report on Unconventional Gas. IEA Publications, Paris, Available from: http://www.worldenergyoutlook.org/media/weowebsite/2012/goldenrules/WEO2012_GoldenRulesReport.pdf.

IRGC–International Risk Governance Council, 2013. Risk Governance Guidelines for Unconventional Gas Development. http://www.irgc.org/wp-content/uploads/2013/12/IRGC-Report-Unconventional-Gas-Development-2013.pdf

Jacquet, J.B., 2013. Risk to Communities from Shale Gas Development. National research council workshop on risks from shale gas development. Washington DC. http://sites.nationalacademies.org/cs/groups/dbassesite/documents/webpage/dbasse_083234.pdf

Jasanoff, S., 2005. Designs on Nature: Science and Democracy in Europe and the United States. Princeton University Press, Princeton, NJ.

Jemielita, T., Gerton, G.L., Neidell, M., Chillrud, S., Yan, B., Stute, M., Howarth, M., Saberi, P., Fausti, N., Penning, T.M., et al.,2015. Unconventional gas and oil drilling is associated with increased hospital utilization rates. PLoS ONE 10 (7), e0131093.

Jenner, S., Lamadrid, A.J., 2013. Shale gas vs. coal: policy implications from environmental impact comparisons of shale gas, conventional gas and coal on air, water, and land in the United States. Energy Policy 53, 442–453.

Jovanovic, A., 2015. Bibliometric Analysis of Fracking Scientific Literature. Scientometrics, 105 (2), 1273–1284.

Jovanovic, A., Klimek, P., Zarea, M., 2012. Monitoring public perception of risks related to unconventional exploitation of gas, in "Think Piece" for the IRGC workshop on "Risk governance guidelines for unconventional gas development". International Risk Governance Council.

Jovanovic, A., Pfau, V., Hahn, R., 2014. Public perception of new energy technologies: Survey on public acceptance of fracking as energy alternative in Germany Discussion paper (Ausgabe 01/2014). (http://www.energy-trans.de/; tp://helmholtz.eu-vri.eu)

Kaplan, T., 2014. Citing Health Risks, Cuomo Bans Fracking in New York State. New York Times December 18, 2014. http://www.nytimes.com/2014/12/18/nyregion/cuomo-to-ban-fracking-in-new-york-state-citing-health-risks.html?

Kasperson, R.E., Renn, O., Slovic, P., Brown, H., Emel, J., Goble, R., Kasperson, J.X., Ratick, S., 1988. The social amplification of risk: a conceptual framework risk analysis. Risk Anal. 8, 177–182.

Kemball-Cook, S., Bar-Ilan, A., Grant, J., Parker, L., Jung, J., Santamaria, W., Mathews, J., Yarwood, G., 2010. Ozone impacts of natural gas development in the Haynesville shale. Environ. Sci. Technol. 44, 9357–9363.

Korfmacher, K., Jones, W.A., Malone, S.L., Vinci, L.F., 2013. Public health and high volume hydraulic fracturing. New Solut. 23, 13–31.

Kovates, S., Depledge, M., Haines, A., Fleming, L.E., Wilkinson, P., Shonkoff, S.B., Scovronick, N., 2014. The health implications of fracking. Lancet 383 (9919), 757–758.

Kriesky, J., Goldstein, B.D., Zell, K., Beach, S., 2013. Differing opinions about natural gas drilling in two adjacent counties with different levels of drilling activity. Energy Policy 58, 228–236.

Krupnick, A., 2012. Risk Matrix for Shale Gas Development. Center for Energy Economics and Policy, Resources for the Future. Washington, DC. http://www.rff.org/centers/energy_economics_and_policy/Pages/Shale-Matrices.aspx

Li, J., Jovanovic, A., Klimek, P., Guo, X., 2014. Bibliographic Analysis and Mapping of Scientific Research Data on Fracking, submitted for publication in Journal of Unconventional Oil and Gas Resources.

Litvak, A. Marcellus Shale waste trips more radioactivity alarms than other products left at landfills. Pittsburgh Post-Gazette August 22, 2013.

Lofsedt, R., 2005. Risk Management in Post Trust Societies. Palgrave Macmillan, London.

Lofstedt, R., 2015. Effective risk communication and CCS: the road to success in Europe. J. Risk Res. 18 (6), 675–691.

Lutz, B.D., Lewis, A.N., Doyle, M.W., 2013. Generation, transport, and disposal of wastewater associated with Marcellus Shale gas development. Water Resour. Res. 49, 647–656.

Mason, K.L., Retzer, K.D., Hill, R., Lincoln, J.M., 2015. Occupational fatalities during the oil and gas boom – Unisted States, 2003-2013. Morbidity Mortality Weekly Report 64(20):551--554.

McKenzie, L.M., Witter, R.Z., Newman, L.S., Adgate, J.L., 2012. Human health risk assessment of air emissions from development of unconventional natural gas resources. Sci. Total Environ. 424, 79–87.

McKenzie, L.M., Guo, R., Witter, R.Z., Savitz, D.A., Newman, L.S., Adgate, J.L., 2014. Birth outcomes and maternal residential proximity to natural gas development in rural Colorado. Environ. Health Perspect. 122 (4), 412–417, DOI:10.1289/ehp.1306722. Available from: http://dx.doi.org/10.1289/ehp.1306722.

Meiners, G.H., Denneborg, M., Müller, F., 2012. Umweltforschungsplan: Umweltauswirkungen von Fracking bei der Aufsuchung und Gewinnung von Erdgas aus unkonventionellen Lagerstätten-Risikobewertung, Handlungsempfehlungen und Evaluierung bestehender rechtlicher Regelungen und Verwaltungsstrukturen. Available from: http://www.bezreg-arnsberg.nrw.de/themen/e/erdgas_rechtlicher_rahmen/gutachten_uba/gutachten_uba_kurz.pdf.

Muehlenbachs, L., Spiller, E., Timmins, C. The Housing Market Impacts of Shale Gas Development; National Bureau of Economic Research, Working Paper 19796, January 2014, http://www.nber.org/papers/w19796

Nash, J., 2013. Assessing the potential for self-regulation in the shale gas industry. In Workshop on Governance of Risks of Shale Gas Development, Washington DC, 2013; Available from: http://sites.nationalacademies.org/DBASSE/BECS/DBASSE_083520

Nelson, A.W., Eitrheim, E.S., Knight, A.W., May, D., Mehrhoff, M.A., Shannon, R., Litman, R., Burnett, W.C., Forbes, T.Z., Schultz, M.K., 2015. Understanding the radioactive ingrowth and decay of naturally occurring radioactive materials in the environment: An analysis of produced fluids from the Marcellus shale. Environ. Health Perspect. 123 (7), 689–696, DOI:10.1289/ehp.1408855.

New York State Department of Health, 2014. A Public Review of High Volume Hydraulic Fracturing for Shale Gas Development, December, 2014. Available from: http://www.health.ny.gov/press/reports/docs/high_volume_hydraulic_fracturing.pdf.

North, D. Warner., Sterm, C. Paul., Webler, Thomas., Field, Patrick., 2014. Public and stakeholder participation for managing and reducing the risks of shale gas development. Environ. Sci. Technol. 48 (15), 8388–8396.

Paulik, L.B., Donald, C.E., Smith, B.W., Tidwell, L.G., Hobbie, K.A., Kincl, L., Haynes, E.N., Anderson, K.A., 2015. Impact of natural gas extraction on PAH levels in ambient air. Environ. Sci. Technol. 49 (8), 5203–5210.

Pennsylvania Department of Environmental Protection, 2015. Online Spud Data Report Database. Available from: http://www.depreportingservices.state.pa.us/ReportServer/Pages/ReportViewer.aspx?/Oil_Gas/Spud_External_Data

Porter, M.E., Gee, D.S., Pope, G.J., 2015. America's Unconventional Energy Opportunity: a win-win plan for the economy, the environment, and a lower-carbon, cleaner energy future. [Report]. pp 1–72. Available from: http://www.hbs.edu/competitiveness/Documents/america-unconventional-energy-opportunity.pdf

Potential Gas Committee, 2015. http://www.frackcheckwv.net/2015/04/10/2015-potential-gas-committee-reports-record-high-natural-gas-resources/

Rabinowitz, P.M., Slizovskiy, I.B., Lamers, V., Trufan, S.J., Holford, T.R., Dziura, J.D., Peduzzi, P.N., Kane, M.J., Reif, J.S., Weiss, T.R., Stowe, M.H., 2015. Proximity to natural gas wells and reported health status: results of a household survey in Washington County, Pennsylvania. Environ. Health Perspect. 123, 21–26.

Renn, O., Benighaus, C., 2013. Perception of technological risk: insights from research and lessons for risk communication and management. J. Risk Res. 16 (3–4), 293–313.

Renn, O., Elliott, E.D., McCright, A.M., 2011. Chemicals. In: Wiener, J.B., Rogers, M.D., Hammitt, J.K., Sand, P.H. (Eds.), The Reality of Precaution. Comparing Risk Regulation in the United States and Europe. Earthscan, London, pp. 223–256.

Renn, O., Schweizer, P., 2014. IRGC's Stakeholder Engagement Resource Guide. http://www.irgc.org/risk-governance/stakeholder-engagement-guide/

Resnikoff, M., 2012. Radioactivity in Marcellus shale challenge for regulators and water treatment plants. Contemporary Technologies for Shale-Gas Water and Environmental Management: in Water Environment Federation, 45–60.

Robinson Township v. Commonwealth of Pennsylvania, 83 A.3d 901.(Pa. 2013).

Rosa, E.A., Renn, O., McCright, A.M., 2014. The Risk Society Revisited: Social Theory Revisited. Temple University Press, Philadelphia, USA.

Royal Society and Royal Academy of Engineering, 2012. Shale gas extraction in the UK: A review of hydraulic fracturing, London. Available from: http://royalsociety.org/uploadedFiles/Royal_Society_Content/policy/projects/shale-gas/2012-06-28-Shale-gas

Rozell, D.J., Reaven, S.J., 2012. Water pollution risk associated with natural gas extraction from the Marcellus shale. Risk Anal. 32 (8), 1382–1393.

Shonkoff, S.B., Hays, J., Finkel, M.L., 2014. Environmental public health dimensions of shale and tight gas development. Environ. Health Perspect. 122 (8), 787–795.

Small, M.J., Stern, P.C., Bomberg, E., Christopherson, S.M., Goldstein, B.D., Israel, A.L., Jackson, R.B., Krupnick, A., Mauter, M.S., Nash, J., et al.,2014. Risks and risk governance in unconventional shale gas development. Environ. Sci. Technol. 48, 8289–8297.

Stacy, S.L., Brink, L.L., Larkin, J.C., Sadovsky, Y., Goldstein, B.D., Pitt, B.R., et al.,2015. Perinatal outcomes and unconventional natural gas operations in southwest Pennsylvania. PLoS ONE 10 (6).

Steinzor, N., Subra, W., Suni, L., 2013. Investigating links between shale gas development and health impacts through a community survey project in Pennsylvania. New Solut. 23 (1), 55–63.

Subra, W., 2010. Community Health Survey Results, Pavillion, Wyoming Residents. Earthworks' Oil and Gas Accountability Project. https://www.earthworksaction.org/files/publications/Pavillion-FINALhealthSurvey-201008.pdf

Vidic, R.D., Brantley, S.L., Vandenbossche, J.M., Yoxtheimer, D., Abad, J.D., 2013. Impact of shale gas development on regional water quality. Science. 340 (6134), 1–9, DOI: 10.1126/science.1235009.

Wachinger, G., Renn, O., Begg, C., Kuhlicke, C., 2013. The risk perception paradox – implications for governance and communication of natural hazards. Risk Anal. 33 (6), 1049–1065.

Werner, A.K., Vink, S., Watt, K., Jagals, P., 2015. Environmental health impacts of unconventional natural gas development: a review of the current strength of the evidence. Sci. Total Environ. 505, 1127–1141.

Wiener, J.B., 2003. Whose precaution after all? A comment on the comparison and evolution of risk regulatory systems. Duke J. Comp. Int. Law 13, 207–261.

Williams, L., Macnaghten, P., Davies, R., Curtis, S., 2015. Framing 'fracking': Exploring public perceptions of hydraulic fracturing in the United Kingdom. Public Underst. Sci., pp 1–17, DOI: 10.1177/0963662515595159

Wilson, J.M., Van Briesen, J.M., 2012. Oil and gas produced water management and surface drinking water sources in Pennsylvania. Environ. Pract. 14 (4), 288–300.

Zucker HA. 2014. New York Department of Health: A Public Review of High Volume Hydraulic Fracturing for Shale Gas Development . Memo to Joseph Martens. Available from: http://www.health.ny.gov/press/reports/docs/high_volume_hydraulic_fracturing.pdf

CHAPTER 9

Transportation of Shale Gas and Oil Resources

Alicia Jaeger Smith, Matthew M. Murphy
BatesCarey LLP, Chicago, IL, USA

Chapter Outline

Environmental and Health Issues in Unconventional Oil and Gashttp://dx.doi.org/10.1016/B978-0-12-804111-6.00009-1

ROUTES AND MODES OF TRANSPORTATION OF SHALE GAS AND OIL

Shale gas and oil resources are located in many parts of the world. However, only certain countries have been able to recover these resources. Once they have been recovered, they need to be transported for refinement and processing, and for use as products. Transportation of shale gas and oil resources primarily occurs via pipeline, tanker, rail, barge, and truck.

Routes of Transportation

SHALE GAS

As of mid-2013, the countries with the most significant technically recoverable shale gas resources were China, Argentina, Algeria, the United States, Canada, Mexico, Australia, and South Africa (EIA, 2013) (see further discussion on International Markets in Chapters 2 and 13). However, out of those countries, the United States and Canada were the only large-scale producers of commercially viable natural gas from shale formations in the world as of that time (EIA, 2013).

Large-scale shale gas production in the United States (US) started around 2000 in the Barnett shale region in Texas, and since then, there has been a dramatic growth in production in eight other regions as well: Marcellus (PA and WV), Bakken (ND), Haynesville (LA and TX), Eagle Ford (TX), Fayetteville (AR), Woodford (OK), Antrim (MI, IN and OH), and Utica (OH, PA and WV) (EIA, 2014) (see fig 1.4 in Chapter 1). Demand for shale gas exports mainly comes from Canada, and increasingly, Mexico, and a future market for exports to Europe and Asia.

SHALE OIL

As of mid-2013, the countries with the most significant technically recoverable shale oil resources were Russia, the United States, China, Argentina, Libya, Australia, Venezuela, and Mexico. Since 2013, the United States has surpassed Russia in shale oil production (EIA, 2013). Most of the shale oil produced in the United States comes from North Dakota, Texas, Oklahoma, New Mexico, Wyoming, Colorado, and Utah.

Almost all oil produced in the US must be transported to refineries before it can be utilized as a product. As of 2012, there were approximately 115 refineries located in the US, that have been organized into districts as delineated by the Petroleum Administration for Defense Districts (PADD) (CRS and EIA, 2012; Conca, 2014).

Modes of Transportation

In the past, the vast majority of US oil and gas (O&G) resources were transported via pipelines and oil tankers, with barge, truck, and rail being the other modes of transport. In fact, the US Department of Transportation's analysis of crude oil and petroleum products transported domestically by mode in 2012 essentially showed that approximately 90% of crude oil, and petroleum products were transported by pipeline and oil tanker (CRS, 2014). However, rail companies have stepped up since 2012 to transport the increasing amount of US shale gas and oil resources, and there has also been a marked increase in transport by barge and truck.

SHALE GAS

Shale gas is mainly transported within the US via pipeline, utilizing existing and new interstate, and intrastate pipeline infrastructure (EIA, 2008). Figure 9.1 presents a map of the US natural gas pipeline network (EIA, 2009). The US liquids pipeline operators totalled 192,396 miles in 2013, of which 60,911 miles were crude oil pipeline. These liquid pipelines transported 14.9 billion barrels of crude oil (8.306 billion) and petroleum products (6.642 billion) in 2013, an increase of 6.2% from 2012 (AOPL and API, 2014).

Notably, pipelines are also the primary method of exporting US-mined natural gas to Mexico. As of 2015, the demand for natural gas has grown substantially since 2010 and is projected to continue increasing due to Mexico's growing

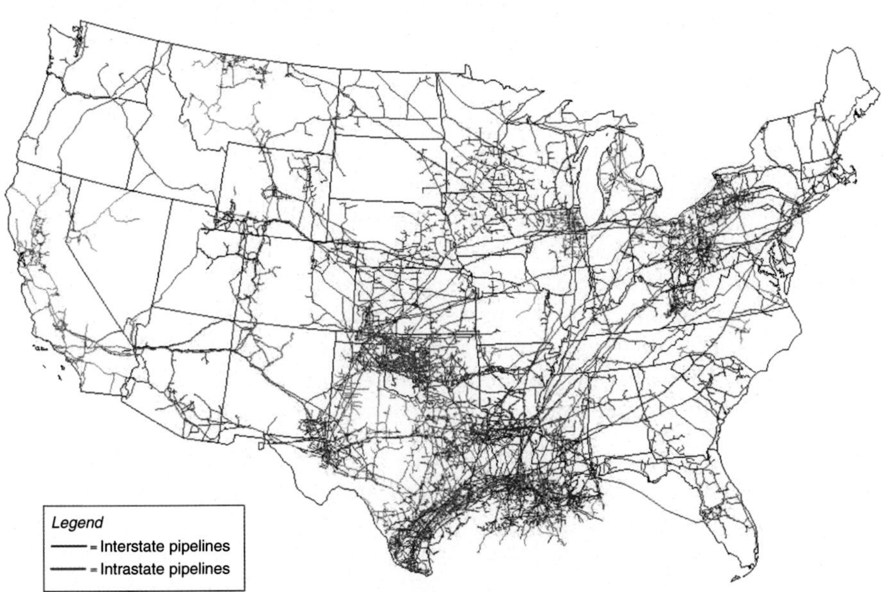

FIGURE 9.1
Source: U.S. Energy Information Administration (2009).

demand for natural gas, primarily for electric power generation (EIA, 2015) (see eg., Chapter 2).

In addition, shale gas is transported via tanker from certain locations on the Atlantic and Gulf coasts (EIA, 2008). Shale gas is also transported in the United States and Canada via railroad and truck.

SHALE OIL

Shale oil is primarily transported via pipeline, utilizing existing and new pipeline infrastructure. US energy companies are continuing to build new pipelines to accommodate the increase in recovery of shale oil resources, particularly from the Bakken region, to market centers in the midwest, and other parts of the United States (Eggleston, 2014).

However, shale oil is increasingly being transported via rail (which has become commonly known as "crude-by-rail") (EIA, 2014). A train unit of 70–120 tank cars can carry anywhere from 50,000–90,000 barrels of oil, depending on the type of crude (CRS, 2014).

Crude-by-rail has increased exponentially: in 2008, the Association of American Railroads (AAR) estimated that member railroads carried approximately 9,500 carloads of crude oil in 2008, compared to 400,000 carloads of crude oil in 2013, further estimated by AAR (AAR, 2013). There are approximately 140,000 miles of private railroad track in the United States (Freight Rail Works and AAR, 2012). Figure 9.2 illustrates the over 93,000 miles of Class I Railroad routes alone.

The number of rail carloads of crude oil has risen since 2012, as production in the Bakken region and other shale regions has grown. In addition to crude-by-rail loading facilities in the Bakken region, new regional rail terminals have been built to transport crude oil produced in the Niobrara shale region (CO and WY) and the Permian Basin (TX and NM) (EIA, 2014). While some refineries are being built or planned in these areas, particularly Bakken, it appears that at least for the foreseeable future, most of the crude oil recovered will continue to be moved out of those regions to be processed at refineries in other parts of the country (particularly on the East and West Coasts). West Coast refineries have ramped up crude-by-rail volumes in recent years, particularly in Washington and California.

While some refineries located on the coasts and near major rivers have onsite rail unloading terminals, others must receive crude oil by barge, after it has been moved by rail to a nearby unloading terminal (MO, AR, IL, NY, VA, PA, and WA). A single river barge can hold 10,000–30,000 barrels of oil (CRS, 2014). Typically, several river barges are tied together in a single tow that allows for the movement of 20,000–90,000 barrels, approximately the same load as a train (CRS, 2014). There are also articulated tug-barges (ATBs), which are coastal tank

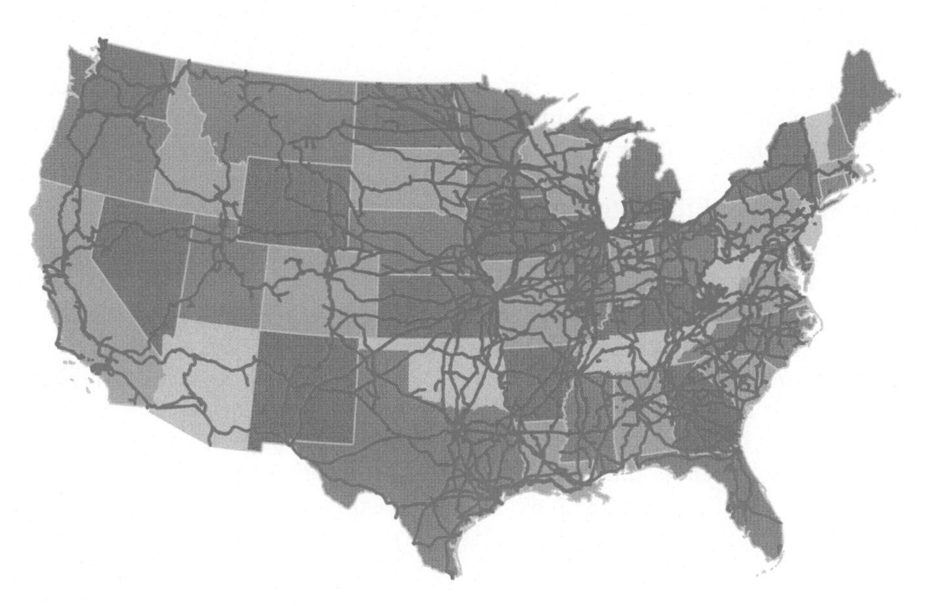

FIGURE 9.2
Source: Association of American Railroads.

barges that travel on the open sea and have a capacity of 50,000–185,000 barrels, while newer ATBs can carry 50,000–185,000 barrels (CRS, 2014).

ISSUES AND CONCERNS ARISING OUT OF THE TRANSPORTATION OF SHALE GAS AND OIL

With the recent boom in the development of shale gas resources, several issues and concerns have arisen with respect to transportation of those resources for refinement, sale, or use. There have been several high-profile incidents in recent years, which have raised environmental and safety concerns about the transportation of O&G, particularly by pipeline and rail. In addition, the need to transport greater amounts of O&G resources within the United States has raised issues regarding the sufficiency of infrastructure, the capacity of the modes of transportation to handle O&G as well as other resources, and territorial and property rights. Finally, this section also touches on insurance considerations attendant to these issues and concerns.

Environmental and Safety Concerns

Transporting shale O&G presents potentially significant harm to the environment and to human health and safety, if there is an accident that compromises the containers. When shale gas is released or oil spills and burns, the affected air, land, and groundwater may be compromised and property and human lives may be damaged or destroyed. The current focus of these concerns is on pipeline and rail.

PIPELINES

With respect to environmental contamination concerns about transportation of oil via pipeline, several recent pipeline incidents have been in the national spotlight. Two high-profile examples are discussed in detail in this section.

Marshall, MI (Enbridge Energy Partners Pipeline)

On July 26, 2010, a 30-in. pipeline ruptured near Marshall, Michigan, spilling an estimated 1 million gallons of heavy crude oil (originating from the Canadian tar sands). The released oil entered Talmadge Creek and flowed into the Kalamazoo River, a Lake Michigan tributary (EPA, 2015). The spill was exacerbated by heavy rains that caused the river to overtop existing dams and carried oil 35 miles downstream on the Kalamazoo River. In mid-May 2015, Enbridge reached a $75 million settlement with the State of Michigan, including cleanup, ongoing monitoring, and restoration and improvement of affected wetlands (Hasemyer, 2015). Concerns have also been expressed about other, older Enbridge pipelines in the area (Lachman, 2015).

Refugio, CA (Plains All American Pipeline)

On May 19, 2015, a pipeline carrying crude oil ruptured along the Gaviota Coast near Santa Barbara, California, resulting in the spill of 100,000 gallons of crude oil (which originated offshore) including 20,000 gallons that flowed into the Pacific Ocean at Refugio Beach. The spill fouled an ecologically sensitive stretch of coast, harmed or killed dozens of marine mammals, and seabirds, shut down a popular state beach park from before Memorial Day to beyond the Fourth of July, and resulted in tar balls washing ashore on distant beaches along the Southern California coast (Herdt, 2015). As of early July 2015, the pipeline owner had already spent at least $100 million on cleanup and response efforts.

In the aftermath of the Refugio spill, federal and state officials are taking a new look at oil pipeline safety. They are asking what can be done to improve an oil-transport system that, "all parties acknowledge is safer than tankers, trains and trucks" but not without safety concerns (Herdt, 2015). They want to ensure that ruptures can be prevented and better contained when they do occur. State officials have opened civil and criminal investigations into the incident, and Santa Barbara's US Congressional representative has written to the Federal Office of Management and Budget, and the US Department of Transportation's Pipeline and Hazardous Materials Safety Administration (PHMSA) demanding a status update on the new rules for implementing 17 pipeline safety measures approved by Congress in 2011 (Herdt, 2015).

With respect to safety of transportation of crude oil by pipeline, pipeline proponents such as American Petroleum Institute (API), the Fraser Institute and the Manhattan Institute for Policy Research have argued that, in terms of human casualties, pipelines are safer than rail, truck, and barge for transporting oil. Specifically, the Fraser Institute stated that while pipelines may leak – trains and trucks can crash (hurting individuals) and barges can sink (Furchtgott-Roth and

Green, 2013). Pipeline proponents have analyzed publicly available data from PHMSA, comparing the safety of transportation of O&G by pipeline, road, and rail in the United States. Data on incident, injury, and fatality rates for these modes of transportation during the 2005–2009 period (the latest data available) demonstrate that road and rail have higher rates of serious incidents, injuries, and fatalities than pipelines, even though more road and rail incidents go unreported. In addition, with respect to its support of the Keystone XL pipeline, the API has stated that the US government's own environmental review indicates that the Keystone XL pipeline would "have a degree of safety over any other" (API, 2015).

In addition, major safety concerns have been raised with respect to the transportation of natural gas by pipeline as a result of rupture, and explosion incidents that have occurred. The most high-profile incident is the Pacific Gas and Electric (PG&E) gas pipeline explosion that occurred in San Bruno, California (near the San Francisco International Airport) on September 9, 2010, killing 8 people, destroying 35 homes, and damaging dozens more (Egelko, 2014). That incident has raised concerns about the age, condition, viability, and maintenance status of the natural gas pipeline infrastructure in the United States.

RAILROADS

Several recent railroad incidents have raised significant environmental and safety concerns about transportation of crude-by-rail. The most high-profile incident is the train derailment and resulting explosion in Lac-Mégantic, Quebec in July 2013. Two other recent crude-by-rail incidents are also discussed in detail. Incidents like these have resulted in crude-by-rail trains being vilified and labelled as "bomb trains" by public media outlets.

Lac-Mégantic, Quebec

On July 6, 2013, an unattended freight train carrying 72 cars of crude oil from the Bakken shale region rolled downhill and derailed in the town center of Lac-Mégantic, Quebec (near the Maine border), resulting in a fire and explosion that killed 47 people and destroyed most of the buildings in the town's center (CBC News, 2015). In addition, most of the remaining buildings in the town's center needed to be demolished due to petroleum contamination. Furthermore, about 26,000 barrels of oil were released into the Chaudière River (Villenueve, 2014).

In addition to the loss of life and property, the Lac-Mégantic incident highlighted many issues associated with the potential environmental impact of a derailment resulting in the release of shale oil. First, the Lac-Mégantic fire department was not aware that they needed to use fire-extinguishing foam to extinguish oil fires. Had they known to do so, the shale oil would have been less likely to migrate to the Chaudière River and other nearby lakes and streams. Moreover, the fire may have been extinguished faster, thereby emitting fewer greenhouse gases to the environment. The actual site of the derailment also was saturated with shale oil. In the aftermath of the incident, government officials debated whether to excavate and replace thousands of cubic yards of tainted soil, or to cordon off the area and relocate buildings and residents to a new downtown area yet to be developed. According to Lac-Mégantic City Councilor Roger Garant, the cost for rebuilding the town could total $2.7 billion over the next decade (Villenueve, 2014). The clean up of the contaminated land, sewer system and nearby bodies of water will cost a minimum of $200 million, but estimates by federal officials have been more than twice that sum (Sharp, 2014).

Aliceville, Alabama

On November 8, 2013, a 90-car freight train operated by Genesee & Wyoming Railroad, carrying 2.7 million gallons of crude oil from the Bakken shale region, derailed near a 60-foot-long wooden trestle in Aliceville, Alabama (Gates and McAllister, 2013). Residents were evacuated from the area immediately following the incident and extensive damage occurred to the adjacent wetland area. Four months later, the Associated Press reported that crude oil continued to leak into the wetland area (AP, 2014). Environmental groups alleged that Genesee & Wyoming Railroad undertook insufficient remediation efforts and left the area once the tracks were repaired. According to the EPA and Alabama Department of Environmental Management, the railroad removed 10,700 gallons of oil from the water and workers collected 203,000 gallons of shale oil from the damaged rail cars and the railroad further excavated some 290 cubic yards of oil-filled dirt (AP, 2014). Due to the greenhouse gas emitting fire, it remained unclear to environmental regulators how much shale oil burned.

Casselton, North Dakota

On December 30, 2013, a westbound BNSF freight train carrying grain derailed just west of Casselton, North Dakota. When it derailed, one of the grain cars in the middle of that train landed on the adjacent track. Less than a minute after the derailment, an eastbound BNSF train carrying petroleum crude oil struck the grain car (NTSB, 2014a). Both head locomotives and 21 cars of the 106-car oil train derailed. While no one was injured, the collision set off a series of explosions and an estimated 18 cars of the oil train were breached and caught fire and burned through the night. In the initial hours after the explosion, authorities advised residents to stay indoors to avoid the smoke (Shaffer, 2014). Later that day, the Cass County Sheriff's' Office urged residents within 5 miles of Casselton to leave until the situation could be evaluated for health hazards (Inforum, 2013). According to Incident Command, roughly 65% of the areas 2500 residents voluntarily evacuated until the next day. According to the National Transportation Safety Board (NTSB) media briefing on January 1, most of the oil burned and there was limited contamination of the surrounding soil (NTSB, 2014b).

As a result of these rail incidents, the safety of DOT-111 tank cars has been called into question. While some new or adapted DOT-111 tank cars have been placed into operation, most of the DOT-111 tank cars currently in use are the older version of the tank cars.

Furthermore, several municipalities in the US have begun a public discussion about preparedness for railroad disasters involving transportation of crude oil.

These municipalities are located on regular routes of transportation of crude oil, including shale oil from the Bakken region in North Dakota. For example, Chicago is considered the nation's rail hub; much of the shale oil coming from North Dakota or the oil sands of Canada must pass through Chicago and its surrounding suburbs on its way to refineries on the East Coast (Wronski, 2014a). This has led to protests and other expressions of concern from local governments and residents.

Other concerns, as noted in a May 25, 2014 article in the *Chicago Tribune*, are that area fire departments and first responders do not have enough firefighting foam or the proper knowledge, training or equipment to appropriately respond to crude-by-rail fires (Wronski, 2014b). In an attempt to increase training and knowledge of first responders, at least one railroad has conducted training seminars whereby first responders are taught emergency response techniques in the event of a derailment. Many towns in the area surrounding Chicago believe they would be better served than other areas to handle a derailment containing hazardous materials, due to the existence of the Mutual Aid Box Alarm System (MABAS) that facilitates numerous municipalities to gather collective resources in the event of an emergency.

In addition, some municipalities have complained about their inability to gain access to lading information to determine whether shale oil is moving through their towns on existing rail lines, which has come into conflict with railroad and governmental concerns about security issues connected to the general public having knowledge of shipment information. In mid-2014, the US Department of Transportation (US DOT) ordered the railroads to disclose to State Departments of Transportation the routes of crude oil movement, against the wishes of the railroads. The railroads initially expressed general safety concerns with the public being aware of the routes of the movement of shale oil. Notwithstanding their collective concern, the US DOT ordered that the State Departments of Transportation be made aware of the rail schedule. The Associated Press subsequently filed FOIA (Freedom of Information Act) requests to the various State Departments of Transportation and published the rail routes of trains carrying shale oil through towns. In May 2015, WGNTV News in Chicago reported that the railroad industry is starting to roll out a new app called "Ask Rail" that will be available to first responders providing information about materials being carried on trains (Jindra, 2015).

SUSPECTED INCREASED VOLATILITY OF SHALE OIL

Leading to concerns about transportation of shale oil, have been several reports suggesting that crude oil extracted from the Bakken shale region in North Dakota has an increased level of volatility, raising issues about the degree of harm that may be caused and the ability to respond to incidents. The increased volatility of shale oil has been suspected, based on fires following numerous recent derailments, including those discussed previously. On November 20, 2013 and January 2, 2014, PHMSA issued safety advisories with regard to the properties of Bakken crude oil (PHMSA and FRA, 2013; PHMSA, 2014). Further studies are being conducted to determine whether the oil from this region is more volatile than other crude oil.

RAIL VERSUS PIPELINE SAFETY DEBATE

While AAR has maintained that carrying shale oil by train provides limited environmental impact, the data from PHMSA found that the majority of the worst railroad spills have occurred in recent years (Doukopil, 2015). The US State Department, in its long awaited report and recommendations for approving the application to build the Keystone XL Pipeline, confirmed this finding, opining that if the Keystone XL was not built, the rail industry would continue to move shale oil at a higher rate and potentially higher environmental impact (US Dept. of State, 2014). More specifically, the State Department concluded that if the Keystone XL Pipeline is not built and existing rail assumed the transport of shale oil from Canada, North Dakota, and the Northeast, it opined that 1,227 barrels of shale oil would be released a year, more than double the number of barrels that would be released if the Keystone XL Pipeline is built. Assuming combined rail and tanker modes of transport, the State Department estimated that 4,633 barrels would be released a year.

The AAR has stated that it vehemently disagrees with the State Department's conclusions (AAR, 2014). The AAR analyzed the data released from PHMSA and concluded that the so-called "spill rate" for railroads from 2002–2012 was approximately 2.2 gallons per million ton miles crude oil generated, but the comparable spill rate for pipelines is approximately 6.3 gallons per million ton miles. Notably, the AAR's data does not take into account the massive increase in transporting Bakken shale oil by rail from the period immediately following the data reported by the PHMSA.

As of the time of writing, the Keystone XL Pipeline has not yet been passed into law. While the US Senate voted 62–35 in favor of the building of the Keystone Pipeline in January 2015, President Obama threatened to veto the bill. Before the vote, Senate Majority Leader Mitch McConnell (R-KY) stated "[c]onstructing Keystone would pump billions into our economy .... And as the President's own state department has indicated, it would do this with minimal environmental impact" (Harder, 2015).

Infrastructure and Capacity Concerns

The increased need to transport shale gas and oil resources from various regions in the US and Canada, has led to a rise in concerns with regard to the ability of existing infrastructure to handle the increase in transportation. These newer supply sources may not be connected to the old refinery infrastructure or may not have the capacity to handle the increased volume. There are also concerns about viability of an aging infrastructure.

PIPELINES

Pipelines have been used for decades in North America to transport crude oil. Furchtgott-Roth and Green (2013) noted that, with the increase in availability of crude oil from the Alberta "oil sands" region in Canada and the Bakken shale region in North Dakota, there is an increased demand for pipeline capacity.

However, industry efforts to install the Keystone XL pipeline in the US have drawn public criticism of pipelines due to concerns over the environment, safety, and territorial and property rights. Furthermore, with respect to natural gas transportation, the 2010 San Bruno pipeline incident (discussed previously) highlights safety concerns with respect to aging natural gas pipelines.

RAILROADS

As the increased amount of shale gas and oil resources being mined in the US and the inability of the existing pipeline infrastructure to handle it all, railroads are being asked to increase their capacity to transport shale gas and oil (CRS, 2014). Moving shale oil by rail has the benefit of an existing rail network. However, capacity is limited by the availability of the appropriate rail cars, the location of the refineries, and increasing output of shale oil. There is little doubt that existing capacity must increase within the next few years in order to take additional product to market. Major rail companies plan to expand rail terminals, purchase new tank cars, utilize longer and heavier trains, and increase speed along their tracks.

One of the most prominent concerns for transportation of crude oil is the ability to provide enough tank cars, and in particular, tank cars that meet safety regulations. In the US, there is currently a reluctance to build new tank cars or retrofit old tank cars before definitive regulations are issued by the FRA. This may create a capacity problem if railroads do not have enough tank cars deemed "acceptable" to handle crude oil to meet the demand for transportation of the oil. In addition, shale oil producers face competition from agricultural producers (particularly corn and soybeans) for availability of tank cars to transport their products.

Furthermore, because of recent crude-by-rail incidents, mandates have been put in place for decreased speed of trains carrying crude oil (50 mph generally, 40 mph in urban areas), which constrict the use of existing railroad infrastructure. Concerns have been expressed about congestion on the rail networks if rail cars are added to provide increased capacity (Rucker, 2014).

Property and Territorial Rights

At this time, there is insufficient pipeline capacity to handle the amount of shale oil coming from Canada, North Dakota, and the Northeast. Building additional pipeline is fraught with complications, including the property and territorial rights of those whose land the pipeline builders would use.

In the case of the Keystone XL pipeline, three Nebraska landowners brought a suit against the governor of the State of Nebraska, on the grounds that state statute L.B. 1161 is unconstitutional in the *Thompson v. Heineman* case (Thompson, 2015). L.B. 1161 is a Nebraska state law that allows major oil pipeline carriers to bypass the regulatory procedures of the Public Service Commission (a constitutional body charged with regulating common carriers) and seek approval from the governor to exercise the power of eminent domain for building a pipeline in Nebraska (State of Nebraska, 2012). In January 2015, the Nebraska Supreme Court held that while the majority of the judges agreed that the constitutionality of L.B. 1161 is questionable, the laws of the State of Nebraska require a supermajority of five judges before the court can strike a state statute (Thompson, 2015). This decision will allow the Keystone XL Pipeline to go forward with eminent domain to procure the land it needs to proceed (if and when the federal government allows it). The example of the *Thompson* case is indicative of the type of legal challenges inherent in building hundreds of miles of additional pipeline.

Insurance Considerations

On August 1, 2014, the US DOT issued a Draft Regulatory Impact Analysis stating, *inter alia* that railroads do not have sufficient insurance to cover even a moderate spill associated with a train derailment carrying shale oil (DOT, 2014a). The US DOT proposed a rule, which may require Class I railroads to carry up to $1 billion of insurance coverage or establish that it has the financial means to pay for a massive spill. At least one major US insurer, AIG, has introduced a $1 billion per occurrence policy after a railroad has incurred more than $1.5 billion of costs in an accident. The policy would cover third-party property damage and bodily injury in the event of a derailment or other accident comparable to the Lac-Mégantic incident. According to Russ Johnston, head of AIG's casualty operations, this is one of the largest policies offered to Class I rail carriers in North America. It is expected that other insurers will explore this market as well.

In Canada, Transport Canada (TC) announced on June 18, 2015 that enhanced insurance requirements have been put into place through the Safe and Accountable Rail Act, which amends the Canada Transportation Act to strengthen the liability and compensation regime for federally regulated railways, and which are set to come into force in June 2016 (TC, 2015). Notably, the Canadian government aims to align these updated railroad regulations with updates being made to the liability and compensation regimes in other sectors of transport, including for marine tankers and oil pipelines (TC, 2015).

REGULATIONS

In the wake of recent incidents involving the transportation of shale gas and oil resources, regulatory bodies in the US and Canada have developed regulations aimed at promoting safety, particularly with respect to transportation of O&G by pipeline and rail. In addition, regulatory agencies have added certain labeling requirements for transportation of shale gas and oil products. A discussion of these regulations is set forth in detail in this section.

Pipelines

PHMSA

The PHMSA is a US DOT agency responsible for developing and enforcing regulations related to pipeline transportation in the US (for interstate pipelines and intrastate when safety standards and practices are not regulated by a state agency or municipality) (49 U.S.C. § 60104(c) and 49 U.S.C. § 60105). The PHMSA's regulatory authority is derived from numerous statutes related to pipeline safety, ranging from the Natural Gas Pipeline Safety Act 1968, to the Pipeline Safety, Regulatory Certainty, and Job Creation Act 2011 (the "Pipeline Safety Act"). The Pipeline Safety Act was passed in partial response to recent pipeline disasters in San Bruno, California, Allentown, Pennsylvania, and Marshall, Michigan. The Pipeline Safety Act and its predecessor statutes authorize the DOT (which in turn delegates its authority to PHMSA) to establish minimum federal safety standards for the transportation of gas and for pipeline facilities (49 U.S.C. § 60102(a)(1)).

Due in part to the pipeline failures referenced previously, as well as the rapid increase in shale gas development, the Pipeline Safety Act charged PHMSA with the task of evaluating the necessity of more stringent safety measures, particularly in PHMSA's integrity management regulations for hazardous liquid pipelines. Most recently, in the wake of the Refugio pipeline spill in California (as well as the Enbridge spill in Michigan) the *Ventura County Star* (Herdt, 2015) reported that the US House of Representatives approved on a bipartisan vote an appropriations measure that would designate a portion of PHMSA's budget to be spent on completing rules to require enhanced automatic shut-off and spill-detection standards.

As of this writing, PHMSA has issued one final rule in response to the 2011 legislation, with reauthorization of the Pipeline Safety Act set for September 2015. Pursuant to 78-FR 58897, which went into effect in October 2013, PHMSA amended its pipeline safety regulations to update the administrative civil penalty maximums for violation of safety standards, updated informal hearing and the adjudication process for pipeline enforcement matters, and made other technical corrections and updates to certain administrative procedures. The amendments did not impose any new operating, maintenance, or other substantive requirements on pipeline owners or operators. In early July 2015, PHMSA made a rulemaking proposal which, among other things, called for a requirement that pipeline operators report spills within an hour of their discovery, in order to help mitigate damage (Graeber, 2015). Congress may pass a "clean" bill in its current form, but will more likely impose new substantive amendments to the statute to address recent incidents (Hunton & Williams LLP, 2015).

OTHER FEDERAL AGENCIES

While PHMSA's focus is geared toward safety and integrity management, numerous federal agencies partake in the pipeline regulatory scheme. For example, the Transportation Security Administration has responsibility for coordinating security for all transportation-related operations, including pipelines. Perhaps more significantly, the Federal Energy Regulatory Commission (FERC) – an independent agency operating in the Department of Energy, has responsibility for energy supplies, refinery operations, and natural gas pipeline regulation. The FERC and the DOT (and thus, PHMSA) entered into a memorandum of understanding in 1993 whereby FERC acknowledges the DOT's exclusive authority to promulgate federal safety standards for facilities used in the transportation of natural gas (DOT and FERC, 1993). Under the same memorandum, the DOT acknowledges that under the Natural Gas Act, FERC exercises the authority over the seating of interstate natural gas transmission facilities and may impose conditions to mitigate the impact of construction or operation on the environment (see Chapter 12 for further discussions on regulatory). The memorandum requires the DOT to promptly notify FERC of safety activities, major accidents, and enforcement actions. In turn, FERC must promptly notify the DOT of safety issues regarding natural gas facilities, planned pipeline construction, and safety issues developed in FERC's environmental assessments.

STATE REGULATIONS

As discussed earlier, PHMSA's regulatory authority, technically does not apply to intrastate pipeline facilities and intrastate transportation of gas when safety standards and practices are regulated by a state agency or municipality (49 U.S.C. § 60105). However, state agencies must submit an annual certification to PHMSA to exercise their regulatory authority over intrastate pipelines. To qualify for certification, a state must adopt the minimum federal regulations and may adopt additional or more rigorous rules as long as they do not conflict with PHMSA's rulemaking. States are also required to develop enforcement sanctions on par with those authorized by the Pipeline Safety Act and other federal pipeline statutes. Most state pipeline safety programs are administered by public utility commissions.

Railroads

FEDERAL REGULATIONS

There is significant and growing regulation of the transportation of O&G by rail by the federal government. In addition to pipeline regulation, PHMSA is also charged with regulating the transportation of hazardous materials (including O&G) by rail. In addition to PHMSA, the Federal Railroad Administration (FRA) – another agency operating in the DOT, provides oversight of railway O&G transport in the US. Outside of the DOT, the Transportation Security Administration, and the NTSB also have a hand in regulating the transportation of O&G by rail. Although the NTSB has no statutory regulatory authority, its recommendations are often agreed with and adopted by regulatory agencies.

Due to an increase in the production of crude oil and recent high-profile rail disasters, federal agencies have looked to adopt more stringent regulations for the transportation of crude-by-rail. In February 2014, the DOT issued an emergency order requiring all shippers to test product from the Bakken region of North Dakota to ensure the proper classification of crude oil before it is transported by rail, while also prohibiting the transportation of crude oil in the lowest-strength packing group (DOT, 2014b). The PHMSA and the FRA have proposed other new rules and sought comments related to transportation of crude oil, but to date, few of the proposals have actually been instituted.

On May 1, 2015, the US DOT announced a long-awaited final rule for the safe transportation of flammable liquids by rail, developed by the PHMSA and the FRA in coordination with Transport Canada (TC) (DOT, 2015). The final rule establishes requirements for any "high-hazard flammable train" (HHFT), which is defined as a train comprised of 20 or more loaded tank cars of Class 3 flammable liquid in a continuous block, or 35 or more loaded tank cars of a Class 3 flammable liquid across the entire train. The rule also defines a "high-hazard flammable unit train" (HHFUT) as a train comprised of 70 or more loaded tank cars containing Class 3 flammable liquids and traveling speeds at greater than 30 mph. The regulatory changes implemented in this final rule are as follows:

- Tank car design: (1) new tank cars constructed after October 1, 2015 are required to meet the enhanced DOT Specification 117 design or performance criteria; (2) existing tank cars must be retrofitted in accordance with the DOT-prescribed retrofit design or performance standard; and (3) retrofits must be completed based on a prescriptive retrofit schedule and a retrofit reporting requirement is triggered if the initial milestone is not achieved. The entities affected by these new requirements are tank car manufacturers, tank car owners, shippers/offerors, and rail carriers. The requirements to retrofit or retire existing tank cars are expected to cost the industry up to $1.747 billion over 20 years, and new tank car construction based on these requirements is expected to cost $34.8 million over 20 years.
- A more accurate classification of unrefined petroleum-based products, including the following: develop and carry out new sampling and testing program directives for all unrefined petroleum-based products, such as crude oil; and certify that program is in place, document the testing and sampling program outcomes; and make information available to DOT personnel upon request. These requirements affect shippers/offerors of unrefined petroleum-based products, and are expected to cost $18.9 million over 20 years.
- Enhanced braking system elements, including: (1) requiring HHFTs to have in place a functioning two-way end-of-train device or a distributed power braking system; (2) requiring trains meeting the definition of a HHFUT to be operated with an electronically controlled pneumatic braking system by January 1, 2021, when transporting one or more tank cars loaded with a Packing Group I flammable liquid; and (3) require trains meeting the definition of a HHFUT be operated with an electronically controlled pneumatic braking system by May 1, 2023, when transporting one or more tank cars loaded with a Packing Group II or III flammable liquid. These requirements affect rail carriers and are expected to cost $492 million over 20 years.
- (1) Rail Routing – Risk Assessment: including performance of a routing analysis that considers, at a minimum, 27 safety and security factors and selects a route based on its findings (as prescribed in 49 CFR § 172.820). (2) Rail Routing – Notification, which ensures that railroads notify State and/or regional fusion centers and State, local, and tribal officials who contact a railroad to discuss routing decisions are provided appropriate contact information for the railroad in order to request information related to the routing of hazardous materials through their jurisdictions. These requirements affect rail carriers and emergency responders, and are expected to cost $8.8 million over 20 years.

- Reduced operating speeds: (1) restrict all HHFTs to 50 mph in all areas and (2) require HHFTs that contain any tank cars not meeting the enhanced tank car standards required by this rule operate at a 40 mph speed restriction in high-threat urban areas. These requirements affect rail carriers, and are expected to cost $180 million over 20 years.

In total, these regulatory requirements are expected to cost the industry nearly $2.5 billion. In comparison, DOT analysis shows that expected damages based on the historical safety record are expected to exceed $4.1 billion (undiscounted) and that damages from high-consequence events could reach $12.6 billion (undiscounted) over a 20-year period in the absence of the rule (DOT, 2015).

STATE REGULATIONS

In addition to the influx of federal regulations discussed earlier, state legislators have also been active in adopting (or attempting to adopt) new regulations related to the transportation of crude oil by rail. As held in the *Tyrrell* case in the Sixth Circuit (Tyrrell et al., 2001) under the federal railroad regulatory scheme, the FRA (and the DOT) is the primary authority over national rail safety policy. Per the *CSX Transp., Inc. v. Plymouth, 2000* out of the Eastern District of Michigan (2000) a state regulation related to railroad safety is permissible where the Secretary of Transportation has not regulated, or where the state regulation is in response to a local, rather than a national, safety concern, so long as the regulation is not in conflict with federal law and does not unduly burden interstate commerce.

In December 2014, the North Dakota Industrial Commission (Order No. 25417) adopted conditioning standards to improve the safety of Bakken crude oil for transport (North Dakota Industrial Commission, 2014). While not directly addressing rail transport, the regulation significantly affects the transportation of Bakken crude oil by rail. The Order requires operators to condition Bakken crude oil to a vapor pressure of no more than 13.7 pounds per square inch (psi) (national standards are 14.7 psi) and imposes a $12,500 fine for producers who fail to comply.

Oju and Hartman (2014) reported that nationally, state legislatures in California, Maine, Minnesota, Oregon, and Pennsylvania have proposed laws that directly address safety issues related to the transportation of O&G. Most of these proposed laws relate to oil spill contingency plans; health and safety studies on the effects of transporting O&G materials; providing funding, personnel, and training for emergency preparedness; enhancing safety standards for tank cars; and conducting comprehensive inspections of railroads.

CANADA

Since July 2013, following the Lac-Mégantic, Quebec incident, TC has implemented numerous safety measures since the Lac-Mégantic derailment (TC, 2015). The following directives and rules have been implemented: requiring that unattended locomotives be secured and establishing the number of crew members required for operating a locomotive carrying dangerous goods; requiring any person who imports or offers for transport crude oil to retest or classify their crude oil prior to shipment, and, in the interim, ship it at the highest packing group level until testing is completed; requiring railway companies to share information with municipalities to support emergency planners, and first responders; requiring railway companies to immediately implement key operating practices, including reducing the speed of trains transporting dangerous goods; ensuring that safety management system (SMS) audits are conducted frequently and in depth and that proper follow-up is performed, and amending the SMS audit cycle to a three-to-five year cycle, and will recruit additional specialized auditors to provide guidance to inspectors on conducting audits, and on the elements of an effective SMS; updating administrative monetary penalty regulations; establishing a standardized minimum for hand break applications, and specific testing requirements, and additional physical defenses for unattended trains; requiring railway companies to develop, and enhance rules on train securement; requiring certain railways (including short lines) to submit training plans to TC for review; conducting audits of short lines with respect to qualification standards of operating crews to determine specific training gaps, and any other issues that arise; requiring railway companies to hold a valid Railway Operating Certificate in order to operate on federally regulated railways in Canada; initiating crude oil sampling as part of a research project to assess properties, behavior, and hazards of crude oil transported in Canada through sampling, testing, and analysis of a variety of crude oils from different regions; requiring Class I and Class II rail carriers to report leading indicator data to TC; and establishing new safety standards for federally regulated grade crossings.

On February 20, 2015, the Transport Canada Minister introduced the Safe and Accountable Rail Act (that came into force on June 19, 2015) that will enhance railway safety and make the rail industry and crude oil shippers more accountable to Canadians (Parliament of Canada, 2015). On May 1, 2015, as discussed earlier, TC announced the TC-117 tank car standard, the next generation of stronger and safer rail tank cars (in conjunction with the US announcement of requirements regarding tank car design going forward). Also, as discussed earlier, enhanced insurance and compensation requirements were passed in June 2015.

Regulations Affecting Additional Modes of Transport

Although the regulatory focus of O&G transportation is pipelines and railways, PHMSA's authority extends to the transportation of hazardous materials by air and sea. Despite holding that authority, there is little regulatory focus on the transportation of O&G by air or sea, perhaps due to the inefficiency of continental shipping using either of those methods.

Several states have taken action targeted at regulating the transportation of O&G by roadways. New York allows municipalities to require trucks transporting hazardous materials to travel on designated routes subject to a road use agreement; in particular, Steuben County requires energy companies to post a $250,000 bond or to pay to upgrade the road and post a $15,000 bond; and Pennsylvania has adopted similar regulations allowing the state to impose "impact fees" on energy companies, a portion of which are dedicated to repairing local roads and bridges (Merrill and Schizer, 2013).

Labeling Requirements

In the United States, rail carriers are required to use proper labeling reflecting the type of crude oil they are carrying (e.g., crude oil from the Bakken region). Following the Lac-Mégantic incident and other crude-by-rail incidents, the US government instituted "Operation Classification", utilizing the PHMSA and the FRA to conduct random inspections and issuing fines to carriers who are found not to be in compliance. Officials have emphasized that proper labeling of the type of crude oil can aid in an appropriate emergency response to any spill incidents (PHMSA, 2014; Mouawad, 2014).

CONCLUSIONS

Transportation of shale gas and oil resources, while significantly established in North America and other global routes prior to the last decade, has been a focus of attention in recent years due to both an increased demand for transportation of such resources and due to high-profile safety and environmental concerns raised by incidents involving transportation of both shale and nonshale O&G resources. The industry and government are working on responses to concerns about capacity and infrastructure, while the public as well as municipal and federal governments (primarily in the United States and Canada) are raising and attempting to address concerns about human health and safety, and the environment through both appeals to the industry and regulations. Going forward, the industry will face substantial logistical and financial challenges to meeting capacity requirements and implementing regulatory requirements, especially as regulations further develop to address ongoing public and industry concerns about transporting shale gas and oil resources.

Acknowledgments

We are grateful for the rail industry knowledge and experience of Scott L. Carey (Partner), and research and writing assistance from Krista C. Sorvino (Special Counsel), I. Jordan Lowe (Associate), and Salli A. Ball (Paralegal) of BatesCarey LLP.

References

49 U.S.C. § 60102(a).(1).

49 U.S.C. § 60104(c).

49 U.S.C. § 60105.

American Petroleum Institute, 2015. Keystone XL Pipeline. Washington, DC. Available from: http://www.api.org/policy-and-issues/policy-items/keystone-xl/keystone-xl-pipeline

Association of American Railroads, 2012. 140,000 Mile Private Rail Network Delivers for America's Economy. AAR, Washington, DC. Available from: http://archive.freightrailworks.org/wp-content/uploads/FRW_Nine_Privat_Rail_Networks8.pdf

Association of American Railroads, 2013. Moving Crude Oil by Rail. AAR, Washington, DC. Available from: http://dot111.info/wp-content/uploads/2014/01/Crude-oil-by-rail.pdf

Association of American Railroads, 2014. Statement from AAR on State Department Environmental Impact Statement on the Keystone XL Pipeline. Available from: https://www.aar.org/newsandevents/Press-Releases/Pages/Statement-from-AAR-on-State-Department-Environmental-Impact-Statement-on-the-Keystone-XL-Pipeline.aspx

Association of Oil Pipelines ("AOPL") and American Petroleum Institute ("API"), 2014. U.S. Liquids Pipeline Usage & Mileage Report. Washington, DC. Available from: http://www.aopl.org/wp-content/uploads/2014/10/U.S. -Liquids-Pipeline-Usage-Mileage-Report-Oct-2014-s.pdf

Canadian Association of Petroleum Producers ("CAPP"), 2015. Canadian and U.S. Crude Oil Pipelines and Refineries. CAPP, Calgary, Alberta, Canada. Available from: http://www.capp.ca/~/media/images/customerportal/page-images/publications-and-statistics/crude-oil-forecast-june-2015/crude-oil-pipeline-and-refinery-map.jpg

CBC News, 2015. Lac-Megantic: Charges Laid for Brake Failure in Train Disaster. CBC News Montreal. Available from: http://www.cbc.ca/news/canada/montreal/lac-m%C3%A9gantic-charges-laid-for-brake-failure-in-train-disaster-1.3122732

Conca, J., 2014. Pick Your Poison for Crude – Pipeline, Rail, Truck or Boat. Forbes Magazine. Available from: http://www.forbes.com/sites/jamesconca/2014/04/26/pick-your-poison-for-crude-pipeline-rail-truck-or-boat/

Congressional Research Service, 2014. U.S. Rail Transportation of Crude Oil: Background and Issues for Congress. Washington, DC. Available from: https://fas.org/sgp/crs/misc/R43390.pdf

Congressional Research Service; Energy Information Administration, 2012. U.S. Refinery Capacity by Petroleum Administration for Defense Districts ("PADD"). Washington, DC.

CSX Transp., Inc. v. City of Plymouth, 92 F. Supp. 2d 643, 649. (E.D. Mich. 2000).

Department of Transportation and Federal Energy Regulatory Commission, 1993 Memorandum of Understanding Between the Department of Transportation and the Federal Energy Regulatory Commission Regarding Natural Gas Transportation Facilities. Available from: http://www.phmsa.dot.gov/staticfiles/PHMSA/DownloadableFiles/1993_DOT_FERC.pdf

Doukopil, T., 2015. Oil Train Spills Hit Record Level in 2014. NBC News. Available from: http://www.nbcnews.com/news/investigations/oil-train-spills-hit-record-level-2014-n293186

Egelko, B., 2014. PG&E Pleads Not Guilty in San Bruno Blast Case. San Francisco Chronicle. Available from: http://www.sfgate.com/crime/article/PG-amp-E-pleads-not-guilty-in-San-Bruno-blast-case-5695990.php

Eggleston, K., 2014. Two New Bakken Crude Oil Pipelines Online by 2016. Bakken Shale. 30 October. Available from: http://bakkenshale.com/pipeline-midstream-news/two-new-bakken-crude-oil-pipelines-online-by-2016/

Energy Information Administration, 2008. U.S. Natural Gas Supply Basins Relative to Major Natural Gas Pipeline Transportation Corridors. U.S. Department of Energy, Washington, DC. Available from: http://www.eia.gov/pub/oil_gas/natural_gas/analysis_publications/ngpipeline/TransportationCorridors.html

Energy Information Administration, 2009. U.S. Natural Gas Pipeline Network, 2009. U.S. Department of Energy, Washington, DC. Available from: http://www.eia.gov/pub/oil_gas/natural_gas/analysis_publications/ngpipeline/ngpipelines_map.html

Energy Information Administration, 2013. Technically Recoverable Shale Oil and Gas Resources: An Assessment of 137 Shale Formation in 41 Countries Outside the United States. U.S. Department of Energy, Washington, DC. Available from: http://www.eia.gov/analysis/studies/worldshalegas/

Energy Information Administration, 2014. Annual Energy Outlook 2014 ("AEO2014"). U.S. Department of Energy, Washington, DC. Available from: http://www.eia.gov/forecasts/aeo/pdf/0383%282014%29.pdf

Energy Information Administration, 2015. Shale in the United States. U.S. Department of Energy, Washington, DC. Available from: http://www.eia.gov/energy_in_brief/article/shale_in_the_united_states.cfm

Furchtgott-Roth, D., Green, K., 2013. Intermodal Safety in the Transport of Oil. Fraser Institute. Available from: http://www.fraserinstitute.org/uploadedFiles/fraser-ca/Content/research-news/research/publications/intermodal-safety-in-the-transport-of-oil.pdf

Gates, V., Mcallister, E., 2013. Crude Oil Tank Cars Ablaze After Train Derails in Alabama. Reuters. Available from: http://www.reuters.com/article/2013/11/09/us-crude-train-explosion-idUSBRE9A70Q920131109

Graeber, T., 2015. U.S. Proposes Tighter Pipeline Spill Rules. UPI. Available from: http://www.upi.com/Business_News/Energy-Resources/2015/07/02/US-proposes-tighter-pipeline-spill-rules/3981435839304/

Harder, A., 2015. Override of Obama's Keystone Veto Fails in Senate. The Wall Street Journal. Available from: http://www.wsj.com/articles/override-of-obamas-keystone-veto-fails-in-senate-1425498369

Hasemyer, D., 2015. Michigan's $75 Million Settlement with Enbridge Draws Praise, Questions. Inside Climate News. Available from: http://insideclimatenews.org/news/13052015/michigan-75-million-settlement-enbridge-kalamazoo-spill-questions

Heineman, T.V., Governor of the State of Nebraska, Case No. S-14-158 (Opinion filed January 9, 2015).

Herdt, T., 2015. Officials Push for Pipeline-Safety Reforms in Wake of Refugio Spill. Ventura County Star. Available from: http://www.vcstar.com/news/state/officials-push-for-pipelinesafety-reforms-in-wake-of-refugio-spill_10981143

Hunton & Williams LLP, 2015. PHMSA Comes Under Scrutiny as Pipeline Safety Act Reauthorization Draws Near. Pipelinelaw.com. Available from: http://www.pipelinelaw.com/2015/07/13/phmsa-comes-scrutiny-pipeline-safety-act-reauthorization-draws-near/

Inforum, 2013. Release from Cass County Sheriff's Office Regarding Casselton Evacuation. Inforum.com. Available from: http://www.inforum.com/content/release-cass-county-sheriffs-office-regarding-casselton-evacuation

Jindra, S., 2015. The Dangers of Crude Oil Trains and the New Rules to Help. WGN-TV News. Available from: http://wgntv.com/2015/05/04/the-dangers-of-crude-oil-trains-and-the-new-rules-to-help/

Lachman, S., 2015. This Aging Oil Pipeline is in Great Lakes' "Worst Possible Place" for a Spill. Huffington Post. Available from: http://www.huffingtonpost.com/2015/05/22/michigan-enbridge-pipeline_n_7308734.html

Merrill, T., Schizer, D., 2013. The Shale Oil and Gas Revolution, Hydraulic Fracturing, and Water Contamination: A Regulatory Strategy, 98 Minn. L. Rev. 145, 264 (2013).

Mouawad, J., 2014. 3 Companies Fined for Mislabeling Crude in Rail Transit. New York Times, NY. Available from: http://www.nytimes.com/2014/02/05/business/energy-environment/3-companies-fined-for-mislabeling-crude-oil-in-rail-transit.html?_r=0

North Dakota Industrial Commission, 2014. Order No. 25417. Bismarck, ND. Available from: https://www.dmr.nd.gov/oilgas/Approved-or25417.pdf

NTSB, 2014a. Member Robert Sumwalt's Last On Scene Briefing on BNSF Train Accident in Casselton, N.D. Available from: https://www.youtube.com/watch?v=_fYz9piUbyQ&feature=youtu.be

NTSB, 2014b. Preliminary Report, Railroad, DCA14MR004. NTSB.gov. Available from: http://www.ntsb.gov/investigations/AccidentReports/Reports/Casselton_ND_Preliminary.pdf

Oju, S., Hartman, K., 2014. Transporting Crude oil by Rail: State and Federal Action. National Conference of State Legislatures. Available from: http://www.ncsl.org/research/energy/transporting-crude-oil-by-rail-state-and-federal-action.aspx

Parliament of Canada, 2015. Statutes of Canada, Chapter 31 (Bill C-52). Available from: http://www.parl.gc.ca/HousePublications/Publication.aspx?Language=E&Mode=1&DocId=8057194

Pipeline and Hazardous Materials Safety Administration ("PHMSA"). PHMSA's Ongoing Bakken Investigation Shows Crude Oil Lacking Proper Testing, Classification. PHMSA, Washington, DC. Available from: http://phmsa.dot.gov/pv_obj_cache/pv_obj_id_2D0F19D85476377CC34AE11384620E21F26E0000/filename/PHMSA%2001-14.pdf

Pipeline and Hazardous Materials Safety Administration and Federal Railroad Administration, 2013. Safety and Security Plans for Class 3 Hazardous Materials Transported by Rail. Federal Register (Doc. 2013-27785), Washington, DC. Available from: https://www.federalregister.gov/articles/2013/11/20/2013-27785/safety-and-security-plans-for-class-3-hazardous-materials-transported-by-rail

Rucker, P., 2014. Oil-By-Rail Traffic Hurts Farmers, Travelers, U.S. Officials Told. Reuters.com. Available from: http://in.reuters.com/article/2014/04/10/usa-railway-congestion-idINL2N0N21SP20140410

Shaffer, D., Casselton, N.D., 2014. Residents Flee Town After Oil Train Explosion. Star Tribune. Available from: http://www.startribune.com/dec-31-casselton-residents-flee-after-oil-train-explosion/238207831/

Sharp, D., 2014. New Railroad Owner Rebuilding after Lac-Mégantic Disaster. Portland Press Herald. Available from: http://www.pressherald.com/2014/12/12/new-railroad-owner-rebuilding-after-quebec-disaster/

State of Nebraska. Legislative Bill 1161, 2012. Available from: http://nebraskalegislature.gov/FloorDocs/102/PDF/Slip/LB1161.pdf

The Associated Press, 2014. Oil Mars West Alabama Swamp Months after Train Crash Near Aliceville. AL.com. Available from: http://blog.al.com/wire/2014/03/oil_mars_west_alabama_swamp_mo.html

The United States Department of State Bureau of Oceans and International Environmental and Scientific Affairs, 2014. Final Supplemental Environmental Impact Statement for the Keystone XL Project: Executive Summary Applicant for Presidential Permit: TransCanada Keystone Pipeline, LP, Washington, DC. Available from: http://keystonepipeline-xl.state.gov/documents/organization/221135.pdf

Transport Canada (TC), 2015. Measures to enhance railway safety and the transportation of dangerous goods. Available from: http://www.tc.gc.ca/eng/mediaroom/infosheets-menu-7564.html

Tyrrell v. Norfolk S. Ry. Co., 248 F.3d 517, 523 (6th Cir. 2001).

U.S. Environmental Protection Agency, 2015. EPA's Response to the Enbridge Oil Spill. Washington, DC. Available from: http://www.epa.gov/enbridgespill/

U.S. Department of State, Bureau of Oceans and International Environmental and Scientific Affairs (U.S. DOS), 2014. Final Supplemental Environmental Impact Statement for the Keystone XL Pipeline Project. U.S. DOS, Washington, DC. Available from: http://keystonepipeline-xl.state.gov/documents/organization/221135.pdf

U.S. Department of Transportation (U.S. DOT), 2014a. Hazardous Materials: Enhanced Tank Car Standards and Operational Controls for High-Hazard Flammable Trains. Available from: https://www.federalregister.gov/articles/2014/08/01/2014-17764/hazardous-materials-enhanced-tank-car-standards-and-operational-controls-for-high-hazard-flammable

U.S. Department of Transportation (U.S. DOT), 2014b. Amended and Restated Emergency Restriction/Prohibition Order. U.S. DOT (Docket No. DOT-OST-2014-0025), Washington, DC. Available from: http://www.transportation.gov/sites/dot.gov/files/docs/Amended%20Emergency%20Order%20030614.pdf

U.S. Department of Transportation (U.S. DOT), 2015. Hazardous Materials: Enhanced Tank Car Standards and Operational Controls for High-Hazard Flammable Trains. U.S. DOT (Docket No. PHMSA-2012-0082 (HM-251)), Washington, DC. Available from: http://www.transportation.gov/sites/dot.gov/files/docs/final-rule-flammable-liquids-by-rail_0.pdf

Villenueve, M., 2014. After "End of the World" Explosion, Lac-Mégantic Aims to Rebuild. Portland Press Herald. Available from: http://www.pressherald.com/2014/04/17/after-end-of-the-world-explosion-lac-megantic-aims-to-rebuild/

Wronski, R. 2014. Chicago at Heart of Crude Oil Shipments, Data Show. Chicago Tribune, Chicago. Available from: http://www.chicagotribune.com/news/local/ct-oil-train-new-data-met-20150403-story.html

Wronski, R., 2014. Area Poorly Prepared for Crude Oil Train Fires. Chicago Tribune, Chicago. Available from: http://articles.chicagotribune.com/2014-05-25/news/ct-railroad-tankers-foam-met-20140525_1_foam-aid-box-alarm-system-fire-chief

CHAPTER 10

An Evaluation of the Hydraulic Fracturing Literature for the Determination of Cause–Effect Relationships and the Analysis of Environmental Risk and Sustainability

MariAnna K. Lane, Wayne G. Landis
Institute of Environmental Toxicology, Huxley College of the Environment, Western Washington University, Bellingham Washington USA

Chapter Outline

INTRODUCTION

Hydraulic fracturing (HF) is a gas extraction technique in which high-pressure water, sand, and chemicals are injected into natural gas wells to dislodge otherwise difficult-to-extract gas, such as that trapped in shale and tight sand deposits. The

Environmental and Health Issues in Unconventional Oil and Gas Development. http://dx.doi.org/10.1016/B978-0-12-804111-6.00010-8

techniques and applications are summarized in the companion chapters. There are 20 shale plays in the continental United States (Brittingham et al., 2014) (see Chapter 1, fig 1.4).

HF is seen as a means of producing domestic energy and a transition fuel from dirtier fuels like coal to renewable resources. While proponents of HF emphasize the potential economic benefits of unconventional shale gas extraction, a variety of associated human and environmental hazards have been identified (see eg., Chapter 8). These hazards include concerns of ground and surface water contamination, air quality, climate change impacts, seismic stability, toxicity of HF chemicals, increased noise and traffic, habitat fragmentation and degradation, and human health concerns. The lack of data is a common refrain in the assessment of these hazards. The environmental effects are the main focus of this study, although the overlap of many of these issues and general lack of information in many cases warrants the inclusion of other related issues.

Sustainability

The original purpose of this chapter was to examine the sustainability of HF. However, it soon became apparent that a systematic evaluation of the cause–effect relationships between the potential sources of stressors and impacts to typical endpoints had not been completed. In a recent review, Burton et al. (2014) concluded that data were not yet sufficient to conduct an ecological risk assessment although no risk assessment was attempted.

Our literature search and the conclusion of Burton et al. (2014) points to the lack of a research program organized around describing causal relationships. We propose a preliminary cause–effect conceptual model to act as a starting point around which to organize future research on the impact and eventual sustainability of HF. The next sections outline this effort.

Conceptual Model and Risk Assessment

Problem formulation is the initial step in an ecological risk assessment (USEPA, 1998). Perhaps the most important piece in this step is the conceptual model. The conceptual model links the stressors or action of interest in a risk assessment by a causal network to the endpoints that drive the decision-making process. The creation of a conceptual model and the evaluation of cause–effect for ecological risk assessment has been a powerful tool in sorting information and in producing an ecological risk assessment to support decision making. This review uses the cause–effect pathway as an organizing principle in collating and integrating the many types of publications generated describing the potential effects of HF. For our purposes we are using the fundamental structure of the relative risk model (Landis and Wiegers, 2005), diagrammed in Figure 10.1.

The initial point is the listing of all the potential sources of stressors for the particular activity being investigated. In this instance we will be taking the entire activity of the process of an HF site, from exploration, drilling, extraction, logistics, and eventual closure. The stressors box represents all the potential stressors

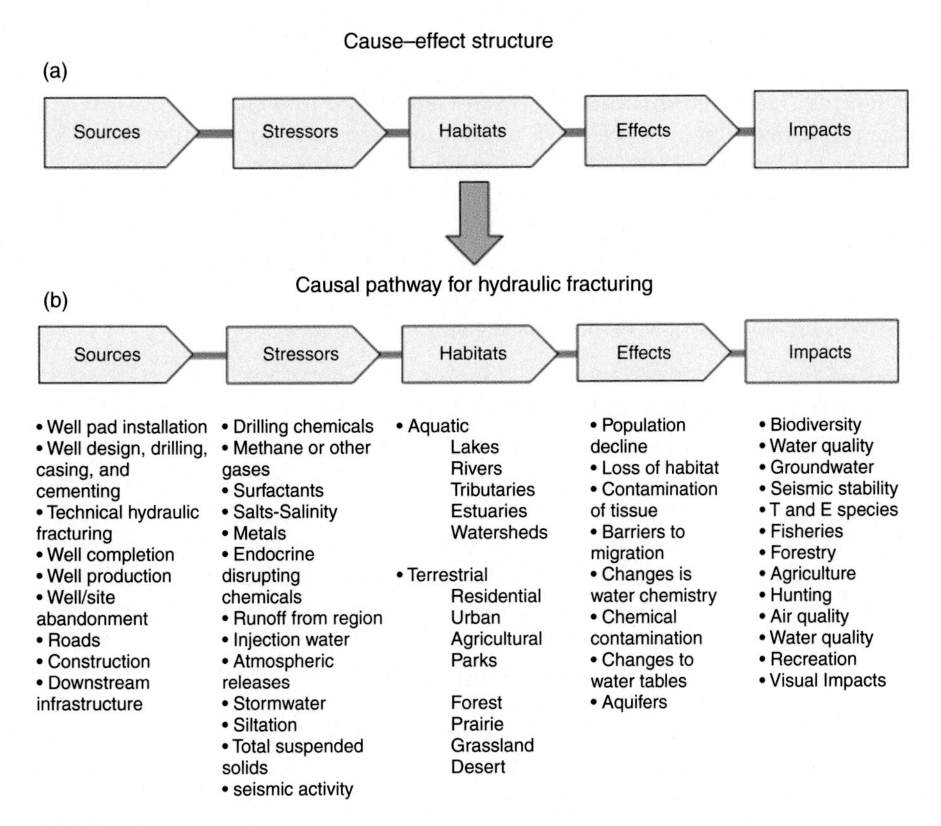

FIGURE 10.1
Causality and the preliminary conceptual model for hydraulic fracturing.
(a) The basic structure of a cause–effect pathway as developed for the relative risk model. (b) list of candidates for each category based on the literature search. Uncertainties exist for many of these factors and these are listed in Table 10.2.

resulting from the process. This item will include not only the drilling materials and chemicals, but also the alteration of the landscape due to having numerous well pads in close proximity to the roads required for support. The habitat box represents both the spatial location of the sources and the exposure of the stressors as well as the types of environments. The habitat segment describes the exposure of the stressors to the endpoints under consideration. Although HF is a land-based operation, it can border a variety of aquatic environments. Groundwater may also become part of the surface water in some locations. Atmospheric releases have the potential for long-range transport, broadening the types of environments under consideration. The effects box can be toxicological in nature, but it can also include fragmentation of the landscape, diversion of water, and increase in total suspended solids or destruction of habitat for a threatened or endangered species. At the end the impact box lists the endpoints being used for the decision-making process and the potential change due to the HF operation. In general, the entire process works best when tied to a specific operation and location and a specific set of regulatory requirements.

The cause–effect model is next populated using data from a number of sources. Site-specific data are preferred for a number of reasons, including reliability and relevance. However, this broader review is not tied to a site and must rely on the published literature. We have made the decision to focus on reviews done during the last few years as the set from which to summarize the state of knowledge regarding risk to HF. Our cutoff date was June 1, 2015. As usual we have screened the literature using criteria such as relevance of the study, quality of the data, tools used during the analysis, and where the study sits in our cause–effect model. However, it has also been demonstrated for a number of controversial issues that another type of analysis is required.

Normative Science and Bending Science

Science becomes normative science when there is an embedded policy preference or goal. Examples of this include concepts like "ecosystem health" and "sustainability," which can be thought of as "stealth policy advocacy" (Lackey, 2004). Taken a step further, science becomes bent science when it is willfully manipulated to serve a predefined policy goal. This bending can take a number of forms, from funding research that already has a specified policy outcome to attacking scientists who publish undesirable results (McGarity and Wagner, 2008). The concepts of bending and normative science have become well established in the realm of science–policy interaction. These issues have been documented extensively in everything from tobacco smoke to climate change. In a discussion on climate change, Oreskes (2013) notes in response to contrarian positions of denial "the cultural and intellectual question becomes, why exactly would someone want to do that?"

In a 2003 paper on the manipulation of science, McGarity (2003) notes that bending is frequently employed by what he calls "risk producing industries" in order to deflect blame and liability. The driving factor in these cases is economic interests. Adding evidence to this idea, in a psychological study Lewandowsky et al. (2013) found a correlation between belief in free markets, conspiratorial thinking, and the rejection of climate change and to a lesser extent other sciences. However, McGarity and Wagner (2008) show bending is not restricted to industry and political conservatives but can also be seen in government agencies, public and environmental interest groups, and trial attorneys. According to McGarity and Wagner (2008), bending is likely to occur if the advocate believes costs of adverse scientific findings are higher than costs spent undermining research and if the advocate has sufficient resources to mount the desired attacks. Given this analysis, one would expect to find instances of science bending on both sides of a contentious policy issue assuming both sides have sufficient resources at their disposal.

Bending science is an obvious issue for scientists because it has the capacity to distort the scientific literature and decrease public trust of science (McGarity and Wagner, 2008). HF makes a good case study to assess the prevalence of bending based on its increasing importance, public visibility, and the number of potential interest groups. To this end, we conducted a literature search of peer-reviewed and nonpeer-reviewed literature on the environmental effects of HF, as well as a specific search for risk assessments to assess the use of decision-making tools.

EVALUATION METHODS

Literature Search

We conducted a literature search, tied the outcomes to the cause–effect pathway and performed an assessment of the prevalence of bending in the peer-reviewed and nonpeer-reviewed literature. Search terms and databases used to find relevant literature are included in Section "Search Term Box." We selected papers based on their relevance to the environmental effects of HF and currency to provide the most current information.

The literature was then sorted as to which part of the causal pathway could be informed by the investigation. Potential endpoints were also derived from this survey and compiled. Many of the papers discussed uncertainties in the datasets and in our understanding of cause–effect relationships and these were also evaluated. The information from these reviews was finally sorted using the bins established from our initial causal diagram.

Criteria for analyzing data were based on the strategies for bending science developed by McGarity and Wagner (2008) and indications of tampering with science based on Oreskes and Conway (2010). The use of literature searches as a methodology for assessing scientific consensus and research was based on Oreskes (2004b). McGarity and Wagner (2008) detail six strategies for bending science summarized later.

1. Shaping science – research commissioned to produce a desired outcome by an outside party with vested interests.
2. Hiding science – preventing undesirable scientific findings from becoming known.
3. Attacking science – illegitimate attacks on undesirable research.
4. Harassing scientists – false allegations of misconduct, subpoenas or depositions, and data sharing requests on researchers with undesirable findings.
5. Packaging science – commissioning review articles or hand-picked panels to generate the image of consensus or communicate findings in the most favorable light.
6. Spinning science – portraying science in a particular way to advance economic or ideological goals rather than to communicate the science accurately.

With the exception of (4), in which a literature search alone was inadequate to assess, we used these strategies to qualitatively analyze each piece of literature surveyed. Unlike Oreskes (2004b), which relied on abstracts, the entire body of text of each piece of literature was closely read, funding sources investigated, and conclusions noted to fully search for signs of bending. Literature that exhibited any of these bending strategies is discussed in Section "Specific Examples of Bending" in the Section "Results."

The narrative of Oreskes and Conway (2010) detailing indications that the science may have been tampered with or bent was summarized as seven discrete criteria and applied to a frame of binary logic to result in a score for each piece of literature, which could be used to compare the literature and assess the likelihood of bending. These criteria are given later phrased as questions with a yes or no answer:

1. Is it peer reviewed? Yes (1), No (0)
2. Is there original research? Yes (1), No (0)
3. Do the funding sources have vested interests? Yes (0), No (1)
4. Does the article have no misleading, or distracting facts, or terminology? Yes (0), No (1)
5. Is there an attack on "undesirable" research or researchers? Yes (0), No (1)
6. Is only one's own research or commentary cited? Yes (0), No (1)
7. Is the author(s) an expert in an unrelated field? Yes (0), No (1)

We assigned a score of either 1 or 0 to the answer to these questions, which is shown next to each question mentioned previously – a score of 1 if the answer did not indicate bending, and 0 if it did. For example, a peer-reviewed article is theoretically less likely to be bent because the peer-review process is designed to prevent poorly designed or corrupt research from being published. So, if an article was peer reviewed it was assigned a score of 1, and if it was not peer reviewed it was assigned a score of 0. It is not guaranteed, however, that a peer-reviewed article will not be bent, so these scores should be viewed as an initial screening of the likelihood of bending and a convenient metric for comparing trends in the literature, not a definitive measure of the amount of bending. An article with a lower total score would be more expected to contain bending than an article with a higher total score. The McGarity and Wagner (2008) bending strategies as described earlier were used to analyze and look for specific examples of bending.

RESULTS

Characterization of Literature Surveyed

A total of 19 peer-reviewed and 10 nonpeer-reviewed articles were included in this study (see Table 10.1). Funding for peer-reviewed articles came from a variety of academic, governmental, industry, and nongovernmental organizational (NGO) sectors.

Endpoints of concern in the literature surveyed included groundwater contamination, surface water contamination, air quality, climate change, ecological impacts, toxicity, seismic stability, and social impacts (Table 10.1). Uncertainties listed included the lack of baseline data, composition of fracturing fluid, toxicity of specific chemicals in the fracturing fluid, chemical fate and transport, mechanisms of groundwater contamination, chronic effects, risks or magnitude of risks,

Table 10.1 Papers Sorted by Relevance to the Cause–Effect Pathway and Then by Endpoint

Cause-Effect Pathway			
Source	**Stressor**	**Exposure**	**Effects**
Molofsky et al. (2013)	Farag and Harper (2014)	Osborn et al. (2011)	Farag and Harper (2014)
Molofsky et al. (2011)	Orem et al. (2014)	Gordalla et al. (2013)	Medical Health Experts (2014)
Darrah et al. (2014)	Barbot et al. (2013)	Barbot et al. (2013)	Vengosh et al. (2014)
Vengosh et al. (2014)	Burton et al. (2014)	Fontenot et al. (2013)	Stringfellow et al. (2014)
Goldstein et al. (2014)	Goldstein et al. (2014)		Farag et al. (2014)
	Stringfellow et al. (2014)		Gordalla et al. (2013)
	Fontenot et al. (2013)		Adams (2011)
			Papoulias and Velasco (2013)
			Kassotis et al. (2014)

Endpoint		
General Endpoint	**Papers**	**Specific Focus**
Groundwater	Barbot et al. (2013)	Water quality parameters
	Bever (2014)	Surfactant toxicity
	Darrah et al. (2014)	Methane contamination
	Fontenot et al. (2013)	Trace metals and total dissolved solids
	Fountain (2014)	Methane contamination
	Goldstein et al. (2014)	Toxicity and contamination pathways
	McHugh et al. (2014)	Comment on Fontenot et al. (2013)
	Concerned Health Professionals of NY (2014)	Drinking water contamination
	Molofsky et al. (2013)	Methane contamination
	Mufson (2014)	Methane contamination

(*Continued*)

Table 10.1 Papers Sorted by Relevance to the Cause–Effect Pathway and Then by Endpoint *(cont.)*

Endpoint		
General Endpoint	**Papers**	**Specific Focus**
	Orem et al. (2013)	Chemistry of organics
	Osborn et al. (2011)	Methane contamination
	Stokstad (2014)	Methane contamination
	Vengosh et al. (2014)	Methane and salt contamination
	Verango (2013)	Methane contamination
Surface water	Adams (2011)	Trees and soil
	Burton et al. (2014)	Hazarads of HFHV operations
	Farag et al. (2014)	Salinity
	Goldstein et al. (2014)	Toxicity and contamination pathways
	Concerned Health Professionals of NY (2014)	Radioactive contamination
	Vengosh et al. (2014)	Organics, salts, metals
Air quality	Goldstein et al. (2014)	Toxicity and contamination pathways
	Concerned Health Professionals of NY (2014)	Direct and indirect pollution
	Rice (2014)	Human health effects
	Small et al. (2014)	Risk and risk governance
Climate change	Small et al. (2014)	Risk and risk governance
	Barbot et al. (2013)	HF as alternative to coal
Ecological	Adams (2011)	Trees and soil
	Burton et al. (2014)	Hazards of HVHF operations
	Papoulias and Velasco (2013)	Fish histopathology
	Small et al. (2014)	Risk and risk governance
Toxicity	Farag and Harper (2014)	Salinity
	Farag et al. (2014)	Salinity
	Goldstein et al. (2014)	Toxicity and contamination pathways

Table 10.1 Papers Sorted by Relevance to the Cause–Effect Pathway and Then by Endpoint *(cont.)*

Endpoint		
General Endpoint	**Papers**	**Specific Focus**
	Gordalla et al. (2013)	HF chemicals in Germany
	Kassotis et al. (2014)	Endocrine disruption in human cell lines
	Stringfellow et al. (2014)	Chemical, physical, and toxicity data
Seismic stability	Concerned Health Professionals of NY (2014)	Deep-well injection as trigger
	Small et al. (2014)	Risk and risk governance
	Vengosh et al. (2014)	Injection of large water volumes
Social	Horn (2013)	Delay of EPA reports
	Concerned Health Professionals of NY (2014)	Boom–bust social dynamics
	Mooney (2014)	NY ban versus MD policies
	Robbins (2013)	HF and the Endangered Species Act
	Small et al. (2014)	Risk and risk governance
	Soraghan (2011)	Use of term "fracking"

Blank, review; gray, nonpeer-reviewed work.

impact of chemicals released into the environment, and public health impacts (Table 10.2). The synthesis of these results is summarized in Figure 10.1b along the lines of the cause–effect model.

The additional risk assessment searches yielded many articles discussing the need to conduct risk assessments and several characterizations of hazards, but were not risk assessments. The exceptions to this were risk assessments conducted by the Maryland Department of the Environment and Department of Natural Resources (2014), a human health risk assessment of air emissions in Colorado (McKenzie et al., 2013), a master's thesis involving groundwater contamination in Pennsylvania (Fletcher, 2012), and water pollution risk in the Marcellus Shale (Rozell and Reaven, 2012). All of these risk assessments were heavily focused on human health and could not be classified as ecological risk assessment, although the Maryland (2014) assessment did have some consideration of risk to the environment and natural resources.

Table 10.2 Uncertainties Discussed in the Peer-Reviewed and Nonpeer-Reviewed Literature

Uncertainty	Peer-Reviewed Literature	Nonpeer-Reviewed Literature
Baseline data	Adams (2011), Burton et al. (2014), Fontenot et al. (2013), Goldstein et al. (2014), Gordalla et al. (2013), Molofsky et al. (2013), Osborn et al. (2011), Vengosh et al. (2014)	Mufson (2014)
Composition of HF fluid	Adams (2011), Barbot et al. (2013), Goldstein et al. (2014), Stringfellow et al. (2014)	None
Toxicity of chemicals in fluid	Burton et al. (2014), Goldstein et al. (2014), Gordalla et al. (2013), Orem et al. (2014), Stringfellow et al. (2014)	Bever (2014)
Chemical fate and transport	Burton et al. (2014), Goldstein et al. (2014), Gordalla et al. (2013), Stringfellow et al. (2014)	None
Mechanism of groundwater contamination	Darrah et al. (2014), Fontenot et al. (2013), Goldstein et al. (2014), McHugh et al. (2014), Molofsky et al. (2013), Orem et al. (2014), Osborn et al. (2011), Stockstad (2014), Vengosh et al. (2014)	Bever (2014), Mufson (2014), Verango (2013)
Chronic effects	Farag and Harper (2014)	Mooney (2014)
Risk/magnitude of risks	Burton et al. (2014)	Concerned Health Professionals of NY (2014), Mooney (2014)
Impact of chemicals released into environment	Adams (2011), Orem et al. (2014), Small et al. (2014)	Concerned Health Professionals of NY (2014)
Public health impacts	Small et al. (2014)	Concerned Health Professionals of NY (2014), Rice (2014)

Given the lack of probabilistic ecological risk assessments and even human health risk assessments to a large degree, we conducted additional research to determine types of analytical tools decision makers were using if not risk assessments. Three specific states were singled out as examples: New York, Maryland, and Michigan. In New York, the recent ban on HF enacted by Governor Cuomo was based in part by a review by the New York State Department of Health, which concluded that the current state of scientific information was insufficient to understand the public health risks associated with HF (New York State Department of Health, 2014). In contrast, Maryland is a state that allows HF. Although Maryland has a much smaller area of potential HF sites, which may have influenced the state's decision (Mooney, 2014), a key difference was also that the decision was based largely on an assessment conducted by the Maryland Department of the Environment and the Maryland Department of Natural Resources. Unlike the New York Department of Health review, this document qualitatively estimates some of the hazards associated with HF.

Michigan is also a state that allows HF. A question and answer for the public on their state website is very firm on the position to not regulate HF, suggesting that the state has been studying HF for decades with no indication of human or environmental impacts (Michigan Department of Environmental Quality, 2015). Recently, the state has commissioned an extensive report on the impact of HF to a wide range of human and environmental issues from the University of Michigan (see Graham Sustainability Institute, 2015).

Cause–Effect Pathways

We were able to find information on a number of factors that would compose a cause–effect pathway for the construction of a suitable conceptual model for performing a risk assessment (Figure 10.1b). The Maryland (2014) assessment was particularly helpful in listing the sources.

There are a large number of potential uncertainties in quantifying the cause–effect pathway in order to produce a quantitative risk assessment. Table 10.2 lists a number of issues regarding the uncertainties for many of these parameters. Some can be eliminated by careful study of fracking sites, their construction, operation, the infrastructure used to build the sites, measurements of downstream emissions, the composition of the stormwater, and the composition of the drilling muds, to name a few examples.

However, many of the stressors that are released, the types of receiving habitats, and the organisms potentially exposed can be reasonably quantified for a specific site. Appropriate geographic information system analysis and a properly designed monitoring program can answer the questions of stressors, habitats, and effects as has been done for a number of contaminated sites. As for chemical-induced effects, many of the chemicals or close analogs found in HF have been tested in the laboratory or there is experience from other contaminated sites. The particular differences in HF are the depths of the drilling and contamination of deep aquifers. However, once those materials find their way to the surface the result is that of a classic contaminated site.

The environmental endpoints are typical of any contaminated or impacted sites. Under that column are the endpoints that comprise evaluations under the Toxic Substance Control Act (TSCA), Resource Conservation and Recovery Act (RCRA), Oil Pollution Act (OPA), or the Clean Water and Clean Air Acts. Specific species, numeric water quality standards, and other site-specific numeric values can be derived depending upon the location of the HF site.

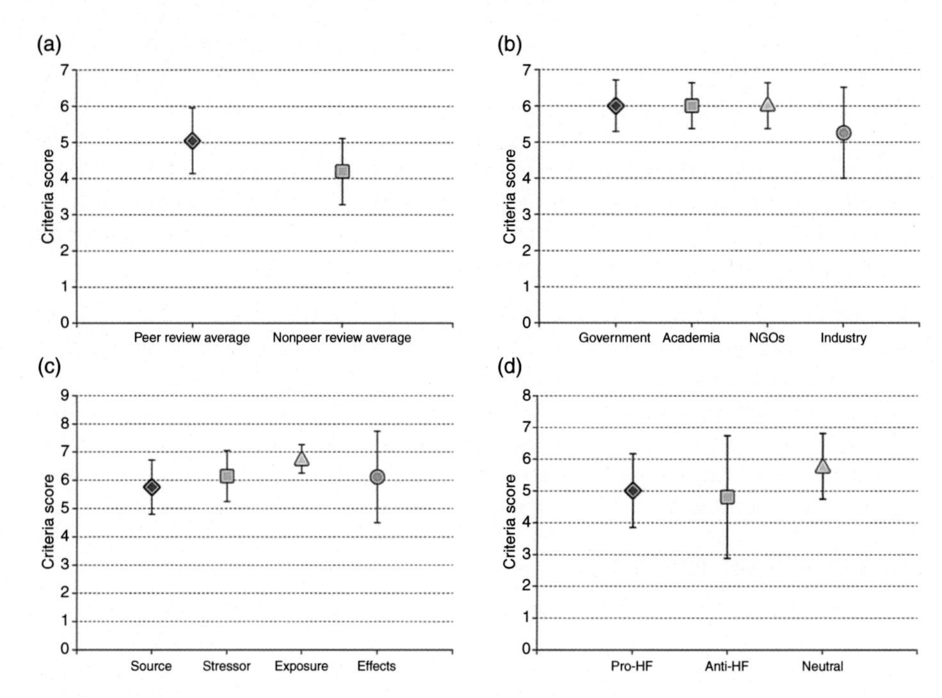

FIGURE 10.2
Results of the bending science analysis showing average criteria scores with error bars showing standard deviation.
The literature is sorted by peer-reviewed and nonpeer-reviewed (a), funding source (b), placement on the cause–effect pathway (c), and position on HF (d) categories.

Bending Science Criteria Analysis

Although some general trends are visible in average criteria scores, none of these trends are significant due to the high standard deviations. Nonpeer-reviewed papers tended to have lower scores than peer-reviewed papers (Figure 10.2a). Peer-reviewed papers with government, academic, or NGOs all had the same average criteria score while papers with industry funding had lower average criteria scores but high standard deviation (Figure 10.2b). There was no discernible trend in the average criteria scores for papers categorized by placement on the cause–effect pathway (Figure 10.2c) or whether the conclusions of the papers leaned pro, anti, or neutral toward HF (Figure 10.2d). The papers dealing with exposure had the highest criteria score and lowest standard deviation, while the papers regarding effects had the highest standard deviation (Figure 10.2c). This is logical considering it is difficult to argue with exposure data – it is typically a yes or no question. Effects, however, are much more up to interpretation and subject to the definition of what constitutes an effect.

RELATIONSHIP BETWEEN CONCLUSION AND SECTOR

In each of the funding categories (government, academic, NGO, and industry) at least 50% of the literature had conclusions that were neutral toward HF. In the government and academic funding categories, the same proportion of papers had pro and anti-HF conclusions. More NGO-funded papers had pro-HF conclusions, although at this sample size this was only a difference of one paper. Similarly, half of industry-funded papers had pro-HF conclusions and none had anti-HF conclusions, but this may be due to a sampling error rather than to an actual trend.

QUALITATIVE ANALYSIS: SPECIFIC EXAMPLES OF BENDING

Shaping. Shaping is when outside interests fund research with a particular goal in mind. Within this set of papers, one potential instance of shaping science was identified. In 2011 and 2013, Molofsky et al. published research aimed at determining the mechanism of methane contamination in groundwater in the Marcellus Shale. Both the researchers and funding for these papers were from GSI Environmental, Inc., an environmental consulting firm, and Cabot Oil & Gas Corporation, which has a stake in HF research as a gas extraction corporation. Interestingly, this research involves the same general location and subject as a group of papers published by a group of researchers at Duke University (Darrah et al., 2014; Osborn et al., 2011). While both sets of researchers ultimately reach the same basic conclusion that gas contamination by upward migration triggered by HF is not occurring, they disagree about whether HF is at all to blame for the methane contamination of groundwater in the region. The GSI Environmental/Cabot Oil & Gas Corporation researchers conclude that methane concentrations are best correlated with topographic and geologic features and therefore natural and unrelated to HF activities. The Duke University research team concluded that where instances of contamination do occur, they are most likely to be the result of well casing failure and so related to the broader gas extraction process but not HF directly.

It is possible that these groups of researchers came to slightly different conclusions simply due to their different expertise and methods. The fact that the GSI Environmental/Cabot Oil & Gas Corporation researchers found conclusions that absolve HF of any responsibility for the methane contamination, however, is increasingly suspect given the additional activities of this research group, which will be explained in Section "Attacking."

Hiding. As McGarity and Wagner (2008) state, hiding is often the most difficult-to-identify science-bending technique because it is only an option for those with something to hide, and by definition it avoids being discovered. With regard to HF, the only indication that hiding was occurring came from a Huffington Post Article (Horn, 2013), which details a mismatch in United States (US) Environmental Protection Agency (EPA) internal documents and reports on a Pennsylvania study, a similar Texas study, and the delay of a report on a Wyoming study. The article suggests industry lobbying is a likely culprit behind the EPA's alleged misdeeds.

Attacking. We identified two potential instances of attacks on HF research in the literature surveyed, both centered on the issue of groundwater contamination and the GSI Environmental/Cabot Oil & Gas Corporation research group. The first instance regards the differences between the aforementioned research group and the Duke University research group introduced in section "Shaping." In an announcement on the website of GSI Environmental about the release of the Molofsky et al. (2011) study, they advertise that "gas development activities in the Marcellus Shale have not caused widespread methane impacts on water wells in northeastern Pennsylvania. This directly counters allegations made in a recent study by Duke University (Osborne et al., 2011) [*sic*]." They further emphasize the "apparent misinterpretation by the Duke study [which] underscores the need for a multiple lines-of-evidence approach for proper characterization" (GSI Environmental Inc, 2013). The tone of this announcement and use of words like "allegations" and "apparent misinterpretation" directly attacks the Osborn et al. (2011) study and Duke researchers.

Disagreements and frank critiques of others' work is a normal part of the scientific and peer-review process. It is often difficult to determine whether such disagreements are simply just that or if they are carefully planned attacks on research whose only real fault is having undesirable conclusions. This point is further illustrated by the second potential attack instance, involving a study of water quality in drinking water wells overlying the Barnett Shale (Fontenot et al., 2013) and a subsequent comment published in the journal regarding that study (McHugh et al., 2014). Fontenot et al. (2013) found levels of arsenic, strontium, selenium, and total dissolved solids exceeding maximum contaminant levels in some wells surveyed. They are careful to note that the source of elevated contaminants is beyond the scope of the study and impossible to determine without measurements from before, after, and during gas extraction activities in the region, but they suggest that mechanical disturbance from drilling is most consistent with their data.

McHugh et al. (2014) attack the conclusions of Fontenot et al. (2013). Their complaints are that (1) Fontenot et al.'s comparison of active and nonactive datasets is not statistically valid or meaningful to establishing causation, (2) the comparison against historical data is flawed because of the increase in detection limits of metals, and (3) the patterns in their data are not consistent with natural gas extraction and they failed to consider mechanisms other than natural gas extraction. On their own, these critiques seem valid. Points (1) and (3) directly attack Fontenot et al.'s conclusion about the mechanism of contamination. However, as stated earlier, Fontenot et al. (2013) were careful to note that they were not able to definitively establish the mechanism or source of contamination. The McHugh et al. (2014) comment makes no mention of this, instead making it seem as if it was the main purpose of Fontenot et al. (2013) to establish the mechanism of the contamination they found. McGarity and Wagner (2008) note that attacks on undesirable scientific research need not be valid to produce the desired damage; simply the presence of a critique and hint of controversy surrounding a piece of research can be enough to dissuade decision makers from seriously considering that research. Given this, it is questionable as to whether the goal of the McHugh et al. (2014) comment was to simply discourage any research that connects gas extraction in any way to groundwater contamination.

Packaging. Packaging involves the use of review articles to create the semblance of scientific consensus surrounding an issue, or present research that best supports the viewpoints of the orchestrator of the review. Two review articles surveyed showed some signs of packaging. The first, Small et al. (2014), was the result of a National Research Council project, which had support from both the Park Foundation and Shell Upstream America. The Park Foundation is an NGO, which has contributed significantly to groups opposing HF (Soraghan, 2012). Shell Upstream America is the US branch of Royal Dutch Shell, a global petrochemical corporation. The resultant text relied heavily on the use of normative language when presenting the issues surrounding HF, noting the potential for "widespread economic benefits" and ability to "considerably" mitigate operational risks. However, despite such instances of normative language, the overall tone of the review remained neither pro nor anti-HF.

The second review Burton et al. (2014) received funding from the Graham Sustainability Institute, Arkansas Game & Fish Commission, National Science Foundation, and the Michigan Society of Fellows. Although the overall portrayal of the issue was neither pro nor anti-HF as in Small et al. (2014), there was a lack of citations for claims and the rather distracting claim that agriculture likely has a larger impact on water resources than HF. Despite these shortcomings, it was not clear in either of these review articles that the packaging had taken place to willfully skew the perception of HF.

Spinning. Spinning is perhaps the easiest of the bending strategies to perform, considering all it requires is that the scientific findings be cast in a light that makes them seem most supportive of the interest group's goals. Spinning was also the most frequently identified category of bending in both the peer-reviewed and nonpeer-reviewed literature. In the peer-reviewed literature the spinning primarily occurred in Section "Discussion and Conclusions," which is logical as this section is typically the most qualitative and open to interpretations by the authors.

The first example of spinning in the peer-reviewed literature occurred in Orem et al. (2014), an article analyzing the chemical components of formation and produced water. In the conclusion, the authors stated "[t]he environmental and human health impacts (if any) of the release of these compounds into surface and groundwater are unclear." This statement implies a familiarity with the literature on the subject of impacts from HF. However, the statement had no citations and there was almost no discussion of toxicity data for those chemicals. Environmental and health impacts would indeed be unclear if you had done little research on the subject. While it could be the case that the authors were familiar with the topic and simply failed to include a citation, the parenthetical "if any" included in the author's claim implies a high probability that there are no impacts. This is quite the claim to make, given the volume of research suggesting the opposite conclusion (see Adams, 2011; Farag and Harper, 2014; Gordalla et al., 2013; Kassotis et al., 2014). Orem et al. (2014) was funded by the US Geological Survey Energy Resources Program and US Department of Energy.

Spinning is also observable in the discrepancies between the research and conclusions of the GSI Environmental/Cabot Oil & Gas Corporation research group and the Duke University research group. As mentioned before, both groups came to the same general conclusion that gas contamination by upward migration triggered by HF is not occurring. In an editorial Stockstad (2014) notes the criticism by Molofsky that the Duke University researchers put too heavy a focus on the few contamination events that have occurred. In their 2013 paper, Molofsky et al. concede that there have in fact been instances of gas well casing malfunctions resulting in stray gas migration but they emphasize that these are isolated, localized issues which do not result in regional scale impacts to water quality. The mainstream media also picked up on this debate, and their reports highlighted either the relatively small number of contamination instances, the fact that contamination is occurring at all, or the possibility of a technological fix – all of which frame the issue very differently (see Fountain, 2014; Mufson, 2014).

In the previous cases the overall conclusion of the mainstream media ultimately remained fairly true to the study they were reporting on, but this was not always true. This is reflected particularly well in the coverage of a Colorado study on the surfactants added to HF fluid. The news media latched onto a claim that these chemicals were no more toxic than household products. One particular *Washington Post* article (Bever, 2014) boasted "Study: Fracking chemicals found in toothpaste and ice cream" as the title. The study also reiterates the findings of Darrah et al. (2014) that water contamination is from well leaks rather than HF directly with an overall effect downplaying the hazards associated with HF.

The final example of spinning comes from a letter by public health professionals to Governor Cuomo and Health Commissioner Zucker of New York (Concerned Health Professionals of New York 2014). As a persuasive piece intended to convince the readers to adopt a moratorium on HF, it serves as a not-so-subtle example of strategies for spinning science. Throughout the document, strong descriptive words like "indisputable" and "unavoidable" emphasize the certainty of science favorable to their goal while pointing out the uncertainties that remain and the need for more research to reduce these uncertainties. Pointing out uncertainties in science is a classic strategy for those hoping to delay action and is facilitated by the standards of proof required by science and the judicial system. As Oreskes (2004a) points out, however, definitive proof has never truly been a requirement for political action.

DISCUSSION AND CONCLUSIONS

The findings of this study suggest that while bending may not be pervasive in the sense that every piece of literature is likely to be bent, it is pervasive in the sense that bending does occur in every category of funding, every side of the debate, and by every bending strategy. This finding is significant because while much literature has focused on the dishonest activities of industry (Oreskes and Conway, 2010) or government under the influence of industry

(Vallianatos and Jenkins, 2014), this review identified bending in peer-reviewed literature of all funding sources and on both sides of the HF debate. This is consistent with the claim by McGarity and Wagner (2008) that an interest group will participate in bending if they have sufficient resources and perceive the cost of adverse findings to be greater than the costs spent undermining research. Additionally, signs of all strategies of bending (with the exclusion of harassing scientists, which was beyond the scope of this study) were found in the literature surveyed.

This finding has a number of implications for the perceived integrity of the peer-reviewed and nonpeer-reviewed literature as well as its ease of identification. While the criteria score rankings were a useful analytical tool, there is no simple checklist that can be used to definitively determine whether or not a piece of literature is bent. Only careful analysis of the text can identify potential instances of bending, and even that alone is often not enough to demonstrate that the science was intentionally bent to further a particular goal. It is difficult to prove bending because that specific crime can only occur with intent, and without some knowledge of internal dynamics it is difficult to know if the potential bending was in fact carefully calculated or simply the result of carelessness, or unintentional bias, on the part of the individual authors.

The differences between New York's and Maryland's decision-making outcomes and supporting reports highlight the impact of the subtle biases of different disciplines. While there is a substantial amount of literature discussing the need for risk assessments on HF and detailing the hazards, only three risk assessments have actually been done. Several of the studies discussing risk assessments have claimed that there is not adequate information to conduct a risk assessment (Adgate et al., 2014; New York State Department of Health, 2014). Burton et al. (2014) also claim that uncertainty is too high for risk assessment. However, the ability to incorporate estimates, expert elicitation, and similar techniques into risk assessments means that a risk assessment can be conducted even when significant data gaps exist. Until a risk assessment is conducted, it is not possible to conclude that the uncertainty would be too high to inform decision making. The New York State Department of Health study (2014) alludes to the fact that a degree of certainty is desirable for decision making, but this is no reason that risk assessments cannot be done. Tools such as sensitivity analysis can point to the variables critical to the outcome of the risk assessment. Given this, the widespread absence of risk assessments (particularly, ecological risk assessments) and the lack of risk assessments in this field may be due to factors other than the uncertainties.

There are a number of reasons to conduct quantitative risk assessments for HF. First, research and monitoring programs need to be based on specific research questions. One of the major outcomes of a risk assessment is the identification of specific hypotheses regarding endpoints at risk and the identification of specific uncertainties. It may be that even with high uncertainty the estimated risk is useful to decision makers in setting at least preliminary policy choices until those uncertainties are resolved. Finally a strong conceptual model can be a framework describing in detail what is known, the uncertainties around each, and provide a quantitative description of the science regarding the risk due to HF. Perhaps the end result would be in making it more difficult to bend the science using the techniques described in this manuscript.

SEARCH TERM BOX

Table A1.1 List of Search Terms and Locations, or Other Means of Location Where Appropriate for Literature Analyzed

Article	Search Term or Other Means of Location	Search Engine	Date
Adams (2011)	Vengosh et al. (2014) citation	N/A	11/15/2014
Barbot et al. (2013)	Hydraulic fracturing and environmental effects	Google Scholar	10/16/2014
Bever (2014)	Fracking	Washington Post	1/26/2015
Burton et al. (2014)	Hydraulic fracturing and toxicity	Web of Science	10/23/2014
Concerned Health Professionals of NY (2014)	Listed as citing Vengosh et al. (2014)	Google Scholar	11/15/2014
Farag and Harper (2014)	Hydraulic fracturing and toxicity	Web of Science	10/23/2014
Farag et al. (2014)	Author search: Farag AM	Web of Science	11/11/2014
Fontenot et al. (2013)	McHugh et al. (2014) citation	N/A	
Goldstein et al. (2014)	Hydraulic fracturing and toxicity	Web of Science	10/23/2014
Gordalla et al. (2013)	Hydraulic fracturing and toxicity	Web of Science	10/23/2014
Jackson et al. (2014)	Darrah et al. (2014) citation	N/A	
Kassotis et al. (2014)	USGS hydraulic fracturing	Google	11/15/2014
McHugh et al. (2014)	Author search: Molofsky LJ	Web of Science	10/19/2014
Mooney (2014)	Fracking	Washington Post	1/26/2015
Molofsky et al. (2013)	Citation in Darrah et al. (2014)	N/A	

Table A1.1 List of Search Terms and Locations, or Other Means of Location Where Appropriate for Literature Analyzed *(cont.)*

Article	Search Term or Other Means of Location	Search Engine	Date
Orem et al. (2014)	Hydraulic fracturing and toxicity	Web of Science	10/23/2014
Osborn et al. (2011)	Darrah et al. (2014) citation	N/A	
Papoulias and Velasco (2013)	Vengosh et al. (2014) citation	N/A	
Rice (2014)	Fracking	USA Today	1/26/2015
Robbins (2013)	USFWS hydraulic fracturing	Google Scholar	11/15/2014
Small et al. (2014)	Jackson et al. (2014) citation	Google Scholar	10/19/2014
Stockstad (2014)	Science article	N/A	10/14/2014
Stringfellow et al. (2014)	Hydraulic fracturing and toxicity	Web of Science	10/23/2014
Vengosh et al. (2014)	Hydraulic fracturing and environmental effects	Web of Science	10/16/2014

Table A1.2 Results of Risk Assessment Searches, Including Summaries of the Most Relevant Results. Searches all Conducted 2/2/2015, Except Those With Asterisks, Which Were Conducted on 2/4/2015

Search Location	Search Term	# Results	# Risk Assessment	Relevant Results
Environmental science and technology	Risk assessment and hydraulic fracturing	43	0	
Environmental toxicology and chemistry (via Wiley Online Library)		2	0	Burton et al. (2014)

(Continued)

Table A1.2 Results of Risk Assessment Searches, Including Summaries of the Most Relevant Results. Searches all Conducted 2/2/2015, Except Those With Asterisks, Which Were Conducted on 2/4/2015 *(cont.)*

Search Location	Search Term	# Results	# Risk Assessment	Relevant Results
Integrated environmental assessment and management (via Wiley Online Library)		3	0	
Human and ecological risk assessment (via Taylor & Francis Online Library)		6	0	
Risk analysis (via Wiley Online Library)		21	1	Rozell and Reaven
Science		8	0	Vidic et al. (2013)
Nature		27	0	
Journal of environmental management (via Springer Link)		18	0	Racicot et al. (2014)
PA department of environmental protection		10	0	
WV department of environmental protection		31	0	
Google	Maryland and hydraulic fracturing and risk assessment		1	Maryland Department of the Environment and Department of Natural Resources (2014)

Table A1.2 Results of Risk Assessment Searches, Including Summaries of the Most Relevant Results. Searches all Conducted 2/2/2015, Except Those With Asterisks, Which Were Conducted on 2/4/2015 *(cont.)*

Search Location	Search Term	# Results	# Risk Assessment	Relevant Results
	*Michigan and hydraulic fracturing and risk assessment		0	University of Michigan (2013)
	*Indiana and hydraulic fracturing and risk assessment		1	Fletcher (2012)
	*Colorado and hydraulic fracturing and risk assessment		1	McKenzie et al. (2012)

References

Adams, M.B., 2011. Land application of hydrofracturing fluids damages a deciduous forest stand in West Virginia. J. Environ. Qual. 40, 1340–1344.

Adgate, J.L., Goldstein, B.D., McKenzie, L.M., 2014. Potential public health hazards, exposures and health effects from unconventional natural gas development. Environ. Sci. Technol. 48 (15), 8307–8320.

Barbot, E., Vidic, N.A., Gregory, K.B., Vidic, R.D., 2013. Spatial and temporal correlation of water quality parameters of produced waters from Devonian-age Shale following hydraulic fracturing. Environ. Sci. Technol. 47, 2562–2569.

Bever, L., 2014. Study: fracking chemicals found in toothpaste and ice cream. The Washington Post (accessed 16.01.2015).

Brittingham, M.C., Maloney, K.O., Farag, A.M., Harper, D.D., Bowen, Z.H., 2014. Ecological risks of shale and oil and gas development to wildlife, aquatic resources and their habitats. Environ. Sci. Technol. 48, 11034–11057.

Broomfield, M., 2014. Shale gas risk assessment for Maryland. Ricardo-AEA Ltd., Oxford, UK.

Burton, G.A., Basu, N., Ellis, B.R., Kapo, K.E., Entrekin, S., Nadelhoffer, K., 2014. Hydraulic "fracking": are surface water impacts an ecological concern? Environ. Toxicol. Chem. 33 (8), 1679–1689.

Concerned Health Professionals of NY, 2014. Letter to Governor Cuomo, May 29 2014. Available from: http://concernedhealthny.org/letters-to-governor-cuomo/ (accessed 26.02.2015)

Darrah, T.H., Vengosh, A., Jackson, R.B., Warner, N.R., Poreda, R.J., 2014. Noble gases identify the mechanisms of fugitive gas contamination in drinking-water wells overlying the Marcellus and Barnett Shales. PNAS 111 (39): 14076–14081.

Farag, A.M., Harper, D.D., 2014. A review of environmental impacts of salts from produced waters on aquatic resources. Int. J. Coal Geol. 126, 157–161.

Farag, A.M., Harper, D.D., Skaar, D., 2014. *In situ* and laboratory toxicity of coalbed natural gas produced waters with elevated sodium bicorbonate. Environ. Toxicol. Chem. 33 (9), 2086–2093.

Fletcher, S.M., 2012. Risk assessment of groundwater contamination from hydraulic fracturing fluid spills in pennsylvania. Master's Thesis. Massachusetts Institute of Technology, Massachusetts.

Fontenot, B.E., Hunt, L.R., Hildenbrand, Z.L., Carlton, D.D., Oka, H., Walton, J.L., Hopkins, D., Osorio, A., Bjorndal, B., Hu, Q.H., Schug, K.A., 2013. An evaluation of water quality in private drinking water wells near natural gas extraction sites in the barnett shale formation. Environ. Sci. Technol. 47, 10032–10040.

Fountain, H., 2014. Well leaks, not fracking, are linked to fouled water. The New York Times (accessed 19.10.2014.).

Goldstein, B.D., Brooks, B.W., Cohen, S.D., Gates, A.E., Honeycutt, M.E., Morris, J.B., Orme-Zavaleta, J., Penning, T.M., Snawder, J., 2014. The role of toxicological science in meeting the challenges and opportunities of hydraulic fracturing. Toxicol. Sci. 139 (2), 271–283.

Gordalla, B.C., Ewers, U., Frimmel, F.H., 2013. Hydraulic fracturing: a toxicological threat for groundwater and drinking-water? Environ. Earth Sci. 70, 3875–3893.

Graham Sustainability Institute, 2015. Hydraulic Fracturing in Michigan. University of Michigan. Available from: http://graham.umich.edu/knowledge/ia/hydraulic-fracturing (accessed 26.02.2015.).

GSI Environmental Inc., 2013. Methane in PA Water Wells Unrelated to Marcellus shale fracturing. Available from: http://gsi-net.com/index.php?option=com_content&view=article&id=197&Itemid=266 (accessed 26.02.2015.).

Horn, S., 2013. Obama EPA censored key Pennsylvania fracking water contamination study. Huffington Post (accessed 19.10.2014.).

Jackson, R.B., Vengosh, A., Carey, J.W., Davies, R.J., Darrah, T.H., O'Sullivan, F., Petron, G., 2014. The environmental costs and benefits of fracking. Ann. Rev. Environ. Resour. 39 (7), 1–36.

Kassotis, C.D., Tillitt, D.J., Davis, J.W., Hormann, A.M., Nagel, S.C., 2014. Estrogen and androgen receptor activities of hydraulic fracturing chemicals and surface and ground water in a drilling-dense region. Gen. Endocrinol. 155 (3), 897–907.

Lackey, R., 2004. Normative science. Fisheries 29 (7), 38–39.

Landis, W.G., Wiegers, J.K. Chapter 2 Introduction to the regional risk assessment using the relative risk model. In: Landis, W.G. (Ed.), Regional Scale Ecological Risk Assessment Using the Relative Risk Model. CRC Press Boca Raton, pp. 11–36.

Lewandowsky, S., Oberauer, K., Gignac, G.E., 2013. NASA faked the moon landing—therefore, (climate) science is a hoax: an anatomy of the motivated rejection of science. Psycholo. Sci. 24 (5), 622–633.

McGarity, T.O., 2003. Our science is sound science and their science is junk science: science-based strategies for avoiding accountability and responsibility for risk-producing products and activities. Kansas Law Rev. 52, 897–937.

McGarity, T.O., Wagner, W.E., 2008. Bending Science: How Special Interest Corrupt Public Health Research. Harvard University Press, Cambridge (MA), USA.

McHugh, T., Molofsky, L., Daus, A., Connor, J., 2014. Comment on "an evaluation of water quality in private drinking water wells near natural gas extraction sites in the Barnett Shale formation. Environ. Sci. Technol. 48, 3595–3596.

McKenzie, L.M., Witter, R.Z., Newman, L.S., Adgate, J.L., 2013. Human health risk assessment of air emissions from development of unconventional natural gas resources. Sci. Total Environ. 424, 79–87.

Michigan Department of Environmental Quality, 2015. Questions and answers about hydraulic fracturing in Michigan. Available from: http://www.michigan.gov/deq/0,4561,7-135-3311_4111_4231-262172--,00.html (accessed 26.02.2015.).

Molofsky, L.J., Connor, J.A., Farhat, S.K., Wylie, A.S., Wagner, T., 2011. Methane in Pennsylvania water wells unrelated to Marcellus shale fracturing. Oil Gas J. 109 (19), 54–67.

Molofsky, L.J., Connoer, J.A., Wylie, A.S., Wagner, T., Farhat, S.K., 2013. Evaluation of methane sources in groundwater in northeastern Pennsylvania. Groundwater 51 (3), 333–349.

Mooney, C., 2014. These two states had the same basic information about fracking. They made very different decisions. The Washington Post (accessed 19.10.2015.).

Mufson, S., 2014. Study: bad fracking techniques let methane flow into drinking water. The Washington Post (accessed 19.10.2014.).

New York State Department of Health, 2014. A public health review of high volume hydraulic fracturing for Shale gas development.

Orem, W., Tatu, C., Varonka, M., Lerch, H., Bates, A., Engle, M., Crosby, L., McIntosh, J., 2014. Organic substances in produced and formation water from unconventional natural gas extraction in coal and shale. Int. J. Coal Geol. 126, 20–31.

Oreskes, N., 2004a. Science and public policy: what's proof got to do with it? Environ. Sci. Policy 7, 369–383.

Oreskes, N., 2004b. The scientific consensus on climate change. Science 306, 1686.

Oreskes, N., 2013. On the "reality" and reality of anthropogenic climate change. Climate Change 119, 559–560.

Oreskes, N., Conway, E.M., 2010. Merchants of Doubt. Bloomsbury Press, New York (NY), USA.

Osborn, S.G., Vengosh, A., Warner, N.R., Jackson, R.B., 2011. Methane contamination of drinking water accompanying gas-well drilling and hydraulic fracturing. PNAS 108 (20), 8172–8176.

Papoulias, D.M., Velasco, A.L., 2013. Histopathological analysis of fish from Acorn Fork Creek, Kentucky, exposed to hydraulic fracturing fluid releases. Southeast. Nat. 12 (4), 92–111.

Racicot, A., Babin-Roussel, V., Dauphinais, J.F., Joly, J.S., Noel, P., Lavoie, C., 2014. A framework to predict the impacts of shale gas infrastructures on the forest fragmentation of an agroforest region. Environ. Manag. 53, 1023–1033.

Rice, D., 2014. Is fracking polluting the air? USA Today (accessed 26.01.2015.).

Robbins, K., 2013. Awakening the Slumbering Giant: How Horizontal Drilling Technology Brought the Endangered Species Act to Bear on Hydraulic Fracturing. Case West. Res. Law Rev. 63 (4), 1143–1166.

Rozell, D.J., Reaven, S.J., 2012. Water pollution risk associated with natural gas extraction from the Marcellus Shale. Risk Anal. 32 (8), 1382–1393.

Small, M.J., Stern, P.C., Bomberg, E., Christopherson, S.M., Goldstein, B.D., Israel, A.L., Jackson, R.B., Krupnick, A., Mauter, M.S., Nash, J., North, D.W., Olmstead, S.M., Prakash, A., Rabe, B., Richardson, N., Tierney, S., Webler, T., Wong-Parodi, G., Zielinska, B., 2014. Risk and risk governance in unconventional shale gas development. Environ. Sci. Technol. 48, 8289–8297.

Soraghan, M., 2011. Baffled About Fracking? You're Not Alone. The New York Times (accessed 19.10.2014.).

Soraghan, M., 2012. Quiet foundation funds the 'anti-fracking' fight. *E&E News.* Available from: http://www.eenews.net/stories/1059961204 (accessed 26.02.2015.).

Stockstad, E., 2014. Will fracking put too much fizz in your water? Science 344 (6191), 1468–1471.

Stringfellow, W.T., Domen, J.K., Camarillo, M.K., Sandelin, W.L., Borglin, S., 2014. Physical, chemical, and biological characteristics of compound used in hydraulic fracturing. J. Hazard. Mater. 275, 37–54.

USEPA, 1998. EPA. Guidelines for Ecological Risk Assessment. EPA/630/R095/002F. Risk Assessment Forum. Washington, DC.

Vallianatos, E.G., Jenkins, M., 2014. Poison Spring: the Secret History of Pollution and the EPA. Bloomsbury Press, New York (NY), USA.

Vengosh, A., Jackson, R.B., Warner, N., Darrah, T.H., Kondash, A., 2014. A critical review of the risks to water resources from unconventional shale gas development and hydraulic fracturing in the United States. Environ. Sci. Technol. 48, 8334–8348.

Verango, D., 2013. Fracking linked to well water methane. USA Today. Available from: http://www.usatoday.com/story/news/nation/2013/06/24/water-fracking-pennsylvania/2452023/

CHAPTER 11
Induced Seismicity

Robert Westaway
School of Engineering, University of Glasgow, Glasgow, Scotland, UK

Chapter Outline

INTRODUCTION

The impact that induced seismicity can have on the energy industry is graphically illustrated by recent events in the United Kingdom (UK), where an exploratory fracking project in the spring of 2011 at Preese Hall in Lancashire (northwest England) caused a sequence of induced earthquakes, the largest of magnitude 2.3 (de Pater and Baisch, 2011; Clarke et al., 2014; Westaway and Younger, 2014). Given the reaction of the local population, of whom maybe two dozen actually felt the ground shake, exacerbated by contributions from environmental activists who have not always bothered to deal in facts and have repeatedly challenged the integrity of research on shale gas by others (Younger and Westaway, 2014), this energy technology has scared the living daylights out of millions of people. Furthermore, the fracking preceded initially with no microseismic monitoring, and the UK seismological community was initially unprepared for the occurrence of induced seismicity; in normal circumstances the British Isles are one of the most aseismic regions in the world (e.g., Ambraseys, 1988; Westaway, 2006), so most of the research effort of this relatively small community involves projects outside the UK. As a result, initial responses were by those who happened to be on hand, rather than those whose qualifications and experience might have best suited them to the task. This led to the recommendation (Green et al., 2012), subsequently incorporated – at

Environmental and Health Issues in Unconventional Oil and Gas Development. http://dx.doi.org/10.1016/B978-0-12-804111-6.00012-1

least for the time being – into government regulation (DECC, 2013), that future fracking projects should be subject to a "red traffic light" regulatory system in which the occurrence of any earthquake above magnitude 0.5 should trigger project shutdown. However, earthquakes as small as this are unlikely to even be felt, let alone cause injuries or damage (Westaway and Younger, 2014). A regulatory system on this basis (which seems to have been motivated by its authors' belief that magnitude 0.5 earthquakes should be regarded as precursors of larger events, when most seismologists are well aware that the Earth typically does not behave in this way, relatively large earthquakes being seldom preceded by smaller foreshocks) therefore makes no sense; until it is changed, its main effect is likely to be to add significantly to the costs of shale gas projects by triggering pointless shutdowns. This legacy, of an inappropriate form of regulation, is a consequence of the 2011 induced seismicity and the associated "panic" reaction; overall, this has not been the finest hour for earthquake seismology in the UK.

The chapter will, first, define induced seismicity, and review its potential causes, both from loading or unloading of the Earth's surface and from fluid injection. It will then present an inventory of case studies of induced seismicity, before reviewing a range of related issues, including how high differential stress affects the character of induced seismicity (an issue of particular significance for the UK), how induced earthquakes scale with volumes of fluid injection, and how the nuisance arising from induced seismicity caused by fracking can be quantified.

DEFINITION OF INDUCED SEISMICITY; RELATION TO HUMAN ACTIVITY

Davis and Frohlich (1993) first proposed a series of questions, to test whether candidate earthquakes qualify as induced. In Box 11.1 these are summarized and the resulting assessment procedure will be applied to the Preese Hall fracking activity. Nonetheless, before proceeding further it is important to be clear as to terminology. The mechanisms whereby human activities can affect seismicity have been widely discussed in recent years (Westaway, 2002, 2006; Seeber et al., 2004; Klose, 2007a,b, 2013; Ellsworth, 2013; Rubinstein and Mahani, 2015; see later). Following Klose (2013), an "anthropogenic earthquake" can be defined as any seismic event for which human activity can reasonably be shown to be the cause, or at least a major influence on timing. Anthropogenic earthquakes can in turn be subdivided into "triggered" and "induced" events; a triggered event is one that would have occurred anyway because the state of stress in the area was tending toward the condition for shear failure, or slip on an active fault, so that the human activity merely brought the earthquake forward in time or "advanced the clock." An earthquake is "induced" if there is no reason to consider that, in the absence of human activity, the state of stress in the area was heading toward the condition for shear failure: in other words, without the human activity the earthquake would never have occurred.

Box 11.1 Davis and Frohlich (1993) criteria for assessing induced seismicity

Davis and Frohlich (1993) proposed that candidate instances of induced seismicity resulting from fluid injection can be confirmed or refuted by considering seven questions, thus:

Q1. Are the events the first known earthquakes of this character in the region?
Q2. Is there a clear correlation between injection and seismicity?
Q3. Are epicenters near (within 5 km of) wells where injection has occurred?
Q4. Do some earthquakes occur at or near injection depths?
Q5. If not, are there known geologic structures that may channel flow to sites of earthquakes?
Q6. Are changes in fluid pressures at well bottoms sufficient to encourage seismicity?
Q7. Are changes in fluid pressures at hypocentral distances sufficient to encourage seismicity?

As regard the 2011 Preese Hall earthquake sequence, Q1 can clearly be answered "yes."

Q2 can also clearly be answered "yes" on the basis of data presented by de Pater and Baisch (2011). The two largest events, of magnitudes 2.3 and 1.5, both occurred ~10 h after the end of injection phases of frack stages, although the physical cause of this time lag has not yet been established.

Q3 and Q4 can clearly also be answered "yes," the epicenters were no more than a few hundred meters from the well. Their precise location is, however, contentious, and is the subject of continuing research. For example, in the view of Clarke et al. (2014) the activity occurred hundreds of meters east of and deeper than the injection. However, in my view (Westaway, in review), their conclusion is an artifact of earthquake location using a simple layered seismic velocity model, which is inappropriate as the local geological structure has significant lateral variations. In my view, the activity took place south of the well and at a depth that is indistinguishable from that of the injection. I note in passing that the association between fracking activity at the Preese Hall site and the resulting induced seismicity was not clear, initially, given that the locations obtained using the UK permanent seismograph network operated by the British Geological Survey (BGS) placed this activity several kilometers from the site, this mislocation being a consequence of the sparseness of this network, there having initially been no local microseismic monitoring. Although the BGS website (BGS, 2015) now states that this association was "immediately suspected," media releases by BGS following the April 1, 2011 event (e.g., BBC, 2011) state the cause as glacioisostasy and/or plate motions and do not mention fracking at all. This matter was only settled after BGS installed seismographs near the site, which enabled more accurate location; as many sources (e.g., DECC, 2012) point out, following the May 27, 2011 event, the regulatory authority then responsible (the UK government Department of Energy & Climate Change; DECC) imposed an immediate moratorium on fracking operations for shale gas in the United Kingdom, pending investigation of the Preese Hall seismicity. However, as the original mislocation was less than the 5 km radius advocated by Davis and Frohlich (1993), the cause-and-effect connection should have been accepted or, at least, considered, at a much earlier stage.

In my view Q5 does not need to be answered because the depth of the activity is indistinguishable from that of the injection. My own (Westaway, in review) analysis of the data gives the local direction of maximum principal stress as N7 ± 3°E–S7 ± 3°W (±2 s), and provides the best estimate of the azimuth of the induced fracture network that was created by the fracking. My working hypothesis is, therefore, that the southward propagation of this fracture network intersected a fault, and leakage of fracking fluid into this fault, near the southern extremity of the fracture network, led to the induced seismicity. On the other hand, Clarke et al. (2014) have argued that the well intersected a fault; fracking fluid thus leaked downward along this fault and induced the seismicity. Whether the fracking fluid was under sufficient pressure to have been able to flow downward by the required distance is far from clear, however (see later); in any case, if the well intersected a permeable fault one would expect the fluid to have flowed upward rather than downward.

As regards Q6 and Q7, measured wellhead pressures during the fracking stages were reported by de Pater and Baisch (2011) and were used by them to calculate bottom hole pressures. Clarke et al. (2014) published a graph labeled "bottom hole pressure" but this is in fact a mistake;

(Continued)

Box 11.1 Davis and Frohlich (1993) criteria for assessing induced seismicity (*cont.*)

it is a copy of the de Pater and Baisch (2011) graph of well head pressure. Furthermore, the input parameter values adopted for the calculation procedure used by de Pater and Baisch (2011) have not been disclosed. I have attempted to "reverse engineer" their calculations, but cannot achieve agreement, so steps are currently in hand to ensure disclosure of their data. *In situ* stress measurements were made before the fracking started, this dataset having since been placed in the public domain (Younger and Westaway, 2014). It indicates, at a representative depth of 2440 m, a maximum horizontal stress of 73.4 MPa, a vertical stress of 62.2 MPa, and a minimum horizontal stress of 43.6 MPa. The lithostatic pressure at this depth is thus 59.7 MPa. The mechanical strength of the Bowland Shale is somewhat variable but typically quite low; Westaway and Younger (2014) adopted representative values of 2 MPa for its tensile strength and 4 MPa for its cohesion. According to de Pater and Baisch (2011) the well head pressure and bottom hole pressure peaked, respectively, at ~7500 psi or ~52 MPa and ~9000 psi or ~62 MPa, during fracking stages. Since the bottom hole pressure exceeded the sum of the minimum principal stress plus cohesion, induced fractures were able to develop. A two-dimensional analysis of the state of stress (see Box 11.3) to test for the possibility of reactivation of vertical strike–slip faults oriented at 45° to the maximum principal stress indicates that such a fault intersecting the well bottom would slip if bottom hole pressure reached 33.6 MPa for a coefficient of friction f of 0.4, increasing to 41.3 MPa for $f = 0.6$. Since, according to de Pater and Baisch (2011), bottom hole pressure reached ~62 MPa, it can be presumed that no such fault intersecting the borehole exists; if it had, it would have slipped before the pressure of the fracking fluid reached such a high value. As regards any nearby fault, if its cohesion C is assumed to be 4 MPa (after Westaway and Younger, 2014) then K_1 (see Box 11.3) would be 0.25 and 0.46 for these two coefficients of friction, indicating that transmission of less than half of the excess of bottom hole pressure over hydrostatic pressure to the fault was required for this fault to slip. Conversely, if $C = 0$, the pressure required for fault slip would reduce to 18.7 and 29.5 MPa for these two coefficients of friction. The first of these values is less than the hydrostatic pressure, indicating that under such high differential stress a cohesionless fault of this orientation would be frictionally unstable even under hydrostatic pressure, whereas $f = 0.6$ would give $K_1 = 0.15$. Three-dimensional calculations, taking account of the actual orientation of the faulting present (Fig. 11.3a), which give similar results, will be published elsewhere. The time delay between the fracking stages and the resulting induced seismicity presumably relates to one or other of the mechanisms, recognized by Davies et al. (2013), for fluid leakage; the geometry suggests that the southward development of the induced fracture network may have intersected this fault, the time lag possibly being determined by the permeability of the fault, which governed the rate at which fluid could leak along it until a sufficient area of fault had been "lubricated" to enable it to slip in an earthquake of the magnitude observed.

McGarr et al. (2002) proposed a different definition, which amounts to specifying that earthquakes can be considered "induced" if they occur in localities where changes to the state of stress caused by human activity are large compared with those caused by natural processes. However, such a definition is not especially helpful; for example, the induced seismicity in northwest England in 2011 was recognized as caused by the fracking because this association was blatantly obvious because earthquakes occurred in close proximity to the fracking site within hours of frack stages (de Pater and Baisch, 2011; Clarke et al., 2014), not because anyone calculated the associated changes in the subsurface state of stress for comparison with those occurring as a result of natural processes (such as erosional isostasy and postglacial rebound; Westaway, in press) in the region. Furthermore, the total mass flux within Britain, caused by human activities such as mining and quarrying, is several hundred million tons per year (Table 11.1), roughly two orders of magnitude larger than natural mass flux

Table 11.1 **Mass Fluxes in the United Kingdom From Human Causes**

Process	Year	Mass Rate (MT/a)	Sources	Note
Waste from construction, demolition, excavation	2012	100.2	DEFRA (2015)	1
Production of crushed rock	2013	94.3	BGS (2014)	
Generation of commercial and industrial waste	2012	47.6	DEFRA (2015)	
Production of onshore sand and gravel	2013	43.4	BGS (2014)	
Production of oil	2013	38.5	BGS (2014)	
Production of natural gas	2013	36.5	BGS (2014)	
Soil erosion caused by agriculture	n.d.	34.1	Van Oost et al. (2009), DEFRA (2009)	2
Generation of domestic waste	2012	26.5	DEFRA (2015)	
Production of marine dredged sand and gravel	2013	14.6	BGS (2014)	
Production of raw materials for cement	2013	11.5	BGS (2014)	3, 4
Excavations for foundations for new housing	2007	~10	NHBC (2007)	5
Production of coal (opencast)	2013	8.6	BGS (2014)	
Production of limestone for industrial use	2013	8.1	BGS (2014)	3, 6
Production of coal (deep mined)	2013	4.1	BGS (2014)	
Production of rock salt	2013	6.6	BGS (2014)	7
Production of raw materials for brick manufacture	2013	4.2	BGS (2014)	8
Production of industrial sand	2013	4.0	BGS (2014)	
Production of limestone for agricultural use	2013	2.1	BGS (2014)	3
Production of gypsum	2013	1.2	BGS (2014)	9
Production of china clay	2013	1.1	BGS (2014)	
Production of other minerals	2013	4.5	BGS (2014)	10
Total onshore UK mineral production	2013	195.8	BGS (2014)	
Total offshore UK mineral production	2013	90.7	BGS (2014)	11
Overall total UK mineral production	2013	286.4	BGS (2014)	

Notes:
1. This value is obtained by multiplying the per capita figure for 2012, 1573 kg, from BGS (2014), by the UK population in 2012, 63.7 million.
2. Van Oost et al. (2009) estimated rates of soil erosion of 16.0 MT/a from the direct effect of ploughing and 18.1 MT/a from associated runoff. However, the sum of these estimates exceeds the quantity reported by the UK government (e.g., 2.2 MT/a; DEFRA, 2009) by more than an order of magnitude.
3. Data exclude sources in Northern Ireland.
4. Limestone, chalk, clay, and shale.
5. In 2007, before the subsequent economic downturn, ~500 new homes used to be completed typically on each of ~250 working days (NHBC, 2007). Assuming that the typical floor area was ~40 m^2 and the foundations for each property involved excavation to a depth of ~1 m in soil of density ~2000 kg/m^3, which was removed by the contractor for reuse elsewhere, the total mass of material excavated during the year was $\sim 500 \times 250 \times 1 \times 40 \times 2000$ or $\sim 10^{10}$ kg.
6. Includes chalk and dolomite.
7. Includes brine.
8. Clay, fireclay, and shale.
9. Excludes desulfo-gypsum, produced by flue gas desulfurization in coal-fired power stations.
10. "Others" include ball clay, peat, potash (potassium chloride), slate, building stone, coal recovered by reprocessing old waste tips, and clay/shale for construction (each between 0.5 and 1.0 MT/a), plus minor amounts of others, primarily calcspar, lead ore, and iron ore.
11. Includes almost all oil and natural gas, plus marine dredged sand and gravel.

resulting from erosion and measurable as sediment transported by Britain's rivers (estimated as ~2 MT/a by Westaway, 2006). The isostatic response to erosion is generally considered to be the main natural mechanism driving deformation of the Earth's crust in Britain at present (Bridgland and Westaway, 2014; Westaway, in press). As a result, if the McGarr et al. (2002) definition were to be adopted, it would follow that all earthquake activity in Britain would automatically be classified as "induced seismicity," irrespective of whether the cause-and-effect mechanism was understood for any particular instance.

Rubinstein and Mahani (2015) have proposed yet another definition of induced seismicity, in which all earthquakes with a human cause are considered as induced even if the human cause merely "advanced the clock." This would make the term "induced earthquake" synonymous with anthropogenic earthquake, making one of these terms superfluous. Furthermore, there is clearly an important distinction between instances such as the magnitude 7.9 Wenchuan (China) earthquake of May 12, 2008, where construction of a reservoir on the seismogenic fault arguably changed the state of stress to "advance the clock," but future earthquakes on this fault will follow the same pattern as those in the past, apart from this one-off "clock advance" (Klose, 2012), and the instances described in this chapter where human activity is pervasively altering the state of stress and, thus, causing earthquakes, in a manner that would not occur if the activity were not taking place. I shall therefore adhere to the standard definition of "induced seismicity."

The study of anthropogenic seismicity began when earthquakes were first felt in Johannesburg (South Africa) in 1894; by 1908 these had been attributed to Witwatersrand gold mining, which had begun in 1886 (McGarr et al., 2002). It is thus appropriate to start the present discussion by considering the scale of mass movement at the Earth's surface as a result of mining and other excavation. Regarding the global picture, Syvitski et al. (2005) estimated that in its natural state, before the Earth's surface experienced significant human alteration, the global flux of sediment transported by rivers to the oceans was ~15 billion tonnes per year (~15 GT/a). In their view, human activity had increased the flux of sediment entering the world's river systems by ~2.3 GT/a, but a sediment flux of ~3.7 GT/a is trapped behind dams, so that, overall, human activity has reduced the sediment flux reaching the oceans by ~1.4 GT/a or ~10%. On the other hand, it is well known that human activity, such as agriculture and mining, has led to dramatic increases in sediment fluxes, for example, in the Mediterranean region starting in antiquity (notably, in the Roman era), and in the eastern United States (US) at the start of settlement by Europeans (Wrench, 1946; Trimble, 1975, 2008; Daniels, 1987; Friedman et al., 2000). Syvitski and Kettner (2011) estimated that the annual flux of mass movement due to human activity is roughly equal to that of the world's rivers, or ~15 GT/a, but offered no calculation, although they noted that very large rates of mass movement are associated with large-scale engineering projects such as iron ore mining (Table 11.2). However, as will become clear later, fluxes of mass movement due to human activity are nowadays far greater than those present in the Earth's natural state. Furthermore, much of this anthropogenic mass movement is over relatively short distances and does not involve transport by rivers; the resulting perturbation to the

Table 11.2 Mass Fluxes Associated With the World's Largest Mines

Name	Location	Mass Rate (MT/a)	Reserves (GT)
Copper			
Escondida mine	El Loa Province, Antofagasta Region (Chile)	~475	104
Iron ore			
Hamersley mine complex	Hamersley Basin, Western Australia	127	1.7
Carajás mine	Carajás Mountains, Pará State (Brazil)	107	7.3
Coal			
North Antelope Rochelle mine	Powder River Basin, Wyoming (USA)	108	>2.3
Black Thunder mine	Powder River Basin, Wyoming (USA)	93	1.5

Copper: BHP Billiton (2012) and International Copper Study Group (2012). Annual mass movement is estimated, as for Table 11.3, from the published daily average figure (1.3 million tonnes) to achieve the annual production of 0.9 MT of copper. Reported reserves consist of 21.7 GT confirmed and 82 GT of probable exploration targets, and are expressed directly as the mass of rock that will have to be processed to extract the copper contained therein.
Iron ore: Mining-technology.com (2014) data for 2012. The scale of the Carajás mining operation has been described as "ore-inspiring" in promotional literature (Bus-ex.com, 2013).
Coal: Mining-technology.com (2013) data for 2012.

state of stress will therefore be rather greater than if the movement was more widely distributed (Westaway, 2002).

Table 11.3 quantifies the global contributions of a number of processes that involve large-magnitude mass movements of human origin. Although Syvitski and Kettner (2011) have argued that these effects are dominated by giant engineering projects, these seem to be too few in number to significantly impact on the inventory of global contributions listed in Table 11.3, many of which represent the cumulative effect of many small operations (see Box 11.2). The total flux from the processes listed exceeds 330 GT/a, and is thus more than 20 times the worldwide 'natural' sediment flux in the global river system. Even so, the resulting total flux is probably a significant underestimate, as movements of mass as a result of excavations, say, for land reclamation (recognized as important by Syvitski and Kettner, 2011), have not been included. The redistribution of water mass, caused by reservoir impoundment, has also been omitted, notwithstanding the significance of reservoir-induced seismicity (Klose, 2013). Furthermore, for production of most of the commodities listed, the annual flux of mass movement has been equated to annual production, so any excavation of overburden has been neglected. The exception is for production of copper, which involves the processing of very large quantities of ore that typically contains only a fraction of ~1% of the metal (Tables 11.2, 11.3).

The largest component of anthropogenic mass movement listed in Table 11.3 is from unsustainable use of water; in many parts of the world, water supplies are drawn from aquifers that are not being replenished, with the resulting wastewater draining into the oceans. Various workers (Wada et al., 2010, 2012; Konikow, 2011, 2013; Pokhrel et al., 2012, 2013) have estimated the magnitude of this component from observed rates of sea level rise; one such estimate is listed in Table 11.3. At present (IPCC, 2013), global sea level is estimated to be

Table 11.3 Worldwide Mass Fluxes From Human Causes

Process	Year	Mass Rate (GT/a)	Source	Notes
Unsustainable use of water	2008	196	Wada et al. (2012)	1
Soil erosion	n.d.	75	Myers (1993), Pimentel (2006)	2
Production of crushed rock	2011	14.2	USGS (2013)	3
Production of sand and gravel for construction	2011	9.8	USGS (2013)	4
Production of copper	2012	9.0	USGS (2013), BHP Billiton (2012)	5
Production of building stone	2011	8.4	USGS (2013)	6
Production of coal	2012	7.86	BP (2013)	
Production of oil	2012	4.12	BP (2013)	
Production of cement	2012	3.60	USGS (2013)	
Production of natural gas	2012	3.03	BP (2013)	
Release of carbon to atmosphere from deforestation	2000	2.1	Houghton (2003, 2005)	7
Production of iron ore	2010	1.82	World Steel Association (2012)	
Production of synthetic oil from tar sand	2010	1.4	Syvitski and Kettner (2011)	8

n.d., not determined – indicates that the estimate is not determined for any particular year.
Notes:
1. This estimate is for unsustainable depletion of aquifers by human activity. The rate of sea level rise from this cause, time-averaged over 1993–2008, from Wada et al. (2012), of 0.54 ± 0.09 mm/a, is multiplied by the percentage of the Earth's surface occupied by the oceans (71% of 5.10×10^8 km^2 of surface area) and by the nominal 1000 kg/m^3 density of water.
2. This long-standing estimate from Myers (1993) is nowadays considered conservative (Pimentel, 2006).
3. The global total is not recorded; it is estimated here by multiplying the US production of crushed stone by the ratio of global to US production of cement, all data being from USGS (2013): 1.16 GT × 842/68.6.
4. The global total is not recorded; it is estimated here by multiplying the US production of sand and gravel by the ratio of global to US production of cement, all data being from USGS (2013): 0.802 GT × 842/68.6.
5. From USGS (2013), the global copper production in 2012 was estimated as 17 MT. From BHP Billiton (2012), the Escondida copper mine in Chile, the world's largest, produced 0.9 MT of copper in 2012 but this required the processing of a typical 1.3 MT of rock per day, so a total of 1.3 × 365 or ~475 MT of rock was thus processed at Escondida in 2012. Assuming the same typical ratio of rock processing to copper production for other mines worldwide, the total rock processed can be estimated as 475 MT × 17/0.9.
6. The global total is not recorded; it is estimated here by multiplying the US production by the ratio of value of US imports to US production, all data being from USGS (2013): 1.71 GT × 1590/323. The resulting quantity is an underestimate of the global total, as use in other countries is not considered; however, USGS (2013) does state that the United States is the principal global consumer of building stone.
7. According to Houghton (2003), rates of biomass loss peaked circa 1990 and have since declined. A wide range of estimates for global biomass loss exist, as detailed in Houghton (2005), reflecting the difficulty in measuring both the reduction of forest area and the biomass remaining in areas that are still forested. Burning of plant biomass will also release water vapor to the atmosphere; the resulting mass loss will be subsumed into the global water budget and so will have already been counted as part of the changed water budget.
8. This estimate is for the Athabasca Tar Sands in Alberta (Canada). Syvitski and Kettner (2011) stated that ~30 GT of tar sand had been processed, taken as meaning by 2010. Assuming a linear increase in production since it began in 1967, the rate of processing for 2010 has been estimated. This is probably an overestimate as in recent years a substantial proportion of the oil from tar sand has been extracted by steam injection rather than by quarrying and processing the tar sand.

rising at ~3 mm/a of which ~1.1 mm/a is caused by thermal expansion of the oceans, ~1.6 mm/a is caused by melting of glaciers (~0.8 mm/a) and ice sheets (Greenland: ~0.5 mm/a; Antarctica: ~0.3 mm/a), as well as the ~0.4 mm/a due to groundwater abstraction. It could indeed be argued that this meltwater component is being caused by anthropogenic global warming and is thus another example of mass movement with a human cause; if so, the resulting rate

Box 11.2 Mass movements associated with large engineering projects

Syvitski and Kettner (2011) suggested that large-scale construction projects make the principal contribution to the global mass flux of human origin, noting as examples Chek Lap Kok airport in Hong Kong and the artificial resort islands off the coast of Dubai. Chek Lap Kok airport required excavation of a mountain and the dredging of marine sediment to create a flat land surface of area ~12.5 km^2, with ~400 million m^3 of material moved in ~31 months (according to TESTRAD, 2012). Taking the density of this rock mass as ~2500 kg/m^3, its mass was ~1 GT and the associated mass flux ~0.4 GT/a. As Syvitski and Kettner (2011) noted, the artificial islands off Dubai had a planned mass of >3 GT. However, the largest of these projects, called Palm Deira, has been scaled back (even so, the resulting land area is ~46 km^2 according to Worldtravellist.com, 2011) and total mass movement that has actually taken place has probably been ~2 GT, with a typical mass flux during the ~10 year construction period of ~0.2 GT/a. For comparison, according to Syvitski and Kettner (2011) construction of the Great Wall of China involved the movement of ~0.4 GT of rock and soil. For another comparison, the largest construction project in Europe at present is the Maasvlakte 2 land reclamation scheme to extend the port of Rotterdam in the Netherlands. The resulting creation of ~20 km^2 of land has involved the emplacement of ~240 million m^3 of sand, dredged from the North Sea, ~7 MT of rock, mostly shipped from Scandinavia, and ~20,000 concrete cubes, each with a mass of ~43 T (Port of Rotterdam, 2013). The total mass moved has thus been ~0.5 GT, indicating a mass flux of ~0.1 GT/a time-averaged during 2009–2013. However, although schemes on this scale will clearly dominate the mass flux in the localities affected, it would seem that there are too few of them to have a significant effect on the overall global total (Table 11.3). Dividing the ~330 GT/a mass flux from Table 11.3 by the land area of the Earth, ~1.5 × 10^8 km^2, gives a mass flux density of ~2.2 kg/m^2/a. Assuming a mean density of the material that has been moved of around ~1500 kg/m^3, as much of it has been inferred to be water rather than rock, one obtains a globally averaged typical rate of anthropogenic surface processes of ~1.5 mm/a.

As regards the United Kingdom, significant contributions to anthropogenic mass movement are listed in Table 11.1. Their breakdown is very different from the global dataset, the largest magnitude contributions being production of crushed rock and generation of waste (Table 11.1). The pattern is very different from in the past; for example, at its peak in 1913–1914 coal production was 287.4 MT/a (Mitchell, 1984, p. 3), comparable in magnitude with all contributions to the present-day mass flux combined. In a developed country such as this with a relatively high population density, large construction projects might be expected to contribute significantly to the total anthropogenic mass movement (Syvitski and Kettner, 2011); however, like for the global comparison, it appears that there are not enough of these to make any substantive contribution, relative to the many small projects whose cumulative effect is estimated approximately in the table. For example, the construction in 2001–2006 of the ~40 km long phase 2 of the Channel Tunnel Rail Link east of London involved ~2.5 million m^3 of excavation (Paul et al., 2002, p. 17), largely as a result of the ~20 km of ~8 m diameter tunnels required for each railway track. Assuming a density of 2000 kg/m^3 for the material excavated, this represents a mass of ~5 MT and a mass flux of ~1 MT/a during each year of construction. The Crossrail scheme, to link the suburban railway networks east and west of London by tunneling beneath the city, involves an estimated 8 million m^3 of excavation (TESTRAD, 2012), equivalent to 16 MT, with an annual mass flux during the ~8 years of construction to 2018 of ~2 MT/a. The United Kingdom's largest recent construction project, for the London Gateway container port at Stanford-le-Hope, Essex, in the estuary of the River Thames, has involved movement of an estimated 30 million m^3 or ~60 MT of silt for dredging the shipping channel and land reclamation, at a mass flux of ~10 MT/a during the 6-year construction period to 2014 (Laing O'Rourke, 2014). The proposal to reclaim ~20 km^2 of land to build the proposed "London Britannia Airport" in the Thames Estuary would involve an estimated 134 million m^3 or ~270 MT of mass movement (TESTRAD, 2012); a ~7-year construction period is envisaged, which would mean a typical mass flux of ~40 MT/a. However, even construction on this scale would add not much more than 10% to the existing anthropogenic mass flux in the United Kingdom (Table 11.1).

of mass movement would exceed the estimate in Table 11.3 by a factor of ~5. Changes in the mass distribution of ice sheets on deglaciation around the end of the Pleistocene are well known as the cause of paleoseismicity, for example, in Scotland (Ringrose, 1989; Stewart et al., 2001) and Scandinavia (Lundqvist and Lagerback, 1976; Mörner et al., 2008; Sutinen et al., 2009; Juhlin et al., 2010). However, the associated changes in mass distribution were much greater than anything being caused by human activity at present. For example, around the Last Glacial Maximum, the British Isles were occupied by an ice sheet with an estimated surface area and volume of ~720,000 km^2 and ~800,000 km^3, which subsequently melted at rates of ~65–260 km^3/a (Clark et al., 2012). Taking the density of ice as ~920 kg/m^3, the resulting mass flux of meltwater from this relatively small ice sheet was in the range ~60–240 GT/a, comparable with total estimated mass flux of human origin at present (Table 11.3).

One additional process involving mass movements of human origin, not hitherto considered, is subsurface injection of water. In the UK, and throughout the European Union (EU), subsurface disposal of wastewater by borehole injection is prohibited, as a result of the EU Water Framework Directive (European Union, 2000), as incorporated into the laws of individual member states. Injection of water into boreholes for purposes other than waste disposal, for example, to maintain the pressure in oilfields or geothermal reservoirs or for hydraulic fracturing, is permitted, subject to appropriate environmental permissions, however. Even before the era of EU regulation, there was reluctance in the UK to engage in borehole disposal of wastewater; for example, even during the crash program in 1950–1952 to produce the plutonium for our first nuclear weapons, the resulting radioactive waste was stored pending eventual treatment, rather than adopting borehole disposal as in contemporaneous efforts in the US and the Soviet Union (Pegg, 2015). The present legal framework indeed specifies the technical definition as a potential "aquifer" of all rocks, in principle down to the center of the Earth, including, for example, rocks containing formation waters that are highly saline brines, which no one would ever contemplate using for water supply. Hydrogeologists have questioned whether such a blanket definition of "aquifers" makes sense (Mather et al., 1998), and it might seem strange to forbid by law the subsurface disposal of water but not the subsurface disposal of other substances, such as carbon dioxide (CO_2), but nonetheless this is the current situation. It has led to some contention in relation to shale gas projects, as objectors have mistakenly presumed that rocks at depths of several kilometers that comply with this legal definition of an "aquifer" are somehow used for water supply. The related issue of whether fluids injected for fracking might escape upward from such depths to contaminate aquifers that are used for water supply, at depths within a few hundred meters of the Earth's surface, is addressed by Younger (in review).

The previously mentioned situation is in marked contrast with the US, where the Environmental Protection Agency (EPA; http://water.epa.gov/type/groundwater/uic/) permits borehole injection of wastewater under many circumstances where, within the European Union, the water would be treated and returned to the environment. The largest scale use of borehole injection comprises EPA Class II disposal, which covers wastewater from hydrocarbon production, including flowback fluid from fracking and formation waters from coalbed methane production. EPA (2012) reported that 144,000 Class II disposal wells were in operation in the US in 2012, injecting water at a rate of ~2 billion gallons per day or ~3.3×10^9 m^3 per year. In the central and eastern US (the conterminous US, excluding the seven western states of Arizona, California, Idaho, Nevada, Oregon, Utah, and Washington), Rubinstein et al. (2015) inventoried >200,000 wells (most of which, 78%, are in Texas and Oklahoma), including >1000 wells which each dispose of >100,000 barrels of water per month, equivalent to >500 m^3 per day or ~200,000 m^3 per year. Individual high-volume disposal wells may inject >1000 m^3 per day; for example, McGarr (2014) mentioned one in Oklahoma, Howard-1, where 1.44×10^6 m^3 of water had been injected within 3 years. The EPA borehole disposal figure, mentioned earlier, equates to a mass flux of ~3.3 GT/a. It is apparent that this is small compared with many other components of the global mass flux of human origin (Table 11.3). Nonetheless, as will become clear, induced seismicity is a particularly prevalent consequence of subsurface wastewater disposal, evidently because of the effect of increased water pressure facilitating slip on faults (Box 11.3) (See, also, Chapter 5).

Box 11.3 Conditions for induced seismicity

The conditions whereby seismicity can be induced by changes to the state of stress have been reported many times (Westaway, 2002, 2006); the standard Mohr–Coulomb frictional analysis in two dimensions, illustrated in Figure 11.1, has also been presented graphically many times, but is also included here for completeness. It should be noted, however, that because faults are typically not oriented parallel to any of the principal stress axes, analysis in three dimensions should be undertaken. This is more difficult to depict graphically and also algebraically more complex, so is not considered here.

The construction utilizes a Mohr Circle, which represents the range of possible combinations of shear stress τ and normal stress σ_N in different directions at a given locality within a rock mass, for a given pair of principal stresses (in two dimensions), σ_1 and σ_2. A frictional failure envelope for the fault is also defined, usually with a constant gradient, or angle of slope ϕ, reflecting the assumption that the coefficient of friction on a fault f, where $\tan(\phi) \equiv f$, is independent of its state of stress. Thus, if C is the cohesion of this fault, the failure envelope has the equation:

$$\tau = C + \sigma_N \tan(\phi) \tag{1}$$

The mean stress in the rock mass, $\sigma_M \equiv (\sigma_1 + \sigma_2)/2$, marks the center of the Mohr Circle (at the point $[\sigma_N = \sigma_M, \tau = 0]$), the diameter of this circle being the differential stress $\Delta\sigma \equiv \sigma_1 - \sigma_2$. In the presence of fluid pressure P, the effective mean stress reduces from σ_M to $\sigma'_M \equiv \sigma_M - P$. From the form of Figure 11.1, the condition that the Mohr Circle touches the failure envelope, meaning that the state of stress enables the fault to slip and thus results in an induced earthquake, can be written, after several algebraic steps, as:

$$P = \sigma_M + C\cot(\phi) - \Delta\sigma\,(\sin(\phi) + \cos(\phi)\cot(\phi)) \tag{2}$$

or

$$P = \sigma_M + \frac{C - \sqrt{(1+f^2)}\Delta\sigma}{f} \tag{3}$$

(Continued)

Box 11.3 Conditions for induced seismicity (*cont.*)

This value can be compared with other measures of pressure, such as the hydrostatic pressure P_o (estimated as the depth at the point of interest × the density of water × the acceleration due to gravity) and the bottom hole pressure of the injected fluid P_B. To facilitate such a comparison, one may calculate ratios such as K_1, K_2, and K_3 where:

$$K_1 \equiv \frac{P - P_o}{P_B - P_o} \tag{4}$$

$$K_2 \equiv \frac{P - P_o}{\sigma_M - P_o} \tag{5}$$

and

$$K_3 \equiv \frac{P - P_o}{P_L - P_o} \tag{6}$$

where P_L is the lithostatic pressure. Once the state of stress has been measured, one may use these metrics in the assessment of the possibility of causing induced seismicity, before proceeding with any project. Thus, for example, if $P \leq P_B$ then a significant likelihood of induced seismicity is apparent; this possibility might be mitigated by reducing P_B or might otherwise be covered; for example, by keeping the volume injected low enough to limit the magnitude of any induced earthquake, by arranging insurance, or by establishing a compensation scheme for anyone affected by induced seismicity. I repeat, however, that analysis of this type should be carried out in three dimensions; the two-dimensional version given here is for illustrative purposes only.

This analysis (and its graphical representation in Figure 11.1) has several important practical consequences for induced seismicity, all of which are apparent from the algebraic form of equation (3). Clearly, the higher the pressure of the injected fluid, the more likely it is to cause an induced earthquake. Second, the "stronger" or "rougher" the rock (i.e., the larger its cohesion or coefficient of friction), the less likely induced seismicity is for a given pressure increase. This is evident from Figure 11.1, since the higher the cohesion the higher the vertical intercept of the failure envelope, whereas the higher the coefficient of friction the steeper the failure envelope; both these factors will therefore act to keep the failure envelope away from the Mohr Circle. Third, the larger the differential stress, the greater the likelihood of induced seismicity. This is apparent because differential stress determines the diameter of the Mohr Circle; the larger this diameter, the more likely (other factors being equal) part of this circle will approach the failure envelope. Conversely, if differential stress is low compared with the cohesion of the rock mass, the Mohr Circle may never intersect the failure envelope, however large the fluid pressure, provided a locality remains in the compressional regime; failure may only occur under the tensile conditions that would result from an even larger fluid pressure, if the left-hand edge of the Mohr Circle were to touch the failure envelope at the point corresponding to the tensile strength of the material (*T* in Figure 11.1). High differential stress is thus a significant factor influencing induced seismicity, albeit one that is often overlooked; for example, it is omitted completely from the discussion by Walters et al. (2015) of risk factors relating to fluid injection. It is evidently the reason why, for example, fracking for shale gas has resulted in induced earthquakes that have been large enough to be felt in regions of high differential stress, such as northern England, Ohio, and British Columbia (see the main text), whereas, for example, in the Barnett Shale – a shale gas province of Texas where differential stress is very low (Gale et al., 2007) – no such event has occurred despite ~15,000 wells having been fracked.

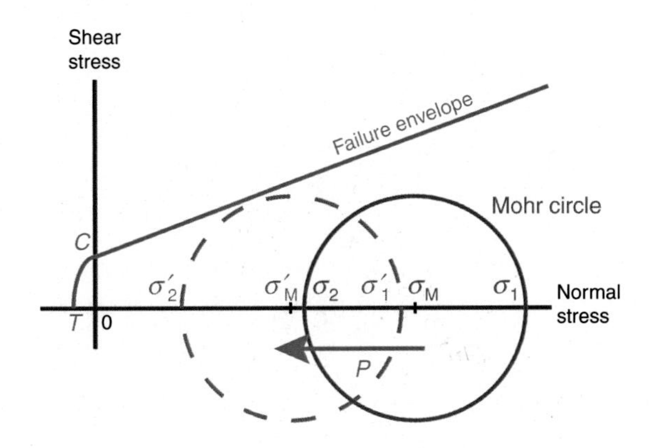

FIGURE 11.1
Schematic illustration of the physical mechanism whereby fluid injection can cause induced seismicity, using a Mohr Circle/failure envelope representation of stress state (Box 11.3). Normal stress (compressional when positive and tensile when negative) is plotted on the horizontal axis, with shear stress on the vertical axis. The maximum and minimum normal stresses acting at a given locality are plotted as σ_1 and σ_2; these lie at opposite ends of a diameter of a circle that is centered at the point on the horizontal axis where the normal stress equals the mean stress σ_M (where $\sigma_M \equiv (\sigma_1 + \sigma_2)/2$) and has radius $\Delta\sigma/2$ where $\Delta\sigma \equiv \sigma_1 - \sigma_2$. This Mohr Circle represents the set of combinations of normal and shear stress that act in different directions at the locality. The failure envelope represents the set of combinations of normal and shear stress that result in rock failing in shear or in tension, or a preexisting fault or tensile fracture being reactivated. The gradient of this line in the compressional regime equals the coefficient of friction of a preexisting fault, or the coefficient of internal friction within a rock mass, *f*. This envelope passes through a point on the horizontal axis where the tensile normal stress equals the tensile strength *T* of the material, and a point on the vertical axis where the shear stress equals its cohesion *C*. Intact rock will have nonzero *C* and *T*; as discussed by Westaway and Younger (2014), preexisting faults or fractures in some rocks, such as shale, might likewise have nonzero *C* and *T*, so this possibility is allowed for here. When fluid pressure *P* is increased, the mean stress is reduced from σ_M to $\sigma'_M \equiv \sigma_M - P$ and the Mohr Circle effectively shifts leftward by a distance *P* (to the dashed circle depicted), resulting in new "effective" maximum and minimum normal stresses $\sigma'_1 \equiv \sigma_1 - P$ and $\sigma'_2 \equiv \sigma_2 - P$. This normal stress reduction effect is known as fault "unclamping"; it has the consequence of bringing the fault nearer to the condition for slip, as can be visualized from the closer proximity of the dashed Mohr Circle to the failure envelope. If the pressure increase is large enough such that the adjusted Mohr Circle touches the failure envelope, the fault will slip (or the rock will fail to create a new fault) and an induced earthquake will result. *Modified from Figure 3 of Rubinstein and Mahani (2015).*

INVENTORY OF INDUCED EARTHQUAKES

Table 11.4 lists an inventory of induced earthquakes, all arguably caused by borehole injection of fluid; the exception is the Newcastle (Australia) event, caused by unloading of the crust as a result of coal mining (Klose, 2007a), which is included for comparison. The physical reason fluid injection can result in induced seismicity is explained in Box 11.3 and illustrated schematically in Figure 11.1. This inventory extends a previous version by McGarr (2014), with additional earthquakes and more information on the events originally included. Figure 11.2 plots, for each of these earthquakes, seismic moment M_o, a standard measure of the elastic strain energy released during an earthquake, which scales with earthquake magnitude M_W, against the volume V of water injected. It is

Table 11.4 Induced Earthquakes Caused by Borehole Injection of Fluid

Name	Date	Time	Cause	Notes	M_w	M_o (Nm)		V_1 (m^3)	V_2 (m^3)	ID
						1	2			
KTB site (Germany)	Dec. 18, 1994	15:26	S	3	1.4	1.43E+11	1.41E+11	200		KTB
Harrison County, Ohio	Oct. 5, 2013	00:16	F SG	4	2.2		2.24E+12	94,175	840	HCO
Preese Hall (England)	April 1, 2011	02:34	F SG	5	2.3	3.20E+12	3.16E+12	4170	2339	BUK
Soultz-sous-Forêts (France)	July 16, 2000	21:41	E	6*	2.4		4.47E+12	23,000		STZ1
Soultz-sous-Forêts (France)	June 10, 2003	22:54	E	6	2.9	2.51E+13	2.51E+13	40,000		STZ2
Eola (Garvin County), OK.	Jan. 18, 2011	03:40	F EPR	7	2.9	3.50E+13	2.51E+13	17,500	8500	GAR
Poland Township, Ohio	Mar. 10, 2014	06:26	F SG	8	3.0		3.55E+13	19,100	1219	PTO
Horn River (Canada)	July 7, 2011	22:46	F SG	9	3.1		5.01E+13	45,000	5000	HRC2
Dallas-Fort Worth, Texas	May 16, 2009	16:24	W	10	3.3	8.90E+13	1.00E+14	370,000		DFW
Basel (Switzerland)	Dec. 8, 2006	16:48	E	6	3.4	1.41E+14	1.41E+14	11,566		BAS
Ashtabula, Ohio	July 13, 1987	05:49	W	11*	3.6	2.82E+14	2.82E+14	60,000		ASH1
Azle, Texas	Dec. 8, 2013	06:10	W	12*	3.6		2.82E+14	2,813,000		ATX
Cooper Basin (Australia)	Dec. 5, 2003	17:45	E	13*	3.7	3.98E+14	3.98E+14	20,000	7300	CBN
Horn River (Canada)	May 19, 2011	13:05	F SG	14	3.8		5.62E+14	63,000	3500	HRC1
Youngstown, Ohio	Dec. 31, 2011	20:05	W	15	3.88	8.30E+14	7.41E+14	78,798		YOH
Ashtabula, Ohio	Jan. 26, 2001	03:03	W	11	3.9	8.00E+14	7.94E+14	340,000		ASH2
Paradox Valley, Colorado	May 27, 2000	21:58	W	16*	4.3	3.16E+15	3.16E+15	2,350,000	490,000	PBN
Berlin (El Salvador)	Sep. 16, 2003	01:20	E	17	4.4		4.47E+15	205,000	136,000	BER
Raton (Trinidad), Colorado	Sep. 5, 2001	10:52	W	18	4.4	4.50E+15	4.47E+15	426,000		RAT1
Cogdell (Snyder), Texas	Sep. 11, 2011	12:27	G EPR	19*	4.41		4.62E+15	325,000		COG
Denver, Colorado	Apr. 10, 1967	19:00	W	20	4.53		7.10E+15	625,000		RMA1
Denver, Colorado	Nov. 27, 1967	05:09	W	20	4.54		7.20E+15	625,000		RMA3
Guy, Arkansas	Feb. 28, 2011	05:00	W	21	4.7	1.20E+16	1.26E+16	3,477,000	612,000	GAK
Painesville, Ohio	Jan. 31, 1986	06:46	W	22	4.8	2.00E+16	1.78E+16	1,190,000		POH
Denver, Colorado	Aug. 9, 1967	13:25	W	20	4.85	2.10E+16	2.10E+16	625,000		RMA2
Timpson, Texas	May 17, 2012	08:12	W	23	4.86	2.21E+16	2.21E+16	3,950,000	1,050,000	TTX
Prague, Oklahoma	Nov. 5, 2011	07:12	W	24	5.0		3.55E+16	12,000,000		POK1

Prague, Oklahoma	Nov. 8, 2011	02:46	W	24	5.0		3.55E+16	12,000,000		POK3
Raton (Trinidad), Colorado	Aug. 23, 2011	05:46	W	25	5.3	1.00E+17	1.00E+17	51,670,000	7,840,000	RAT2
Newcastle (Australia)	Dec. 27, 1989	23:26	C	26	5.6		2.82E+17	778,000,000		NEW
Prague, Oklahoma	Nov. 6, 2011	03:53	W	24	5.7	3.92E+17	3.98E+17	12,000,000		POK2

Earthquakes are listed in order of increasing magnitude M_W and seismic moment M_o. Causes, or processes considered responsible for each event, comprise: C, coal mining; E, hydraulic fracturing for enhanced geothermal systems; F EPR, fracking for enhanced petroleum recovery; F SG, fracking for shale gas; G EPR, carbon dioxide injection for enhanced petroleum recovery; S, a scientific experiment; and W, wastewater disposal. V_1 denotes the measured or estimated volume of fluid injected, V_2 (where specified) denoting significantly smaller estimates that might also apply in some cases. ID codes identify individual earthquakes in Figure 11.2. KTB denotes the German Continental Deep Drilling Program site (Kontinentales TiefBohrprogramm in German), near Windischeschenbach in Bavaria. In the "Notes" column, * indicates that I had to search the online catalog of the International Seismological Centre (http://www.isc.ac.uk/iscbulletin/search/catalogue/) for source parameters that were not provided by the other information sources listed. Other notes indicate sources of information:

1. This column lists M_o values as reported by McGarr (2014).
2. This column lists M_o values determined in this study from M_W using the standard Hanks and Kanamori (1979) equation ($\log_{10}(M_o/\text{Nm}) = 9.05 + 1.5M_W$). For some of the earthquakes the listed M_W values were obtained first, by applying the inverse of this equation to measured values of M_o.
3. Data are from Dahlheim et al. (1997) and Zoback and Harjes (1997).
4. Data are from Friberg et al. (2014). These authors reported that three laterals were fracked: Ryser 2, 20 frack stages during September 7–13, 24,500 m^3 total volume; Ryser 4, 24 frack stages during September 14–26, 30,175 m^3 total volume; and Ryser 3, 47 frack stages during September 19–October 6, including some after the induced event, 39,500 m^3 total volume. V_1 is the sum of these three total volumes; V_2 is the mean volume for a frack stage for Ryser 3.
5. Data are from de Pater and Baisch (2011). V_1, used by McGarr (2014), is total fluid volume injected prior to the earthquake, V_2 being the volume in the fracking stage immediately before the earthquake.
6. Data are from Majer et al. (2007).
7. Magnitude, from Holland (2013), was 2.9, not 3.0 as stated by McGarr (2014). About 9000 m^3 of the fluid volume reported by McGarr (2014) was injected during an earlier fracking stage, and so might have had no causal effect for this particular induced earthquake, hence the difference between V_2 and V_1.
8. Data are from Skoumal et al. (2015), who reported that the induced seismicity adjoined a well with six laterals that had undergone a total of 94 frack stages. The volume of fluid thus injected has not been disclosed (Beiersdorfer, 2014). However, Ridlington and Rumpler (2013) reported that up to the end of 2012, 1.4 billion gallons of water had been used to frack 334 wells in Ohio. This works out at an average per well of 19,100 m^3, which I have taken as V_1. For comparison, the median volume of water used in the seven wells in Mahoning County, Ohio, which on August 11, 2015, were documented on the FracFocus website (https://www.fracfocusdata.org/DisclosureSearch/SearchResults.aspx), was 4,469,118 gallons or 20,317 m^3. For V_2 I have divided this mean value by the number of frack stages, then multiplied the answer by 6, as Skoumal et al. (2015) noted that six of the frack stages resulted in induced seismicity, so one might infer that the sum of the fluid volumes injected during these six frack stages contributed to "lubricating" the patch of fault that eventually slipped in the largest induced event.
9. Data are from BCOGC (2012). Unlike its larger predecessor, 3 weeks earlier (see later) at the time of this event the induced seismicity was being monitored by a local seismograph network, so more detail is evident as to what occurred. This event adjoined the horizontal part of well G of pad d-1-D, in the Eshto area of the Horn River basin, and can be presumed induced by fracking this well. BCOGC (2012) report that the 7 wells drilled from this pad used an average of 138,005 m^3 of fluid and had an average of 27 frack stages, indicating typical use of 5111 m^3 of fluid per frack, which I round to 5000 m^3. This earthquake occurred after 9 frack stages had taken place in well G; I thus take V_1 as 9×5000 m^3 and V_2 as 5000 m^3.
10. Data are from Frohlich et al. (2011). McGarr (2014) derived an injected volume of 282,000 m^3 from this source. However, to the best of my understanding, the Frohlich et al. (2011) dataset indicates that the nearest injection well operated for 246 days before this earthquake with a typical daily injected volume of ~9500 barrels, giving a total volume of ~370,000 m^3.
11. Data are from Seeber et al. (2004), who reported local magnitudes of 3.8 and 4.3, rather than the values listed here, which are from McGarr (2014). The fluid volume listed for the second event includes the volume already "assigned" to the first.
12. Data are from Hornbach et al. (2015). I have determined V_1 by summing the stated figures for the two adjacent injection wells: well 1, 44,000 m^3 per month for 53 months; well 2, 13,000 m^3 per month for 37 months. Although its epicenter falls within the Barnett Shale, this relatively large induced earthquake occurred much deeper, within basement beneath the Ellenberger Formation (highly permeable, karstified, Ordovician limestone; the injection target). It is therefore not a counterexample to the view that induced earthquakes within the Barnett Shale, which are large enough to be felt, have not been reported.

(Continued)

Table 11.4 Induced Earthquakes Caused by Borehole Injection of Fluid (*cont.*)

13. V_1, from McGarr (2014), is the total volume of fluid injected, after Baisch et al. (2006), including fluid injected after the earthquake. V_2 being my estimate, also from Baisch et al. (2006), of the volume injected up to the time of the earthquake.
14. The magnitude listed is as reported by BCOGC (2012); Ellsworth (2013) reported M_W 3.6. BCOGC (2012) report that this event occurred during fracking from pad c-34-L, also in the Eshto area, but little is known of its geometrical relation to the fracking as no microseismic monitoring was in place. BCOGC (2012) also report that the nine wells of pad c-34-L used an average of 63,000 m^3 of fluid (which I adopt as V_1) and had an average of 18 frack stages, indicating typical use of 3500 m^3 of fluid per frack (which I adopt as V_2).
15. Data are from Kim (2013), who reported $M_W = 3.88$ rather than the value of 4.0 stated by McGarr (2014). Likewise, the volume of fluid injected was reported by Kim (2013) as 78,798 m^3 rather than the value of 83,400 m^3 stated by McGarr (2014).
16. McGarr (2014) estimated an injected volume of 3,287,000 m^3, citing Ake et al. (2005). However, according to this reference, injection involved two phases. Thus, phase 1 spanned July 22, 1996, to July 25, 1999, or 1100 days, less ~100 days for shutdowns. The injection rate was 1290 L/min or ~1860 m^3/day, making the volume injected ~1,860,000 m^3. Phase 2 spanned July 26, 1999, to June 23, 2000, or 332 days, less 40 days for planned shutdowns and a 28-day shutdown after the magnitude 4.3 earthquake, so 264 days net. The injection rate was the same as before, making the volume injected ~490,000 m^3. The sum of these two values gives V_1; the latter value gives an estimate of V_2, since injection phase 1 induced many smaller earthquakes.
17. Data are from Bommer et al. (2006). V_1 is the volume injected before the earthquake; V_2 is the volume injected during the most recent injection phase.
18. Data are from Meremonte et al. (2002), who reported a magnitude of 4.6 rather than the value 4.4 adopted by McGarr (2014). Meremonte et al. (2002) did not give any figure for the volume injected, but provided a link to the Colorado Oil & Gas Conservation Commission website, which they said provided "more information." However, this link did not work when I tried it; I am therefore unable to verify the value quoted by McGarr (2014) for the fluid volume injected.
19. Data are from Gan and Frohlich (2013).
20. This is the "classic" case study of induced seismicity caused by borehole disposal of wastewater, from weapon manufacture at the Rocky Mountain Arsenal, Denver, Colorado (Healy et al., 1968). Earthquake source parameters are from Herrmann et al. (1981). The volume of fluid injected is from Hsieh and Bredehoeft (1981); McGarr (2014) appears to have determined this from their graphical depiction of the injection history.
21. Data are from Horton (2012), who reported that there were eight injection wells in this vicinity, which he numbered 1–8. Taking into account their individual injection rates and dates of operation, prior to this induced earthquake, I estimate injection volumes of 486,135, 1,215,505, 479,948, 397,612, 126,156, 201,477, 510,874, and 59,708 m^3, respectively. I take V_1 as the sum of these values. I take V_2 as the sum of the values from wells 1 and 5, as these were shut down following the occurrence of induced seismicity, leading to its gradual cessation, indicating a causal connection. McGarr (2014) stated a volume of 629,000 m^3, similar to my V_2, suggesting that he carried out a similar calculation.
22. Data are from Nicholson et al. (1988), who reported the magnitude as 5.0, the value listed in the table being from McGarr (2014).
23. Data are from Frohlich et al. (2014). McGarr (2014) stated that this source gave the volume injected as 991,000 m^3, but it actually says that the volumes injected in two nearby wells were 1,050,000 and 2,900,000 m^3. I have taken V_1 as the sum of these values and V_2 as the smaller of the values.
24. Keranen et al. (2013) documented the three large events in the Prague, Oklahoma, area and associated them with ~170,000 m^3 of fluid injection in two nearby wells. However, McGarr (2014) noted that other nearby higher volume injection wells make the relevant volume for fluid injected much higher, ~12,000,000 m^3.
25. Data are from Rubinstein et al. (2014). These authors reported that wastewater injection preceding the largest induced event amounted to ~185 million barrels in that part of the Raton Basin in Colorado and ~140 million barrels in its part in New Mexico. Using the standard conversion factor between units, these values equate to ~29,410,000 and ~22,260,000 m^3. On the contrary, McGarr (2014) has reported the injected volume as 7,840,000 m^3, which equates to ~49,310,000 barrels. It is unclear where this number comes from; possibly he has summed the injected volumes (which were not reported for individual wells by Rubinstein et al., 2014), or maybe there is a mistake in his working (e.g., over conversion of units). Notwithstanding these vagaries, I take the total volume injected for both parts of the Raton Basin as V_1 and the volume stated by McGarr (2014) as V_2.
26. Data are from Klose (2007a). This earthquake did not involve fluid injection; it was induced instead by unloading of the crust as a result of mass removal by coal mining and associated dewatering. The mass removed was estimated by Klose (2007a) as 778,000,000 tonnes and is converted here to an equivalent volume as if the mass were entirely composed of water. Strictly speaking, in this situation the right-hand side of equation (11.1) should contain an additional numerical factor, which can take values in the range 1/2 to 4/3 depending on the geometry of the strain resulting from the unloading in relation to the stress field (McGarr, 1976); however, such detail is omitted from this first-order comparison.

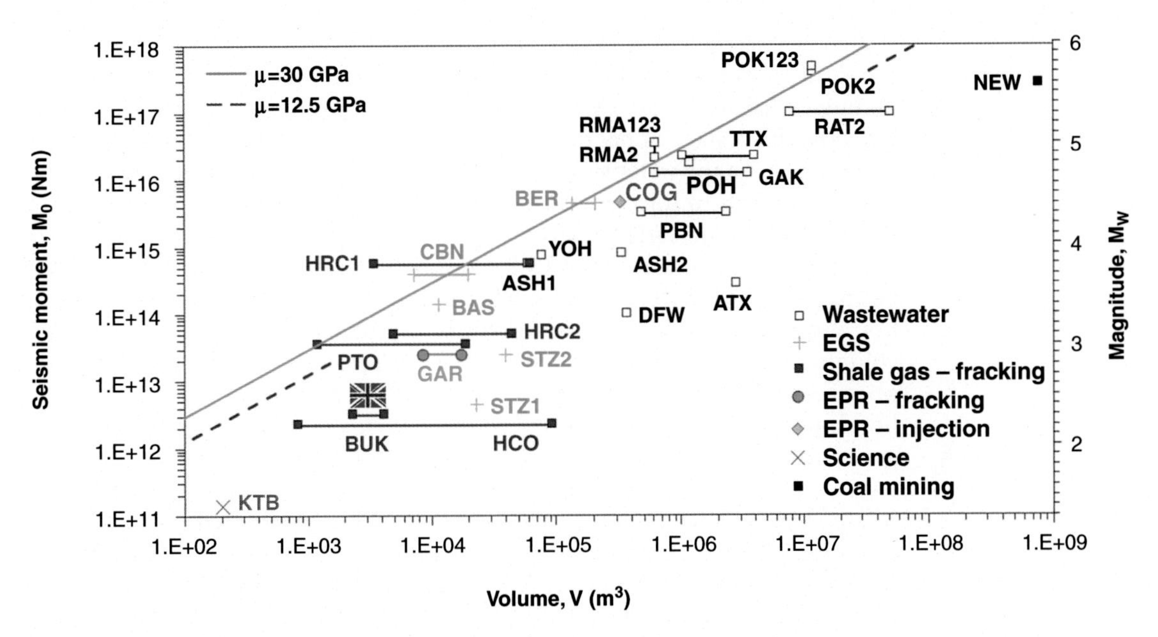

FIGURE 11.2
Graph of seismic moment M_o against water volume V for the population of induced earthquakes listed in Table 11.4. Sloping line denotes the prediction of the limiting value of M_o for a given V, using equation (11.1) for two representative values of shear modulus μ, the prediction line for μ = 12.5 GPa being only shown near the edges to avoid cluttering the figure. See text for discussion.

ornamented to indicate the six causal mechanisms for this population of fluid injection-related earthquakes: scientific experimentation; hydraulic fracturing for enhanced geothermal systems (EGS), shale gas, and enhanced petroleum recovery (EPR); injection of CO_2 for EPR; and wastewater disposal. As was noted earlier, much of the wastewater injection that occurs in the US is of spent fracking fluid or otherwise related to unconventional hydrocarbon development; thus, each of the last four groups of induced earthquake listed falls within the scope of the present study.

This inventory is not exhaustive; indeed, Rubinstein and Mahani (2015) have noted that many hundreds of induced earthquakes of $M_W \geq 3$ ($M_o \geq \sim 4 \times 10^{13}$ N·m) have occurred in recent years in the US, in excess of this country's natural seismicity, although most of these have not been subjected to detailed study; others (Westaway, 2002, 2006; Klose, 2007b, 2013) have documented other earthquakes that may well also have been induced as a result of mining or other human activity. Furthermore, in many cases, including some listed in Table 11.4, the volume of water injection that is considered directly causal to the earthquake is a matter of considerable uncertainty. This is in part because there is no obligation on the part of shale gas developers in some US jurisdictions to disclose volumes of fluid used in fracking, this being considered a matter of commercial confidentiality. An example is the M_W 3.0 Poland Township, Ohio, earthquake, of March 10, 2014 (Skoumal et al., 2015), currently the largest earthquake to have been caused by fracking for shale gas on US territory. To a UK observer, this is another aspect of US custom and practice that seems extraordinary; the environmental permit for each UK shale gas well will specify the volume of water that can be injected per frack stage or per day, this being a matter of legitimate public concern as it impacts on the water supply available to other consumers. Second, in some cases no records of injected volumes have been kept, an example being for the M_W 5.3 Raton Basin earthquake of August 23, 2011 (Rubinstein et al., 2014). This earthquake occurred near the Colorado–New Mexico boundary in the Raton Basin, where coalbed methane extraction has released large volumes of formation water that is injected into boreholes. Systematic records exist of injected volumes for boreholes in Colorado, but not for those in New Mexico where, prior to June 2006, this information was not recorded. Third, it can be unclear how much of the water that has been injected in a given locality has influenced any resulting induced seismicity. It is unclear, for example, whether one should sum all the water injected for all previous fracking projects in a given locality, or all previous frack stages for the well that was being fracked when an earthquake occurred, or only consider the volume injected during the most recent frack stage. Such uncertainty results in multiple possibilities, hence the large margins of uncertainty depicted in Figure 11.2 for volumes of water associated with induced earthquakes caused by fracking. Finally, there have been instances (Table 11.4) of multiple induced earthquakes in regions with multiple injection boreholes; it is thus difficult to apportion the volume injected between the earthquakes, especially since some instances of induced seismicity occurred some distance from the boreholes and some of the injected volume may well have flowed through the subsurface in other directions and may thus have had no effect on the seismicity.

As Figure 11.2 indicates, the largest earthquake to have been induced by fracking for shale gas is the M_W 3.6 or 3.8 Horn River event of May 19, 2011, in British Columbia (Canada). In seismic moment terms, this was larger than the Poland Township, Ohio, event by a factor of ~16 and larger than the Preese Hall, Lancashire, event by a factor of ~180 (Table 11.4). However, nuisance caused by earthquakes in this magnitude range, expressed as peak ground velocity (PGV), is relatively insensitive to M_o (Westaway and Younger, 2014); the implications for how to regulate this form of nuisance in an appropriate manner are therefore quite complex (mentioned later). On the other hand, much larger induced earthquakes have occurred in response to water injection, the largest currently on record (Table 11.4) being the M_W 5.7 Prague, Oklahoma, event of November 6, 2011; in seismic moment terms this was ~700 times larger than the Horn River event and ~130,000 times larger than the Preese Hall event.

DISCUSSION

Walters et al. (2015) have proposed a risk assessment workflow for projects involving hydraulic fracturing. This emphasizes the need for a wide range of inputs, from geoscientists, regulators, and project operators, including inputs that require the latter group to disclose data. In general, their comments seem reasonable, especially their emphasis on the need for operator disclosure of data. However, their comments on UK issues, such as "it is clear that the stakeholders involved in hydraulic fracturing operations in the UK have a very low tolerance for risk ..." indicate limited familiarity with this topic. The issue facing the UK, regarding both induced seismicity and the other environmental issues discussed by Westaway et al. (2015), is not risk – it is nuisance (see later). The recent lack of public acceptance in the UK of nuisance resulting from induced seismicity caused by fracking for shale gas is, indeed, comparable with the lack of public acceptance in previous years in France and Switzerland of the nuisance resulting from induced seismicity caused by hydraulic fracturing for EGS (Majer et al., 2007); it is not something unique to the UK (see e.g., Chapter 13). Particular issues affecting shale gas development in the UK include, first, the presence of high differential stress (see Box 11.3), which has long been apparent, for example, from *in situ* stress measurements in coal mines (Cartwright, 1997). In summary, *in situ* stress measurements in the Preese Hall-1 well indicate, at a representative depth of 2440 m, maximum and minimum horizontal stresses of 73.4 and 43.6 MPa and a vertical stress of 62.2 MPa (Younger and Westaway, 2014), the maximum horizontal stress being oriented N7 ± 3°E–S7 ± 3°W (±2 s) (Westaway, in review). Second, the stratigraphy is pervasively faulted, notably with normal faults inherited from Early Carboniferous crustal extension (British Coal Corporation, 1997; Kirby et al., 2000). Many of these ancient normal faults happen to be favorably oriented for roughly strike–slip reactivation under the present day stress field, for which the maximum principal stress is N–S and the minimum principal stress is E–W (Evans and Brereton, 1990; Baptie, 2010). This slip sense is reflected in the focal mechanism of the 2011 induced seismicity (Figure 11.3a–c), the extent to which only a modest perturbation to the state of stress is needed to cause faults to slip being evident from Figure 11.3d (Box 11.3). A second general area of difficulty in the UK concerns the quality of geological information available. Although some quadrangles have been remapped recently, many have not been. Many, indeed, have not been revisited for a century or so; when they were originally mapped it was using old topographic base maps that were not prepared photogrammetrically, so the topography was not accurate. The existing maps also generally do not distinguish data from interpretation; in particular, some of the faults included are notional structural devices to reconcile data, as in an undergraduate structural geology exercise, rather than being based on evidence. It is a familiar occurrence at UK shale gas conferences (which are currently far more numerous than shale gas wells) for an employee of the British Geological Survey to present sophisticated computer visualizations of three-dimensional geology; this is fine, except its availability is the exception rather than the rule. Until recently, these difficulties did not matter because they did not bear upon any issue at hand. Shale gas development in a high-differential-stress, faulted environment such as this requires a level of detail of geological knowledge that is way beyond the capabilities of existing datasets. It therefore requires a strategy to deal with uncertainty in the underlying information, in particular for dealing with the presence of faults. Does one, for example, invest in expensive exploration (3-D seismic reflection?)

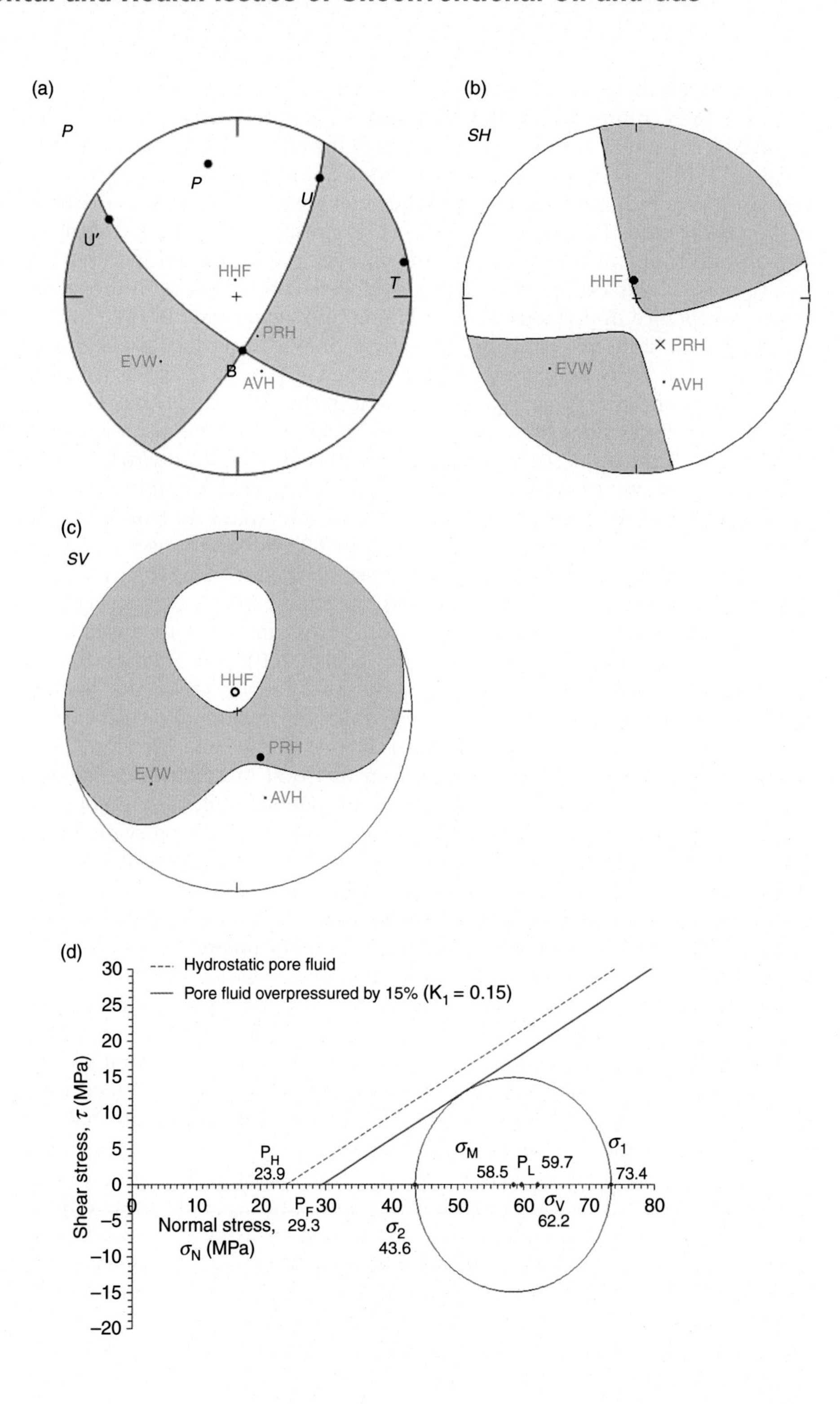
(a)
P
P
U
U′
HHF
T
PRH
EVW
B
AVH
(b)
SH
HHF
× PRH
EVW
AVH
(c)
SV
HHF
PRH
EVW
AVH
(d)
Hydrostatic pore fluid
Pore fluid overpressured by 15% (K1 = 0.15)
Shear stress, τ (MPa)
Normal stress, σN (MPa)
PH 23.9
PF 29.3
σ2 43.6
σM 58.5
PL 59.7
σV 62.2
σ1 73.4

FIGURE 11.3
Illustrations of the deformation sense and stress field for the Preese Hall induced seismicity.
(a–c) Preferred focal mechanism for the earthquake of August 2, 2011 (strike 030°, dip 75°, rake −20°, *P*-axis azimuth 347° with plunge 25°, and *T*-axis azimuth 087° with plunge 3°). This small event was better recorded as a result of improved local seismograph coverage compared with the earlier, larger events in the sequence; the character of seismograms indicates that the focal mechanism orientation stayed the same throughout the sequence (see Westaway, in review, for full details). All diagrams are equal area projections of the lower focal hemisphere, with compressional quadrants (for *P*-waves) and positive-polarity quadrants (for *S*-waves) shaded, for a hypocenter south of the Preese Hall-1 borehole (again, see Westaway, in review, for full details). (a) *P*-wave radiation pattern showing stations marked to indicate no clear polarity picks, although the first motions appear dilatational at HHF and AVH and compressional at PRH. (b) Corresponding *SH*-wave radiation pattern. (c) Corresponding *SV*-wave radiation pattern. Solid and open symbols in (b) and (c) denote signals of positive and negative polarity; cross in (b) denotes an unclear (?nodal) signal. (d) Simple conceptual model for the state of stress in relation to Preese Hall induced seismicity. This modified Mohr Circle construction illustrates the state of stress at 2440 m depth in the Preese Hall-1 borehole relative to the condition for shear failure on an optimally oriented vertical strike–slip fault. σ_1, σ_2 and σ_V denote the measured maximum and minimum horizontal stresses and vertical stress, 73.4, 43.6, and 62.2 MPa, respectively. σ_M, the mean of σ_1 and σ_2, is 58.5 MPa; P_L, the lithostatic pressure $(\sigma_1 + \sigma_2 + \sigma_V)/3$, is 59.7 MPa; and P_H, the hydrostatic pressure, is 23.9 MPa. Dashed sloping line illustrates the frictional condition for slip, for a fault with $f = 0.6$. The bold sloping line is constructed assuming the same coefficient of friction but that injection of fracking fluid raises its pressure within the fault above P_H by 15% of the difference between P_H and P_L, or to $P_F = 29.3$ MPa (i.e., $K_3 = 0.15$). This line now touches the Mohr Circle, making the fault frictionally unstable and thus able to slip in an induced earthquake. This is essentially the same construction as for Figure 11.1 except zero cohesion has been assumed for the fault, and its failure envelope has been adjusted to the right rather than moving the Mohr Circle left. The optimum fault orientation to which this calculation applies would involve strike at 45° to both horizontal principal stresses (i.e., at azimuth 052° for left-lateral slip or 322° for right-lateral slip on vertical faults). Corresponding calculations for non-optimally-oriented faults (such as for activation of the ESE-dipping nodal plane in (a) with the slip sense indicated) are beyond the scope of the present study and will be presented elsewhere; they require slightly greater overpressures. *Modified from Westaway (in review).*

to identify and avoid all faults, thereby "sterilizing" much of the resource? Or does one attempt to mitigate the induced seismicity nuisance by regulating volumes of fracking fluid to limit the seismic moment of induced events (given later)? This is a policy decision, but so far the required political leadership has not been provided.

The rest of this section will concentrate on three specific topics: how high differential stress affects the character of induced seismicity; how induced earthquakes scale with volumes of fluid injection; and how the nuisance resulting from induced seismicity caused by fracking can be quantified.

Issues Relating to Differential Stress

As already noted, the high differential stress in northern England and the near-optimum orientation for reactivation of faults given the orientation of the present-day stress field combine to make induced seismicity in response to fluid injection a relatively likely occurrence in this region (Figure 11.4). The conclusion of de Pater and Baisch (2011) that the 2011 occurrence of "nuisance" (i.e., felt) induced seismicity was the result of such a unique combination of conditions that the probability of recurrence is very low (they estimated this probability as ~0.0001 for each future shale gas well) therefore seems unwarranted, and is indeed a hostage to fortune.

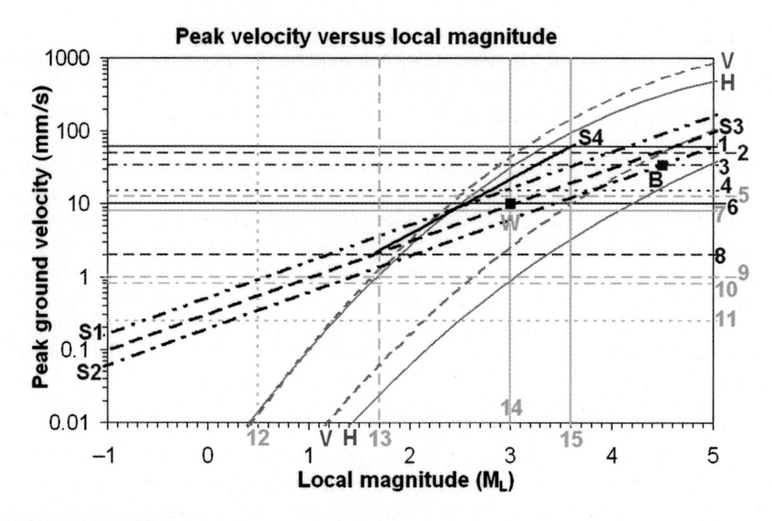

Prediction equations:

V - - - Peak vertical velocity; Bragato & Slejko (2005)

H —— Peak horizontal velocity; Bragato & Slejko (2005)

S —— Prediction using spectral technique

Velocity thresholds:

1 —— Threshold for major damage (BS7385-2): 60 mm/s

2 - - - Threshold for plaster cracking (Calder, 1977): 50 mm/s

3 ·−·− Threshold for minor damage at MMI 5 (Wald et al., 1999): 34 mm/s

4 ····· Threshold for cosmetic damage (BS7385-2): 15 mm/s

5 - - - 'Safe' limit (Siskind et al., 1980): 12.7 mm/s

6 —— Upper limit for quarry blasting during working day (BS6472-2): 10 mm/s

7 —— Limit for jumping onto a wooden floor (Stagg et al., 1980): 8 mm/s

8 - - - Limit for quarry blasting at night (BS6472-2): 2 mm/s

9 - - - Threshold for minimum felt effect at MMI 2 (Wald et al., 1999): 1 mm/s

10 ·−·− Limit for walking on a wooden floor (Stagg et al., 1980): 0.8 mm/s

11 ····· Threshold for minimum felt effect by a very sensitive person (Oriard, 2002): 0.25 mm/s

Magnitude thresholds:

12 ····· Threshold to suspend fracking (Green et al., 2012): M_L 0.5

13 - - - Threshold to suspend fracking (de Pater & Baisch, 2011): M_L 1.7

14 —— Upper limit for earthquakes induced by fracking in the UK (de P & B): M_L 3.0

15 —— Earthquake associated with development of a 600 m long fracture: M_L 3.6

Regulatory recommendations:

B ■ Bull (2013): PGV 34 mm/s; M_L 4.5

W ■ This study: PGV 10 mm/s; M_L 3.0

◀ **FIGURE 11.4**
Comparison of regulations for peak ground velocity at residential property from quarry blasting, applicable in the United Kingdom, with felt effects, proposed magnitude thresholds (expressed in terms of local magnitude M_L, which both Westaway and Younger (2014) and the present study equate with M_W) for regulating fracking in the United Kingdom, estimates of PGV from other forms of environmental nuisance, and predictions of vertical and horizontal PGV after Bragato and Slejko (2005) and from the spectral method developed by Westaway and Younger (2014), where the four alternative predictions S1–S4 are fully explained.
All predictions are for points on the Earth's surface directly above fracking at a depth of 2.5 km; the two sets of predictions after Bragato and Slejko (2005) correspond to different adaptations of their prediction equations to deal with very shallow seismicity, as discussed by Westaway and Younger (2014). British Standard (BS) 7385-2, specifying permitted levels of ground vibration affecting buildings in the UK, is described by BSI (1993). *Modified from Figure 1 of Westaway and Younger (2014).*

This situation is rather different from the innocuous consequences of induced fractures intersecting natural fractures in many US shale gas provinces, such as the Barnett Shale in Texas (Gale et al., 2007), where horizontal stresses are similar in magnitude, and where induced earthquakes caused by fracking for shale gas and strong enough to be felt are unknown, even though >15,000 shale gas wells have been fracked. Nonetheless, two earthquakes large enough to have been felt have now been induced in the US as a result of fracking for shale gas: the M_W 2.2 Harrison County and M_W 3.0 Poland Township events (Table 11.4; HCO and PTO in Figure 11.2), both in Ohio (Friberg et al., 2014; Skoumal et al., 2015). There has also been the M_W 2.9 Eola, Oklahoma, event, which was caused by hydraulic fracturing for EPR (Holland, 2013). An indication of the state of stress in the vicinity of the Ohio events is provided by *in situ* measurements at Norton Mine, a former limestone mine at ~670 m depth, under assessment as a future compressed air energy storage facility. Bauer et al. (2005) reported recent principal stress measurements of 36.7 MPa (E–W), 28.2 MPa (N–S), and 22.5 MPa (vertical), as well as compilations of historical and recent measurements of 42.1 MPa, 25.0 MPa, and 20.9 MPa. High differential stress is thus indicated in this part of Ohio, which would be expected to increase when extrapolated to the depths of the fracking (>2 km).

The Horn River earthquake of May 19, 2011 (HRC1 in Fig. 11.2), provides another pertinent example. This event, and many others in the same locality, in the magnitude ~2–3 range, likewise occurred in a formation at relatively high differential stress (Roche et al., 2015). Roche et al. (2015) indeed noted a correlation between the state of stress and the size distribution for the resulting earthquakes induced by fracking in different shale formations in the Horn River region. Earthquake populations are typically characterized by "*b*-values," where *b* is the gradient of a plot of the logarithm of the cumulative frequency of occurrence *N* of the earthquakes in a given region, during a given span of time, which are above magnitude *M* (i.e., $b = -d\log_{10}(N)/dM$). Populations of earthquakes induced by fracking often have high *b*-values, ~2–3, so their frequency of occurrence tails off abruptly above a particular size threshold that is often at quite a small magnitude; in contrast, naturally occurring earthquake populations typically have $b \sim 1$. However, Roche et al. (2015) showed that b-values for different populations of induced earthquakes in northwest Canada are inversely correlated with local differential stress, with $b \sim 1$ in the localities with the highest differential stress. The Harrison County and Poland Township induced earthquake populations also had relatively low b-values, 0.88 ± 0.08 and ~ 0.89, respectively (Friberg et al., 2014; Skoumal et al., 2015). Decades earlier, Scholz (1968) showed that laboratory rock mechanics experiments demonstrate a negative correlation between differential stress and b-values, and suggested that variability in differential stress is thus the main cause of variability in b-values between different earthquake populations. It now seems evident that this deduction is of significance for populations of earthquakes induced by fracking; in regions of high differential stress, including the earlier examples, b-values can be expected to be low and populations of induced earthquakes can thus be expected to extend across more of the magnitude range, thus giving rise to a significantly higher probability of occurrence of events that are large enough to be felt. The Preese Hall induced earthquake population evidently fits this general pattern, given high local differential stress, although the absence of systematic microseismic monitoring makes it impossible to confirm a low b-value directly.

Scaling Dependence of Induced Earthquakes on Injected Volumes

Figure 11.2 illustrates the general correlation between magnitude and seismic moment of induced earthquakes with injected volume, as others (McGarr, 2014) have previously noted. McGarr (2014) derived an equation linking these quantities:

$$M_o = \mu V, \tag{11.1}$$

where M_o is the seismic moment of the largest possible induced earthquake that can occur for an injected volume V in rocks of shear modulus μ. However, the derivation of this equation envisages that the injected volume pervades a permeable rock mass. This assumption is likely to be valid for fluid injection into permeable rocks, but less so for the other settings, especially during fracking for shale gas where – rather than pervading a permeable rock mass – the water is taken up within faults or fractures in impermeable rock. However, using standard fracture-mechanical theory from Eshelby (1957), Westaway and Younger (2014) established that an equation of the form of (11.1) is also followed for earthquakes induced by water injected under pressure into a preexisting fault, opening up or "unclamping" the fault and causing it to slip. Thus, for example, if a circular fault of radius a is unclamped by fluid pressure while maintaining constant shear stress S in its surroundings, the resulting induced earthquake will have a seismic moment M_o where:

$$M_o = \int_0^a \frac{8\,S(1-\nu)}{\pi(2-\nu)}\sqrt{(a^2-r^2)} \times 2\pi r\, dr = \frac{16\,S(1-\nu)a^3}{3(2-\nu)}, \tag{11.2}$$

where r is distance from the center of the fault and v is the Poisson ratio of the rock. The corresponding volume of fluid required to enter the fault to achieve this unclamping effect is V where:

$$V = \int_0^a \frac{8S(1-v)}{\pi\mu(2-v)}\sqrt{(a^2-r^2)} \times 2\pi r\,\mathrm{d}r = \frac{16S(1-v)a^3}{3\mu(2-v)}; \quad (11.3)$$

taking the ratio of these two equations gives equation (11.1).

Crustal basement rocks will have $\mu \sim 30$ GPa; μ will be less in other rocks (Varga et al., 2012, recommend 12.5 GPa for the Barnett Shale of Texas). Lines illustrating equation (11.1) for these values of μ are shown in Figure 11.2. In two instances where the comparison is for basement rocks, the Prague, Oklahoma (POK), and Denver, Colorado (RMA), earthquakes, the overall seismic moment for all the induced earthquakes slightly exceeds the expectation for the injected volume. The Cooper Basin (Australia) (CBN) earthquake exceeds the expected upper bound for M_o unless the injected volume was near its own upper bound. Likewise, in two instances where the comparison is for shale, the larger Horn River event (HRC1) and the Poland Township, Ohio (PTO), event, M_o exceeds the expected upper bound unless, once again, the injected volume was near its own upper bound.

As regards the UK, using equation (11.1) with the value of μ for shale, the cumulative V before the time of occurrence of the M_W 2.3 Preese Hall event (Table 11.4) gives $M_o = \sim 5.2 \times 10^{13}$ Nm ($M_W \sim 3.1$), whereas V for the preceding frack stage gives $M_o = \sim 2.9 \times 10^{13}$ Nm ($M_W \sim 2.9$), as the largest possible earthquake that might have been induced at Preese Hall in 2011. The largest induced earthquake to occur was rather smaller than these predictions, suggesting that only a small proportion of the fracking fluid reached the seismogenic fault, consistent with this fault having been breached near an extremity of the induced fracture network. de Pater and Baisch (2011) estimated $M_W \sim 3$ as the upper limit for induced earthquakes from fracking in the UK, but this was based on historical experience of mining-induced seismicity; its agreement with my estimates is fortuitous. Current regulatory practice in England is to limit V for any individual frack stage to 750 m^3, primarily for water supply considerations; from equation (11.1) this will limit M_o to $\sim 9.4 \times 10^{12}$ Nm and M_W to ~ 2.6. Limiting V in fracking also has the effect of limiting the vertical and horizontal extents of the induced fracture network (Fisher and Warpinski, 2012; Westaway and Younger, 2014), avoiding the possibility of growth of these networks beyond the intended shale formations, which would be undesirable on multiple grounds.

Weingarten et al. (2015) have recently argued that the probability of an induced earthquake does not depend on the cumulative volume injected into a well. This may be so, but the magnitude and seismic moment of any possible induced earthquake clearly depend on fluid volume (Figure 11.2). Weingarten et al. (2015) argued instead that the probability of an induced earthquake correlates with the injection rate in a well, this being because increasing the injection rate increases the pressure rise in the water around the well (Keranen et al., 2014; Figure 11.1). Such observations reflect earlier experience in EGS; for example, Cuenot et al. (2008) and Dorbath et al. (2009) noted that once the injection rate in a well exceeds a threshold, the pressure in the surroundings increases; also, once the injected fluid enters a permeable fault, the overall response of the injection system changes. Such knowledge is potentially transferrable to fracking for shale gas, and might conceivably result in the development of real time control systems to mitigate induced seismicity. Likewise, time delays evident in these systems, such as the ~10 h delay between injection and induced seismicity at Preese Hall (de Pater and Baisch, 2011), require explanation. As, for example, Davies et al. (2013) have noted, such delays might in principle relate either to properties of faults (e.g., storage and transmissibility characteristics) or to properties of the adjoining rock mass (e.g., poroelasticity, or the time required for the transmission of fluid pressure by pressure diffusion). Indeed, it is currently unclear whether any measurement made during or after injection can indicate that such a time delay is in progress and can establish a course of action to prevent the impending earthquake, or whether once the injection has been completed the earthquake is inevitable.

As regards comparison between scaling of earthquakes induced by fluid injection and by surface loading or unloading effects, it is noteworthy (Figure 11.2 and Table 11.4) that the Newcastle (Australia) and Prague, Oklahoma, earthquakes were of similar size but the former required almost two orders of magnitude more mass/volume change than the latter. This illustrates the essential point that surface loading/unloading effects are inherently much less "efficient" as causes of induced earthquakes than fluid injection. The latter process can act directly on the effective normal stress in a fault (Box 11.3), whereas elastic modeling (Westaway, 2002; Klose, 2007a) demonstrates that the surface loading/unloading is largely absorbed by the elastic response of rock layers between the Earth's surface and the depths at which earthquakes nucleate, so a given load has a much smaller effect on the state of stress on the fault at the depth of nucleation. The need for such elastic modeling to link cause and effect in turn makes it relatively difficult to establish instances of induced seismicity caused by surface loading/unloading, whereas instances caused by borehole injection can generally be established simply by association (Davis and Frohlich, 1993; Box 11.1). Nonetheless, the changes to surface loading/unloading that are occurring worldwide are many orders of magnitude larger (Table 11.3) than those caused by fluid injection, and are likely to increase as a result of continued global economic growth. It is thus possible that sooner or later these surface effects will begin to impact on overall levels of seismicity, in a manner analogous to what happened in the US circa 2009, for fluid injection–induced seismicity, as a result of the continued growth in rates and volumes of fluid injection (Rubinstein and Mahani, 2015).

The Nuisance Caused by Induced Seismicity

As Table 11.4 and Figure. 11.2 indicate, wastewater injection in the US has resulted in induced earthquakes approaching magnitude 6. Many instances are known, worldwide, where earthquakes in this size range have caused injuries, damage to property, and even fatalities - indeed, the Newcastle (Australia) earthquake caused 13 fatalities and US$5 billion of damage, after adjustment for inflation (Klose, 2007b). An obvious way to mitigate such a possibility would be to require shale gas operators in the US to treat their wastewater in lieu of borehole disposal, this being one of many issues over which environmental practices relating to shale gas, considered acceptable in the US, will not be permitted in other countries, including the UK (Westaway et al., 2015) (refer to Chapter 5). It seems evident that if the US trend for ever-increasing rates and volumes of wastewater injection continues, then sooner or later the associated induced seismicity will result in fatalities or, at the very least, in costly damage to property. The resulting possibility of multimillion dollar lawsuits may prompt reassessment of the merits of water treatment as an alternative to borehole injection of wastewater.

Nonetheless, as already noted, subsurface injection of CO_2 is permitted in the UK and European Union and is being developed as part of the strategy to meet international targets for reduction of greenhouse gas emissions. One such project involves CO_2 injection in the Sleipner Field of the North Sea, utilizing the high permeability of the Cenozoic Utsira Sand Formation (Chadwick et al., 2004; Bickle et al., 2007). Although located in the Norwegian sector of the North Sea, the Sleipner Field is connected by pipeline to northern England, providing a significant proportion of the UK gas supply. In future, the flow in this pipeline might be reversed, to sequester CO_2 that would otherwise be emitted from the UK. By late 2011 ~13 MT of CO_2 had been injected into the Utsira Sand, the amount increasing by ~1 MT per year (Verdon et al., 2013). The developing record of induced seismicity from fluid injection (Table 11.4) now includes the first significant event associated with CO_2 injection (Cogdell, Texas: M_W 4.41; Gan and Frohlich, 2013; Table 11.4). Read in the light of this experience, an environmental report on the Sleipner project when in its early stages (Solomon, 2007) now seems somewhat complacent, phrases such as "readily applicable methods exist to assess and control induced fracturing or fault activation ...," "there have been no significant seismic effects attributed to CO_2-EOR ...," and "the fact that only a few individual seismic events associated with deep-well injection have been recorded suggests that the risks are low" being more hostages to fortune. Currently, no monitoring of the Sleipner project for induced seismicity is in place (Verdon, 2014); however because the Utsira Sand is so extensive and so permeable, its deformation in response to current rates of injection has been reported as minimal (Verdon et al., 2013).

Setting aside the issue of relatively large induced earthquakes from wastewater injection, induced seismicity is evidently a matter of public nuisance rather than risk or hazard. Westaway and Younger (2014) suggested that a satisfactory regulatory framework for induced seismicity arising from fracking for shale gas in the UK, from consideration of felt effects, might be based on the long-standing and uncontroversial regulatory framework for ground vibrations (quantified in terms of peak ground velocity or PGV) from quarry blasting, currently provided by British Standard 6472 part 2 (BSI, 2008). This recommends that PGV in the seismic wavefield incident on any residential building should not exceed 10 mm/s during the working day (8 a.m. to 6 p.m. on Mondays to Fridays or 8 a.m. to 1 p.m. on Saturdays), 2 mm/s at night (11 p.m. to 7 a.m.), or 4.5 mm/s at other times, these guidelines being for avoidance of disturbance to occupants rather than considerations of damage. An alternative lower limit of 6 mm/s during the working day was also recommended, with PGV between 6 and 10 mm/s allowable if justifiable on a case-by-case basis.

Figure 11.4 compares these guidelines with other thresholds of PGV estimated to cause damage to buildings or various forms of environmental nuisance and with a range of predictions of PGV by Westaway and Younger (2014) for earthquakes at a depth of 2.5 km. This indicates that the BS6472-2 10 mm/s upper limit to exposure to PGV from quarry blasting during the working day roughly matches the central prediction (S3) for the upper bound to PGV expected for a microearthquake of M_W 3 at 2.5 km depth. Likewise, the 6 mm/s exposure recommendation during the working day roughly matches the predicted upper bound to PGV expected for a microearthquake of M_W 2.6 at this depth. M_W 2.6 is the largest expected on the basis of current permitting arrangements for water supply (given earlier); it is thus to be expected that the largest induced earthquakes that might occur, under current regulations for water supply, will (assuming, of course, that they occur during the working day) produce ground vibrations that would be considered acceptable were the source a quarry blast rather than an earthquake. Moreover, ground vibrations at this level are comparable with those produced by many other forms of environmental nuisance that are accepted, such as from traffic (Westaway and Younger, 2014; Figure 11.4).

The notion that environmental nuisance resulting from seismicity should be determined on the basis of PGV has a long-standing pedigree. For example, Siskind et al. (1980) recommended that PGV ~ 12.7 mm/s (i.e., 0.5 in. per second) is an appropriate limit for avoidance of even cosmetic damage to buildings in the US. The Swiss "Basel Deep Heat Mining" EGS project adopted a regulatory limit for PGV from induced seismicity of 5 mm/s (Majer et al., 2007). The Walters et al. (2015) workflow for assessing induced seismicity is likewise based on thresholds determined by felt effects, which might arguably be based on PGV, while leaving the setting of thresholds to individual jurisdictions. Indeed, I am unaware of any precedent, worldwide, other than Green et al. (2012), for basing such an assessment on earthquake size (i.e., on magnitude), rather than on felt effects.

The question also arises as to what action to take if such thresholds for PGV are exceeded. For fracking for shale gas in the UK, regulatory limits on water volume use per frack stage will dictate that any such exceedance is likely to be marginal and not indicative that significantly larger induced earthquakes might occur later. Rather than being seen as a "red traffic light" situation requiring project shutdown, a more appropriate response might therefore be for the operator to reduce the water volume used in future frack stages. As Westaway and Younger (2014) suggested, a system for compensation of people affected by PGV exceeding any threshold might also be put in place; such a scheme could readily be validated using instrumental monitoring of the induced seismicity, which is another UK regulatory requirement.

CONCLUSIONS

This chapter has set out to provide a snapshot of current knowledge regarding induced seismicity, with emphasis on issues affecting the US and the UK and with focus on topics relating to unconventional petroleum development. In the US the principal current issue is the increasing occurrence of moderate-to-large induced earthquakes caused by borehole disposal of wastewater, much of it spent fracking fluid from shale gas development. Borehole injection of wastewater is occurring in the US on a very large and increasing scale; induced earthquakes that result reach the size limit, in terms of magnitude and seismic moment, predicted from the volumes injected. Borehole injection is thus a particularly effective process for inducing earthquakes, much more so than surface loading and unloading effects, which require much larger movements of mass to cause induced seismicity of a given magnitude. The US experience of induced seismicity from borehole injection of wastewater can shed light on the induced seismicity to be expected from future CO_2 sequestration efforts. In the UK, where borehole disposal of wastewater is illegal, the principal current issues are, first, the need to incorporate induced seismicity resulting from fracking for shale gas into an appropriate regulatory framework, to contribute to creating the conditions whereby shale gas might have a future. The second key issue concerns the implications of the combination of conditions prevailing in the UK, combining high differential stress and pervasively faulted rocks, with the faulting close to the optimum orientation for reactivation by injection of fracking fluid. The occurrence of one induced sequence of earthquakes strong enough to have been felt from the fracking of just one shale gas well has not been "bad luck," it has been a predictable consequence of fracking in this difficult geological setting.

Acknowledgments

I thank many people for helpful discussions on this topic, in particular Jennifer Roberts, Zoe Shipton and Paul Younger.

References

Ake, J., Mahrer, K., O'Connell, D., Block, L., 2005. Deep injection and closely monitored induced seismicity at Paradox Valley, Colorado. Bull. Seismol. Soc. Am. 95, 664–683.

Ambraseys, N.N., 1988. Engineering seismology. Earthquake Eng. Struct. Dynam. 17, 1–105.

Baisch, S., Weidler, R., Voros, R., Wyborn, D., de Graaf, L., 2006. Induced seismicity during the stimulation of a geothermal HFR reservoir in the Cooper Basin, Australia. Bull. Seismol. Soc. Am. 96, 2242–2256.

Baptie, B., 2010. Seismogenesis and state of stress in the UK. Tectonophysics 482, 150–159.

Bauer, S.J., Munson, D.E., Hardy, M.P., Barrix, J., McGunegle, B., 2005. *In situ* stress measurements and their implications in a deep Ohio mine. In: Proceedings of Alaska Rocks 2005, the 40th U.S. Symposium on Rock Mechanics, Anchorage, Alaska, 25-29 June 2005. American Rock Mechanics Association paper ARMA-05-804. Available from: https://www.onepetro.org/conference-paper/ARMA-05-804 (accessed 16.08.2015.).

BBC, 2011. Small earthquake hits Blackpool. BBC News; England; April 1, 2011. Available from: http://www.bbc.co.uk/news/uk-england-12930915 (accessed 25.11.2015.).

BCOGC, 2012. Investigation of observed seismicity in the Horn River Basin. British Columbia Oil and Gas Commission, Fort St John, British Columbia, 29 pp. Available from: http://www.bcogc.ca/node/8046/download?documentID=1270 (accessed 11.08.2015.).

Beiersdorfer, R., 2014. Mahoning County earthquakes and ODNR deja vu all over again. Columbus Free Press, Columbus, Ohio, Available from: http://columbusfreepress.com/article/mahoning-county-earthquakes-and-odnr-deja-vu-all-over-again (accessed 11.08.2015.).

BGS, 2014. Minerals produced in the United Kingdom in 2013. British Geological Survey. Available from: http://www.bgs.ac.uk/mineralsuk/statistics/downloads/MineralsProducedInTheUnitedKingdom.pdf (accessed 13.08.2015.).

BGS, 2015. Fracking and Earthquake Hazard. British Geological Survey, Nottingham. Available from: http://earthquakes.bgs.ac.uk/research/earthquake_hazard_shale_gas.html (accessed 25.11.2015.).

BHP Billiton, 2012. Escondida site tour, 32 pp. Available from: http://www.bhpbilliton.com/~/media/bhp/documents/investors/reports/2012/121001_escondida-site-visit-presentation.pdf?la=en (accessed 13.08.2015.).

Bickle, M., Chadwick, A., Huppert, H.E., Hallworth, M., Lyle, S., 2007. Modelling carbon dioxide accumulation at Sleipner: implications for underground carbon storage. Earth Planetary Sci. Lett. 255, 164–176.

Bommer, J.J., Oates, S.J., Cepeda, M., Lindholm, C., Bird, J., Torres, R., Marroquín, G., Rivas, J., 2006. Control of hazard due to seismicity induced by a hot fractured rock geothermal project. Eng. Geol. 83, 287–306.

BP, 2013. Statistical review of world energy. Available from: http://www.bp.com/content/dam/bp/excel/Statistical-Review/statistical_review_of_world_energy_2013_workbook.xlsx (accessed 13.08.2015.).

Bragato, P.L., Slejko, D., 2005. Empirical ground-motion attenuation relations for the Eastern Alps in magnitude range 2.5–6.3. Bull. Seismol. Soc. Am. 95, 252–276.

Bridgland, D.R., Westaway, R., 2014. Quaternary fluvial archives and landscape evolution: a global synthesis. Proc. Geol. Assoc. 125, 600–629.

British Coal Corporation, 1997. Three-dimensional seismic surveying to investigate the geological structure of shear zones within the Selby coalfield. European Union, Directorate-General Energy, report EUR 17161 EN. Office for Official Publications of the European Communities, Luxembourg, 122 pp.

BSI, 1993. Evaluation and measurement for vibration in buildings — Part 2: Guide to damage levels from groundborne vibration. BS 7385-2: 1993. British Standards Institution, London, 16 pp.

BSI, 2008. Guide to evaluation of human exposure to vibration in buildings – Part 2: Blast-induced vibration. BS 6472-2: 2008. British Standards Institution, London, 24 pp.

Bull, J., 2013. Induced seismicity and the O&G Industry. Ground Water Protection Council, 2013 Underground Injection Control Conference, Sarasota, Florida, pp. 22–24. Available from: http://www.gwpc.org/sites/default/files/event-sessions/Bull_Jeff.pdf (accessed 16.08.2015.).

Bus-ex.com, 2013. Vale Brazil – Carajás Iron Ore Mine. http://www.bus-ex.com/article/vale-brazil-%E2%80%93-caraj%C3%A1s-iron-ore-mine (accessed 13.08.2015.).

Calder, P.N., 1977. Perimeter blasting. In: Pit Slope Manual, Chapter 7, CANMET Report 77-14. Canadian Center for Mineral and Energy Technology, Ottawa, Canada, 82 pp.

Cartwright, P.B., 1997. A review of recent *in-situ* stress measurements in United Kingdom Coal Measures strata. In: Sugawara, K., Obara, Y. (Eds.), Rock Stress: Proceedings of the International Symposium on Rock Stress, Kumamoto, Japan. Balkema, Rotterdam, pp. 469–474.

Chadwick, R.A., Zweigel, P., Gregersen, U., Kirby, G.A., Holloway, S., Johannessen, P.N., 2004. Geological reservoir characterization of a CO_2 storage site: The Utsira Sand, Sleipner, northern North Sea. Energy, 29, 1371–1381.

Clark, C.D., Hughes, A.L.C., Greenwood, S.L., Jordan, C., Sejrup, H.P., 2012. Pattern and timing of retreat of the last British-Irish Ice Sheet. Quat. Sci. Rev. 44, 112–146.

Clarke, H., Eisner, L., Styles, P., Turner, P., 2014. Felt seismicity associated with shale gas hydraulic fracturing: The first documented example in Europe. Geophys. Res. Lett. 41, 8308–8314.

Cuenot, N., Dorbath, C., Dorbath, L., 2008. Analysis of the microseismicity induced by fluid injections at the EGS site of Soultz-sous-Forêts (Alsace, France): Implications for the characterization of the geothermal reservoir properties. Pure Appl. Geophys. 165, 797–828.

Dahlheim, H.-A., Gebrande, H., Schmedes, E., Soffel, H., 1997. Seismicity and stress field in the vicinity of the KTB location. J. Geophys. Res. 102, 18,493-18,506.

Daniels, R.B., 1987. Soil erosion and degradation in the southern piedmont of the USA. In: Wolman, M.G., Fournier, F.G.A. (Eds.), Land Transformation in Agriculture. John Wiley & Sons Ltd, Chichester, England, pp. 407–428.

Davies, R., Foulger, G., Bindley, A., Styles, P., 2013. Induced seismicity and hydraulic fracturing for the recovery of hydrocarbons. Mar. Petroleum Geol. 45, 171–185.

Davis, S.D., Frohlich, C., 1993. Did (or will) fluid injection cause earthquakes? Criteria for a rational assessment. Seismol. Res. Lett. 64, 207–224.

DECC, 2012. Written Ministerial Statement by Edward Davey: Exploration for shale gas. UK Government Department of Energy & Climate Change. Available online: https://www.gov.uk/government/news/written-ministerial-statement-by-edward-davey-exploration-for-shale-gas (accessed 25.11.2015.).

DECC, 2013. Onshore oil and gas exploration in the UK: Regulation and best practice. Department of Energy and Climate Change, London, 49 pp. Available from: https://www.gov.uk/government/uploads/system/uploads/attachment_data/file/265988/Onshore_UK_oil_and_gas_exploration_England_Dec13_contents.pdf (accessed 13.08.2015.).

DEFRA, 2009. Safeguarding our soils: a strategy for England. UK Government Department for Environment, Food and Rural Affairs, 48 pp. Available from: http://webarchive.nationalarchives.gov.uk/20130123162956/http://archive.defra.gov.uk/environment/quality/land/soil/documents/soil-strategy.pdf (accessed 13.08.2015.).

DEFRA, 2015. Digest of waste and resource statistics – 2015 Edition. UK Government Department for Environment, Food & Rural Affairs. Available from: https://www.gov.uk/government/uploads/system/uploads/attachment_data/file/422618/Digest_of_waste_England_-_finalv2.pdf (accessed 13.08.2015.).

de Pater, C.J., Baisch, S., 2011. Geomechanical study of Bowland Shale seismicity: synthesis report. Cuadrilla Resources Ltd., Lichfield, 71 pp. Available online: http://www.rijksoverheid.nl/bestanden/documenten-en-publicaties/rapporten/2011/11/04/rapport-geomechanical-study-of-bowland-shale-seismicity/rapport-geomechanical-study-of-bowland-shale-seismicity.pdf (accessed 05.08.2015.).

Dorbath, L., Cuenot, N., Genter, A., Frogneux, M., 2009. Seismic response of the fractured and faulted granite of Soultz-sous-Forêts (France) to 5 km deep massive water injections. Geophys. J. Int. 177, 653–675.

Ellsworth, W.L., 2013. Injection-induced earthquakes. Science 341, 8. doi: 10.1126/science.1225942.

EPA, 2012. Class II Wells - Oil and Gas Related Injection Wells (Class II). United States Environmental Protection Agency, Washington, D.C. Available from: http://water.epa.gov/type/groundwater/uic/class2/ (accessed 05.08.2015.).

Eshelby, J.D., 1957. The determination of the elastic field of an ellipsoidal inclusion, and related problems. Proc. Royal Soc. Lond. 241, 376–396.

European Union, 2000. Directive 2000/60/EC of the European Parliament and of the Council establishing a framework for the Community action in the field of water policy. Available from:

http://eur-lex.europa.eu/resource.html?uri=cellar:5c835afb-2ec6-4577-bdf8-756d3d694eeb.0004.02/DOC_1&format=PDF (accessed 12.08.2015.).

Evans, C.J., Brereton, N.R., 1990. *In situ* crustal stress in the United Kingdom from borehole breakouts. In: Hurst, A., Lovell, M.A., Morton, A.C., (Eds.), Geological Applications of Wireline Logs. Geological Society, London, Special Publications, 48, pp. 327–338.

Fisher, K., Warpinski, N., 2012. Hydraulic-fracture-height growth: Real data. Soc. Petroleum Eng., Prod. Operat. J. 27, 8–19.

Friberg, P.A., Besana-Ostman, G.M., Dricker, I., 2014. Characterization of an earthquake sequence triggered by hydraulic fracturing in Harrison County, Ohio. Seismol. Res. Lett. 85, 1295–1307.

Friedman, E.S., Sato, Y., Alatas¸, A., Johnson, C.E., Wilkinson, T.J., Yener, K.A., Lai, B., Jennings, G., Mini, S.M., Alp, E.E., 2000. An X-ray fluorescence study of lake sediments from ancient Turkey using synchrotron radiation. In: Proceedings of the 47th Annual Denver X-ray Conference, Colorado Springs, Colorado, 3–7 August 1998. Advances in X-ray Analysis, 42, pp. 151–160.

Frohlich, C., Ellsworth, W., Brown, W.A., Brunt, M., Luetgert, J., MacDonald, T., Walter, S., 2014. The 17 May 2012 M 4.8 earthquake near Timpson, east Texas: An event possibly triggered by fluid injection. J. Geophys. Res. 119, 581–593.

Frohlich, C., Hayward, C., Stump, B., Potter, E., 2011. The Dallas-Fort Worth earthquake sequence: October 2008 through May 2009. Bull. Seismol. Soc. Am. 101, 327–340.

Gale, J.F.W., Reed, R.M., Holder, J., 2007. Natural fractures in the Barnett Shale and their importance for hydraulic fracture treatments. Am. Assoc. Petroleum Geol Bull. 91, 603–622.

Gan Wei, Frohlich, C., 2013. Gas injection may have triggered earthquakes in the Cogdell oil field, Texas. Proc. Natl. Acad. Sci. 110, 18,786-18,791.

Green, C.A., Styles, P., Baptie, B.J., 2012. Preese Hall shale gas fracturing: Review and recommendations for induced seismic mitigation. UK Government Department of Energy and Climate Change, London, 26 pp. Available from: https://www.gov.uk/government/uploads/system/uploads/attachment_data/file/48330/5055-preese-hall-shale-gas-fracturing-review-and-recomm.pdf (accessed 13.08.2015.).

Hanks, T.C., Kanamori, H., 1979. A moment magnitude scale. J. Geophys. Res. 84, 2348–2350.

Healy, J.H., Rubey, W.W., Griggs, D.T., Raleigh, C.B., 1968. The Denver earthquakes. Science 161, 1301–1309.

Herrmann, R.B., Park, S.-K., Wang, C.-Y., 1981. The Denver earthquakes of 1967–1968. Bull. Seismol. Soc. Am. 71, 731–745.

Holland, A., 2013. Earthquakes triggered by hydraulic fracturing in south-central Oklahoma. Bull. Seismol. Soc. Am. 103, 1784–1792.

Hornbach, M.J., DeShon, H.R., Ellsworth, W.L., Stump, B.W., Hayward, C., Frohlich, C., Oldham, H.R., Olson, J.E., Magnani, M.B., Brokaw, C., Luetgert, J.H., 2015. Causal factors for seismicity near Azle, Texas. Nature Commun. 7728, 11. doi: 10.1038/ncomms7728.

Horton, S., 2012. Disposal of hydrofracking waste fluid by injection into subsurface aquifers triggers earthquake swarm in central Arkansas with potential for damaging earthquake. Seismol. Res. Lett. 83, 250–260.

Houghton, R.A., 2003. Revised estimates of the annual net flux of carbon to the atmosphere from changes in land use and land management 1850–2000. Tellus 55, 378–390.

Houghton, R.A., 2005. Aboveground forest biomass and the global carbon balance. Global Change Biol. 11, 945–958.

Hsieh, P.A., Bredehoeft, J.D., 1981. A reservoir analysis of the Denver earthquakes: A case of induced seismicity. J. Geophys. Res. 86, 903–920.

International Copper Study Group, 2012. The world copper factbook 2012, pp. 24. Available from: http://www.slideshare.net/PresentacionesVantaz/the-world-copper-factbook-2012-16401361 (accessed 13.08.2015.).

IPCC, 2013. Climate Change 2013 The Physical Science Basis. Intergovernmental Panel on Climate Change, pp. 2216. Available online: http://www.climatechange2013.org/images/uploads/WGIAR5_WGI-12Doc2b_FinalDraft_All.pdf (accessed 16.08.2015.).

Juhlin, C., Dehghannejad, M., Lund, B., Malehmir, A., Pratt, G., 2010. Reflection seismic imaging of the end-glacial Pärvie Fault system, northern Sweden. J. Appl. Geophys. 70, 307–316.

Keranen, K.M., Savage, H.M., Abers, G.A., Cochran, E.S., 2013. Potentially induced earthquakes in Oklahoma, USA: Links between wastewater injection and the 2011 M_w 5.7 earthquake sequence. Geology 41, 699–702.

Keranen, K.M., Weingarten, M., Abers, G.A., Bekins, B.A., Ge, S., 2014. Sharp increase in central Oklahoma seismicity since 2008 induced by massive wastewater injection. Science 345, 448–451.

Kirby, G.A., Baily, H.E., Chadwick, R.A., Evans, D.J., Holliday, D.W., Holloway, S., Hulbert, A.G., Pharaoh, T.C., Smith, N.J.P., Aitkenhead, N., Birch, B., 2000. The structure and evolution of the Craven Basin and adjacent areas. Subsurface Memoir. The Stationery Office, London, pp.130.

Kim Won Young, 2013. Induced seismicity associated with fluid injection into a deep well in Youngstown, Ohio. J. Geophys. Res. 118, 3506–3518.

Klose, C.D., 2007a. Geomechanical modeling of the nucleation process of Australia's 1989 M5.6 Newcastle earthquake. Earth Planetary Sci. Lett. 256, 547–553.

Klose, C.D., 2007b. Mine water discharge and flooding: a cause of severe earthquakes. Mine Water Environ. 26, 172–180.

Klose, C.D., 2012. Evidence for anthropogenic surface loading as trigger mechanism of the 2008 Wenchuan earthquake. Environ. Earth Sci. 66, 1439–1447.

Klose, C.D., 2013. Mechanical and statistical evidence of the causality of human-made mass shifts on the Earth's upper crust and the occurrence of earthquakes. J. Seismol. 17, 109–135.

Konikow, L.F., 2011. Contribution of global groundwater depletion since 1900 to sea-level rise. Geophys. Res. Lett. 38, 5. doi: 10.1029/2011GL048604, L17401.

Konikow, L.F., 2013. Comment on "Model estimates of sea-level change due to anthropogenic impacts on terrestrial water storage" by Pokhrel et al. Nature Geosci. 6, 2.

Laing O'Rourke, 2014. London Gateway Port, London, UK. Laing O'Rourke, Ltd., Dartford, England. Available from: http://www.laingorourke.com/our-work/all-projects/london-gateway-port.aspx (accessed 16.08.2015.).

Lundqvist, J., Lagerback, R., 1976. The Parve Fault: A late-glacial fault in the Precambrian of Swedish Lapland. Geologiska Föreningens I Stockholm Förhandlingar 98, 45–51.

McGarr, A., 1976. Seismic moments and volume changes. J. Geophys. Res. 81, 1487–1494.

McGarr, A., 2014. Maximum magnitude earthquakes induced by fluid injection. J. Geophys. Res. 119, 1008–1019.

McGarr, A., Simpson, D., Seeber, L., 2002. Case histories of induced and triggered seismicity. Int. Handbook Earthquake Eng. Seismol. 81A, 647–661.

Majer, E., Baria, R., Stark, M., Oates, S., Bommer, J., Smith, B., Asanuma, H., 2007. Induced seismicity associated with enhanced geothermal systems. Geothermics 36, 185–222.

Mather, J., Halliday, D., Joseph, J.B., 1998. Is all the groundwater worth protecting? The example of the Kellaways Sand. In: Groundwater Pollution, Aquifer Recharge, Vulnerability, Robins, N.S., (Eds.), Geological Society, London, Special Publications, 130, pp. 211–217.

Meremonte, M.E., Lahr, J.C., Frankel, A.D., Dewey, J.W., Crone, A.J., Overturf, D.E., Carver, D.L., Bice, W.T., 2002. Investigation of an earthquake swarm near Trinidad, Colorado, August-October 2001. U.S. Geological Survey Open-File Report 02-0073, 32 pp. Available from: http://pubs.usgs.gov/of/2002/ofr-02-0073/ofr-02-0073.html (accessed 12.08.2015.).

Mining-technology.com, 2013. The 10 biggest coal mines in the world. http://www.mining-technology.com/features/feature-the-10-biggest-coal-mines-in-the-world/ (accessed 13.08.2015.).

Mining-technology.com, 2014. The world's biggest iron ore mines. http://www.mining-technology.com/features/featurethe-worlds-11-biggest-iron-ore-mines-4180663/ (accessed 13.08.2015.).

Mitchell, B.R., 1984. Economic Development of the British Coal Industry, 1800–1914. Cambridge University Press, Cambridge, 360 pp.

Mörner, N.-A., Sjöberg, R., Audemard, F., Dawson, S., Sun Guangyu, 2008. Paleoseismicity and uplift of Sweden. Field guide, for International Geological Congress excursion no. 11, July–August 2008, 109 pp. Available from: http://www.iugs.org/33igc/fileshare/filArkivRoot/coco/FieldGuides/No 11 Palaeoseismisity.pdf (accessed 16.08.2015.).

Myers, N., 1993. Gaia: an atlas of planet management. Anchor/Doubleday, Garden City, New York.

NHBC, 2007. Housing completions show moderate rise. National House-Building Council. Available from: http://www.nhbc.co.uk/NewsandComment/UKnewhouse-buildingstatistics/Year2007/Name,32187,en.html (accessed 13.08.2015.).

Nicholson, C., Roeloffs, E., Wesson, R.L., 1988. The northeastern Ohio earthquake of 31 January 1986: was it induced? Bull. Seismol. Soc. Am. 78, 188–217.

Paul, T., Chow, F., Kjekstad, O., 2002. Hidden Aspects of Urban Planning: Surface and Underground Development. Thomas Telford, London, 85 pp..

Pegg, I.L., 2015. Turning nuclear waste into glass. Phys. Today 68 (2), 33–39.

Pimentel, D., 2006. Soil erosion: a food and environmental threat. Environ. Dev. Sustainability 8, 119–137.

Pokhrel, Y.N., Hanasaki, N., Yeh, P.J.F., Yamada, T.J., Kanae, S., Oki, T., 2012. Model estimates of sea-level change due to anthropogenic impacts on terrestrial water storage. Nature Geosci. 5, 389–392.

Pokhrel, Y.N., Hanasaki, N., Yeh, P.J.F., Yamada, T.J., Kanae, S., Oki, T., 2013. Overestimated water storage. Reply. Nature Geosci. 6, 2–3.

Port of Rotterdam, 2013. Projectorganisatie Maasvlakte 2 Construction, from plan to execution. Available from: https://www.maasvlakte2.com/en/index/show/id/198/Construction (accessed 16.08.2015.).

Ridlington, E., Rumpler, J., 2013. Fracking by the numbers: Key impacts of dirty drilling at the state and national level. Environment America Research & Policy Center, Boston, Massachusetts, 47 pp. Available from: http://www.environmentamerica.org/sites/environment/files/reports/EA_FrackingNumbers_scrn.pdf (accessed 11.08.2015.).

Ringrose, P.S., 1989. Recent fault movement and palaeoseismicity in western Scotland. Tectonophysics 163, 315–321.

Roche, V., Grob, M., Eyre, T., Van Der Baan, M., 2015. Statistical characteristics of microseismic events and *in-situ* stress in the Horn River Basin. In: Proceedings of GeoConvention 2015, Calgary, Canada, 4–8 May 2015, 5 pp. Available from: http://www.geoconvention.com/uploads/2015abstracts/080_GC2015_Statistical_characteristics_of_microseismic_events.pdf (accessed 16.08.2015).

Rubinstein, J.L., Ellsworth, W.L., McGarr, A., Benz, H.M., 2014. The 2001–present induced earthquake sequence in the Raton Basin of northern New Mexico and southern Colorado. Bull. Seismol. Soc. Am. 104, 2162–2181.

Rubinstein, J.L., Mahani, A.B., 2015. Myths and facts on waste water injection, hydraulic fracturing, enhanced oil recovery, and induced seismicity. Seismol. Res. Lett. 86, 1060–1067.

Scholz, C.H., 1968. The frequency-magnitude relation of microfracturing in rock and its relation to earthquakes. Bull. Seismol. Soc. Am. 58, 399–415.

Seeber, L., Armbruster, J., Kim, W.-Y., 2004. A fluid-injection-triggered earthquake sequence in Ashtabula, Ohio: Implications for seismogenesis in stable continental regions. Bull. Seismol. Soc. Am. 94, 76–87.

Siskind, D.E., Stagg, M.S., Kopp, J.W., Dowding, C.H., 1980. Structure response and damage produced by ground vibration from surface mine blasting. United States Bureau of Mines, Report of Investigations No. 8507.

Skoumal, R.J., Brudzinski, M.R., Currie, B.S., 2015. Earthquakes induced by hydraulic fracturing in Poland Township, Ohio. Bull. Seismol. Soc. Am. 105, 189–197.

Solomon, S., 2007. Carbon Dioxide Storage: Geological Security and Environmental Issues – Case Study on the Sleipner Gas field in Norway. The Bellona Foundation, Oslo, Norway, 128 pp. Available from: http://bellona.org/filearchive/fil_CO2_storage_Rep_Final.pdf (accessed 15.08.2015.).

Stagg, M.S., Siskind, D.E., Stevens, M.G., Dowding, C.H., 1980. Effects of repeated blasting on a wood frame house. United States Bureau of Mines, Report of Investigations No. 8896.

Stewart, I., Firth, C., Rust, D., Collins, P., Firth, J., 2001. Postglacial fault movement and palaeoseismicity in western Scotland: a reappraisal of the Kinloch Hourn fault, Kintail. J. Seismol. 5, 307–328.

Sutinen, R., Piekkari, M., Middleton, M., 2009. Glacial geomorphology in Utsjoki, Finnish Lapland proposes Younger Dryas fault-instability. Global Planetary Change 69, 16–28.

Syvitski, J.P.M., Kettner, A.J., 2011. Sediment flux and the Anthropocene. Philos. Transact. Royal Soc. Lond. 369, 957–975.

Syvitski, J.P.M., Vörösmarty, C.J., Kettner, A.J., Green, P., 2005. Impact of humans on the flux of terrestrial sediment to the global coastal ocean. Science 308, 376–380.

TESTRAD, 2012. Thames Estuary Airport: feasibility review. The Thames Estuary Research And Development Company, London, 64 pp. Available from: http://testrad.co.uk/wp-content/uploads/2012/08/TEAFRreport.pdf (accessed 15.08.2015.).

Trimble, S.W., 1975. A Volumetric Estimate of Man-induced Erosion on the Southern Piedmont. U.S. Department of Agriculture, Agricultural Research Service Publication S40, pp. 142–145.

Trimble, S.W., 2008. Man-induced soil erosion on the Southern Piedmont: 1700-1970 (enhanced 2nd Edition). Soil and Water Conservation Society, Ankeny, Indiana, 80 pp.

USGS, 2013. Mineral commodity summaries 2013. U.S. Geological Survey, Reston, Virginia, 198 pp. Available from: http://minerals.usgs.gov/minerals/pubs/mcs/2013/mcs2013.pdf (accessed 13.08.2015.).

Van Oost, K., Cerdan, O., Quine, T.A., 2009. Accelerated sediment fluxes by water and tillage erosion on European agricultural land. Earth Surf. Processes Landforms 34, 1625–1634.

Varga, R., Pachos, A., Holden, T., Pendrel, J., Lotti, R., Marini, I., Spadafora, E., 2012. Seismic inversion in the Barnett Shale successfully pinpoints sweet spots to optimize wellbore placement and reduce drilling risks. Proceedings of the 2012 Society of Exploration Geophysicists Annual Meeting, 4–9 November 2012, Las Vegas, Nevada, paper SEG-2012-1266. Curran Associates, Red Hook, New York, pp. 4023--4027.

Verdon, J.P., 2014. Significance for secure CO_2 storage of earthquakes induced by fluid injection. Environ. Res. Lett. 9, 064022.

Verdon, J.P., Kendall, J.-M., Stork, A.L., Chadwick, R.A., White, D.J., Bissell, R.C., 2013. A comparison of geomechanical deformation induced by 'megatonne' scale CO_2 storage at Sleipner, Weyburn and In Salah. Proc. Natl. Acad. Sci. 110, E2762–E2771.

Wada, Y., van Beek, L.P.H., van Kempen, C.M., Reckman, J.W.T.M., Vasak, S., Bierkens, M.F.P., 2010. Global depletion of groundwater resources. Geophys. Res. Lett. 37, 5, L20402.

Wada, Y., van Beek, L.P.H., Weiland, F.C.S., Chao, B.F., Wu, Y.H., Bierkens, M.F.P., 2012. Past and future contribution of global groundwater depletion to sea-level rise. Geophys. Res. Lett. 39, 6, L09402.

Wald, D.J., Quitoriano, V., Heaton, T.H., Kanamori, H., 1999. Relationships between peak ground acceleration, peak ground velocity, and modified Mercalli intensity in California. Earthquake Spectra 15, 557–564.

Walters, R.J., Zoback, M.D., Baker, J.W., Beroza, G.C., 2015. Characterizing and Responding to Seismic Risk Associated with Earthquakes Potentially Triggered by Fluid Disposal and Hydraulic Fracturing. Seismol. Res. Lett. 86, 1110–1118.

Weingarten, M., Ge, S., Godt, J.W., Bekins, B.A., Rubinstein, J.L., 2015. High-rate injection is associated with the increase in U.S. mid-continent seismicity. Science 348, 1336–1340.

Westaway, R., 2002. Seasonal seismicity of northern California before the great 1906 earthquake. Pure Appl. Geophys. 159, 7–62.

Westaway, R., 2006. Investigation of coupling between surface processes and induced flow in the lower continental crust as a cause of intraplate seismicity. Earth Surf. Processes Landforms 31, 1480–1509.

Westaway R., in press. Isostatic compensation of Quaternary vertical crustal motions: coupling between uplift of Britain and subsidence beneath the North Sea. J. Quat. Sci.

Westaway R., in review. The importance of characterizing uncertainty in controversial geoscience applications: induced seismicity associated with hydraulic fracturing for shale gas in northwest England. Proc. Geol. Assoc.

Westaway, R., Younger, P.L., 2014. Quantification of potential macroseismic effects of the induced seismicity that might result from hydraulic fracturing for shale gas exploitation in the UK. Quarterly J. Eng. Geol. Hydrogeol. 47, 333–350.

Westaway, R., Younger, P.L., Cornelius, C., 2015. Comment on 'Life cycle environmental impacts of UK shale gas' by L. Stamford and A. Azapagic. Appl. Energy, 134, 506–518, 2014. Appl. Energy, 148, pp. 489–495 (with 2015 corrigendum: Appl. Energy 155, 949).

World Steel Association, 2012. World steel in figures 2012, 15 pp. Available from: http://www.worldsteel.org/dms/internetDocumentList/bookshop/WSIF_2012/document/World Steel in Figures 2012.pdf (accessed 13.08.2015.).

Worldtravellist.com, 2011. Palm Deira of Dubai is the largest man made island in the world. Available from: http://trip.worldtravellist.com/2011/06/palm-deira-dubai/ (accessed 16.08.2015.).

Wrench, G.T., 1946. Reconstruction by Way of the Soil. Faber and Faber, London, 262 pp.

Younger, P.L., in review. How can we be sure fracking will not pollute aquifers? Lessons from a major longwall coal mining analogue (Selby, Yorkshire, UK). Earth and Environmental Sciences, Transactions of the Royal Society of Edinburgh.

Younger, P.L., Westaway, R., 2014. Review of the Inputs of Professor David Smythe in Relation to Planning Applications for Shale Gas Development in Lancashire (Planning Applications LCC/2014/0096/0097/0101 and /0102) and Associated Recommendations. Report to Lancashire County Council, 12 pp. + 1 p. preface. University of Glasgow Research Reports Available from: http://eprints.gla.ac.uk/108343/

Zoback, M.D., Harjes, H.-P., 1997. Injection-induced earthquakes and crustal stress at 9 km depth at the KTB deep drilling site, Germany. J. Geophys. Res. 102, 18,477–18,491.

CHAPTER 12
State and Federal Oil and Gas Regulations

Karen J. Anspaugh
Indiana University Robert McKinney School of Law and Surrett & Anspaugh, Traverse City, MI, USA

Chapter Outline

WHAT IS THE PURPOSE OF OIL AND GAS REGULATIONS?

Affordable, abundant, and readily available fossil fuels impact almost every aspect of the lives of Americans. Oil and Gas (O&G) enable the United States (US) standard of living and power national and global economies. In each O&G–producing state, a body of rules and regulations arose to protect the state's natural resources, to encourage the effective production of fossil fuels, to make production available for public use, to safeguard the rights of mineral interest owners, to protect the environment, to enhance the well being of society, to promote economic growth, and to enhance national security by encouraging energy independence.

Reliance on fossil fuels is a reality, making domestic production necessary. In 2014, the US used energy as follows: petroleum (35%), natural gas (28%), coal (18%), renewables (10%), and nuclear (8%) (US Energy Information Administration, 2014, www.eia.gov). Yet, production occurs within a global economy.

Environmental and Health Issues in Unconventional Oil and Gas Development. http://dx.doi.org/10.1016/B978-0-12-804111-6.00012-1

Energy supply and energy independence are global issues (see eg., Chapter 13). Natural gas and crude oil benchmark prices are driven by global supply and demand and the price of crude oil is driven largely by the Organization of the Petroleum Exporting Countries (OPEC; www.opec.org) (see economics discussion in Chapter 2). Unlike the US, foreign producers rarely encounter the delays and cost increases inherent in regulation. It is estimated that the Islamic State generates between one million and two million dollars a week in oil revenue and competes for market share (Bronstein et al., 2014; www.cnn.com). What can the US do to compete?

Counterproductive regulations and overregulation (perhaps motivated by politics rather than founded upon extensive knowledge of geologic formations and production techniques) do not achieve the appropriate goals of regulation. Overregulation diminishes the availability of affordable energy, weakens the job market by reducing the profitability of businesses, and drives up the price of goods and services that promote public health.

It is well documented that unconventional O&G development using hydraulic fracturing (fracking) has been very beneficial to the economy of the US, even during the recent years of economic downturn. It is well documented that O&G industry employment outpaced job growth in all other areas of the economy by a significant margin and that areas of the country experiencing shale booms enjoyed dynamic economic growth and low unemployment.

It is well documented that fracking can be conducted safely given sufficient regulation (including water testing and disclosure of the compounds utilized), fully funded inspection and verification, and the best management practices of operators (see eg., Chapter 4). The US Environmental Protection Agency (US EPA) released draft findings of a 5-year study on June 4, 2015, which reports that spills are very infrequent and that fracking has not resulted in any widespread or systemic impact on drinking water resources (US EPA, 2015) (see eg., Chapter 6).

WHAT O&G ISSUES DO US STATE AGENCIES REGULATE?

State O&G conservation laws are enforced by state regulatory agencies (e.g., the Department of Natural Resources, the Department of Environmental Quality, the Office of Oil, Gas and Minerals, or a variety of other names). Such agencies are granted authority by state legislation and are tasked with the following responsibilities: to insure the efficient and economic development of natural resources, to supervise the leasing of state-owned lands, to protect natural resources from wasteful and harmful practices, to protect groundwater, to prevent adverse environmental events and impacts, to enforce well density and unit size requirements, to limit the proximity of wells to neighboring land, to ensure the equitable sharing of O&G via integrated pooling, to permit wells (oil, gas, coalbed methane, disposal, enhanced recovery, noncommercial gas, water supply, gas storage, and geological structure test wells), to collect the fees and bonds required to drill wells, to enforce well construction standards, to inspect wells, to enforce the filing of completion reports and other well records, to enforce the timely reporting and remediation of spills, and to supervise the plugging and abandoning of wells (see eg., Chapters 1 and 4).

State O&G regulatory agencies insure that well design, construction, and integrity meet state specifications, and monitor the migration and containment of fluids, water availability, the storage of fracking fluids, the storage of additives, the storage of flowback fluids and solids, and the movement, management, and disposal of fracking waste (see eg., Chapter 5). Since the enactment of O&G conservation laws (most were adopted more than 50 years ago), state agencies have successfully regulated fracking by requiring that operators use multiple impermeable layers of well casing and cement to protect the ground water (see eg., Chapters 1 and 4). Well construction specifications are reviewed and approved by state regulators, who are generally trained geologists, as part of the well-permitting process. Thereafter, state inspection offices investigate and assess all well sites and continually monitor drilling practices.

Detailed information, such as the volume and types of fluids used and the nature and amounts of proppants used, must be provided in well completion reports that become part of the permanent well record. For example, in Wyoming, operators must report all chemical additives, the compound type, and concentrations and mixture rates utilized (http://soswy.state.wy.us/Rules/RULES/7928.pdf). In Louisiana, to insure that ground water is available for public use, the injection of fracking fluids into the Haynesville Shale formation is limited (Memorandum WH-1, Requirements for Water Use in E&P Operations, Louisiana DNR, September 2009). In Michigan, ground water is protected by requiring that surface casing be set to 100 feet below the base of glacial deposits, into competent bedrock, and 100 ft. below all fresh water strata. The casing must be sealed by circulating cement up to the surface (http://www.legislature.mi.gov/(S(cckjiglylu5vb2jbqs2yp5cz))/mileg.aspx?page=getObject&objectName=mcl-451-1994-III-3-2-615).

WHAT O&G ISSUES DO US FEDERAL AGENCIES REGULATE?

Federal regulations pertaining to waste and ground water are generally enforced at the state level. Air quality is subject to a combination of state and federal requirements and water discharge is enforced at the federal level. The following federal environmental statutes, enacted by the US Congress, pertain specifically to the regulation of O&G:

1. *Clean Air Act (1970)* (http://www2.epa.gov/laws-regulations/summary-clean-air-act). This act identifies various O&G activities as potential sources of air emissions, such as engine emissions from drill rigs, fracking equipment and onsite power generators, fugitive emissions from hydrocarbons in flowback liquids, emissions from venting and flaring of gas during flowback, separators, storage vessels, pneumatic controls, glycol dehydrators, compressors, and desulfurization units. The Act

created a New Source Performance Standard (NSPS; 40 C.F.R. Part 60, Subpart OOOO), which was the first federal air standard to apply to fracked wells. The NSPS was enacted to reduce emissions of volatile organic compounds and sulfur dioxide from O&G operations, including natural gas wells, storage tanks, and other equipment and processing units and applies to all subject facilities constructed after August 23, 2011, and to facilities modified or reconstructed thereafter. Existing facilities were required to comply within 60 days of August 16, 2012. Effective January 1, 2015, wells that have been fracked (other than wildcat wells, for which modified rules apply) must control flowback emissions using Reduced Emission Completion Technology (known as "green completions").

2. *Clean Water Act (1977)* (http://www2.epa.gov/laws-regulations/summary-clean-water-act). This Act establishes water use guidelines pertaining to O&G operations and to the discharge of oil into navigable waterways. The Oil Pollution Act (http://www2.epa.gov/laws-regulations/summary-oil-pollution-act), which amends the Clean Water Act, pertains to clean up and damage assessments of large oil spills into navigable waters and shorelines. The National Pollutant Discharge Elimination System (NPDES; http://water.epa.gov/polwaste/npdes/) regulates the discharge of pollutants into waters of the US. Section 404 of the Clean Water Act (http://water.epa.gov/lawsregs/guidance/cwa/dredgdis/) regulates and requires a permit for the discharge of dredged or fill material into waters of the US, which applies to any discharge of a pollutant into navigable waters, including wetlands. Permits are approved by the US Army Corps of Engineers. Examples of fill materials are rock, sand, soil, concrete, riprap, culverts, sewer and utility lines. Note that fracking is exempted from coverage by the Safe Drinking Water Act (1974), which is part of the Clean Water Act and which protects drinking water and above and below ground water sources that may be used for human consumption. The Safe Drinking Water Act requires that the EPA establish minimum regulations for Class II injection wells, which are governed by state underground injection control programs. Class II wells are utilized for enhanced oil recovery or to dispose of produced water and brine into deep formations located at depths below the base of fresh water. Of significant import is the fact that the 2005 Energy Policy Act (https://www.congress.gov/bill/109th-congress/house-bill/6) amended the Clean Water Act to exclude (1) the underground injection of natural gas for purposes of storage; and (2) the underground injection of fluids or propping agents (other than diesel fuels) used in fracking operations.
3. *Endangered Species Act (1973)* (http://www2.epa.gov/laws-regulations/summary-endangered-species-act). This Act grants the US Fish and Wildlife Service the authority to prohibit activities that may harm the habitats and ecosystems of threatened and endangered species (such as darter fish, prairie chickens, and sagebrush lizards).

WHAT ISSUES DO THE VARIOUS FEDERAL AGENCIES GOVERN?

The federal agencies of the US described below serve as regulatory authorities with jurisdiction over O&G production:

1. *Army Corps of Engineers (A Division of the Department of Defense) (USACE)*: Protects the nation's aquatic resources using engineering, design, and construction management, and evaluates permit applications for construction activities in Waters of the U.S., including wetlands.
2. *Bureau of Indian Affairs (BIA)*: Regulates oil development of native Indian lands along with the Bureau of Land Management.
3. *Bureau of Land Management (BLM)*: Regulates oil, gas, and coal operations, development, exploration, and production on federal onshore properties. The Bureau manages 700 million acres of subsurface mineral estates, which constitutes approximately 29% of the land in the United States.
4. *Bureau of Ocean Energy Management (BOEM)*: Manages the exploration and development of the nation's offshore resources.
5. *Bureau of Safety and Environmental Enforcement (BSEE)*: Promotes safety, protects the environment, and conserves offshore resources.
6. *Department of Energy (DOE)*: Manages the Strategic Petroleum Reserve, conducts energy research, and gathers and analyzes energy industry data.
7. *Department of Interior (DOI)*: Regulates the extraction of O&G from federal lands.
8. *Environmental Protection Agency (US EPA)*: Provides oversight for environmental, health, and safety issues and has primary responsibility for enforcing many of the environmental statutes and regulations of the US. The EPA was created in 1970 and currently employs approximately 16,000 people in 10 regional offices and 27 laboratories.
9. *Federal Energy Regulatory Commission (FERC)*: Regulates interstate pipelines and the interstate transmission of electricity, natural gas, and oil. The agency reviews proposals to build liquefied natural gas terminals and interstate natural gas pipelines, and licences hydropower projects.
10. *Federal Occupational Safety and Health Administration (A Division of the Department of Labor) (DOL-OSHA)*: Promotes the safety of employees.
11. *Fish and Wildlife Service (A Division of the Department of the Interior) (DOI-FWS)*: Administers and enforces the Endangered Species Act of 1973, and lists endangered and threatened species and designates critical habitat.
12. *Office of Natural Resources Revenue (ONRR)*: Collects royalties owed to the government for onshore and offshore production.
13. *Pipeline and Hazardous Materials Safety Administration (PHMSA)*: Regulates the risks inherent in transportation of hazardous materials, including O&G and enforces the Natural Gas Pipeline Safety Act of 1968 (P.L. 90-481).

WHAT STATES CURRENTLY ALLOW FRACKING?

The states that produce O&G and that allow fracking are listed in Table 12.1. To promote transparency in fracking procedures, the Groundwater Protection Council (GWPC, www.gwpc.org) and the Interstate Oil and Gas Compact Commission (iogcc.publishpath.com; a voluntary association of regulators from O&G–producing states) established FracFocus (Groundworks, 2011). Individual states participate voluntarily by furnishing information and FracFocus publishes the additives utilized by operators.

WHAT STATES CURRENTLY BAN FRACKING?

The following states have banned fracking or have a moratorium in place:

1. *Vermont*: In 2012, Vermont became the first state to ban fracking, although there has been no O&G production in the state (http://www.leg.state.vt.us/docs/2012/bills/Intro/H-464.pdf).
2. *New York*: In 2014, New York banned all fracking within the state, after 6 years of study (http://www.nysenate.gov/legislation/bills/2011/S4220). A moratorium had been in place since 2010. New York officials explained they were aware there is no conclusive research that fracking is harmful; however, they believe enough uncertainty exists regarding the potential risks to public health and the environment to warrant the ban on fracking. What are the financial ramifications? If only 10% of the gas reserves within the state reached production in the next 15 years, at US$4/MMbtu New York would profit by US$57 billion in gross value, which equates to US$7 billion in lost revenue to landowners (Joy et al., 2015).
3. *Maryland*: In May of 2015, Maryland approved a 2.5-year moratorium on fracking (http://www.dontfrackmd.org/gov-hogan-quietly-passes-fracking-moratorium/).

WHAT IS THE WEIGHT OF EVIDENCE ON FRACKING RISKS?

The weight of scientific study and research evidences that the New York fracking ban is not based upon supported safety, environmental, or human health concerns.

Studies by the US EPA, state regulators (note that O&G development is the exclusive regulatory domain of state agencies unless the activity takes place on federal lands), industry groups, and academia have all determined that hydraulic fracturing does not pose adverse environmental risks. Even the New York Department of Environmental Conservation (DEC), during the Cuomo administration, determined that hydraulic fracturing would not adversely affect the environment in New York.

A comprehensive study by GWPC in 1998 that surveyed state agencies responsible for regulating hydraulic fracturing found no evidence that public health is at risk as a result of hydraulic fracturing.

Table 12.1 States That Produce O&G and Allow Fracking

1.	Alabama	FracFocus disclosure of fracking solution is required
2.	Alaska	FracFocus disclosure of fracking solution is required
3.	Arkansas	Disclosure of fracking solution to state is required
4.	California	FracFocus disclosure of fracking solution is required
5.	Colorado	FracFocus disclosure of fracking solution is required
6.	Florida	Reporting to FracFocus being considered
7.	Idaho	FracFocus disclosure of fracking solution is required
8.	Indiana	Disclosure of fracking solution to state is required
9.	Illinois	Disclosure of fracking solution to state is required
10.	Kansas	FracFocus disclosure of fracking solution is required
11.	Kentucky	FracFocus disclosure of fracking solution is required
12.	Louisiana	FracFocus disclosure of fracking solution is required
13.	Maryland	No wells have been fracked
14.	Michigan	FracFocus disclosure of fracking solution is required
15.	Mississippi	FracFocus disclosure of fracking solution is required
16.	Montana	FracFocus disclosure of fracking solution is required
17.	Nebraska	FracFocus disclosure of fracking solution is required
18.	Nevada	FracFocus disclosure of fracking solution is required
19.	New Mexico	Disclosure of fracking solution to state is required
20.	North Carolina	FracFocus disclosure of fracking solution is required
21.	North Dakota	FracFocus disclosure of fracking solution is required
22.	Ohio	FracFocus disclosure of fracking solution is required
23.	Oklahoma	FracFocus disclosure of fracking solution is required
24.	Oregon	No wells have been fracked
25.	Pennsylvania	FracFocus disclosure of fracking solution is required
26.	South Dakota	FracFocus disclosure of fracking solution is required
27.	Tennessee	FracFocus disclosure of fracking solution is required
28.	Texas	FracFocus disclosure of fracking solution is required
29.	Utah	FracFocus disclosure of fracking solution is required
30.	Virginia	Reporting to FracFocus being considered
31.	West Virginia	FracFocus disclosure of fracking solution is required
32.	Wyoming	Disclosure of fracking solution to state is required

Note: Arizona, Connecticut, Delaware, Georgia, Hawaii, Idaho, Iowa, Massachusetts, Minnesota, Missouri, New Hampshire, New Jersey, Oregon, Rhode Island, South Carolina, South Dakota, Vermont, and Washington have no economically feasible O&G reserves.
Sources: FracFocus, 2015.

The comprehensive Interstate Oil and Gas Compact Commission (a multistate agency made up of regulators from O&G producing states, including New York) did a survey in 2002 that determined that while approximately 1 million wells had been hydraulically fractured since the 1940s, there were no substantiated claims of ground water contamination from hydraulic fracturing (see eg., Chapters 4 and 6).

The Massachusetts Institute of Technology (MIT) released a study in 2011 on the potential risks of hydraulic fracturing to ground water aquifers that found "no incidents of direct invasion of shallow water zones by fracture fluids during the fracturing process have been recorded."

In January, 2013 the US Geological Survey (USGS) released a study of ground water samples from a substantial part of the Fayetteville Shale region and again found no regional effects on ground water contamination from activities related to gas production. The Fayetteville Shale is similar to the Marcellus and Utica Shales in New York in that it is a low-permeability, low-porosity shale formation that must be hydraulically fractured to be economically productive.

Regulators from Arkansas, Colorado, Louisiana, North Dakota, Ohio, Oklahoma, Pennsylvania, and Texas advised the US Government Accountability Office in 2012 that, based on state investigations, hydraulic fracturing has not caused ground water contamination in any of those states, despite significant volumes of oil and natural gas production resulting from the process of hydraulic fracturing. The Alaska Oil and Gas Conservation Commission stated in 2011 that "[i]n over fifty years of oil and gas production, Alaska has yet to suffer a single documented instance of subsurface damage to an underground source of drinking water."

Federal government regulators working in the administration of President Barak Obama have also acknowledged that there is no evidence that hydraulic fracturing has adversely impacted ground water. On May 24, 2011, US EPA Administrator Lisa Jackson appeared before the House Committee on Oversight and Government Reform and testified that EPA was "not aware of any water contamination associated with the recent drilling" in the Marcellus Shale. Similarly, Robert Abbey, Director of the Bureau of Land Management (BLM), stated in Congressional Testimony in 2011 that he had "never seen any evidence of impacts to ground water from the use of fracing [*sic*] technology on wells that have been approved by" the BLM.

After extensive study and analysis, the New York Department of Environmental Conservation (DEC) released a draft generic environmental impact statement in 2009, concluding that the physical process of hydraulic fracturing of shale formations in New York does not pose any risk to ground water. The DEC conclusions were based in part on evidence gathered from comparable state regulatory agencies in Colorado, New Mexico, Pennsylvania, Ohio, Texas, and Wyoming that all concluded that hydraulic fracturing operations did not pose a risk to ground water contamination (Joy et al., 2015).

SHOULD STATE OR FEDERAL AGENCIES REGULATE FRACKING?

State O&G regulators (experts in protecting groundwater) have not raised the issue of whether additional federal fracking regulations are needed, which evidences their confidence and success in regulating fracking. Yet the issue of the risks associated with fracking is taken up again and again and considered by parties who are not familiar with state geologic characteristics and who have no expertise in the O&G industry or with fracking. Underground reservoirs, rock composition, stratigraphic formations, and the depth of water aquifers vary dramatically from state to state. Correspondingly, the volume of water and additives utilized in fracking vary from state to state. State regulatory agencies are familiar with unique characteristics within their state and thus are logically the best suited to enforce fracking regulations. States can also respond to operators more rapidly, which is of significant benefit. State regulators, GWPC, and the Interstate Oil and Gas Compact Commission have expressed concern that additional federal regulation would be redundant and that the procedural elements of any federal involvement would overburden state officials, who are presently fully engaged in the protection of ground water.

The best suited regulators of hydraulic fracturing are the individual states, which are the source of the common law of the domestic O&G industry because of their unique position and their collective expertise on matters concerning the industry. US states have adopted comprehensive laws and regulations to provide for safe operations and to protect the nation's drinking water sources, and have trained personnel to effectively regulate O&G exploration and production. Hydraulic fracturing is currently, and has been for decades, a common operation used in exploration and production by the O&G industry in all gas-producing states. Because the unique position of the states and their collective expertise on matters concerning the O&G industry, regulation of hydraulic fracturing should remain the responsibility of the states. The states have as much of a vested interest in the protection of groundwater as the federal government and, as such, will continue to regulate the process effectively and efficiently, taking into account the particulars of the geology and hydrology within their boundaries. There is not a "one size fits all" approach to effective regulation (Groundworks, 2015).

DO REGULATIONS ADEQUATELY CONSIDER GLOBAL DYNAMICS?

How is the US performing in the global market (see global discussion in Chapter 13)? It is competing for part of the global market share; therefore, the regulations must allow production in a climate that is not rapidly changing, where regulatory bodies do not cause significant delays in approval processes, and where taxes and fees and the costs of doing business in compliance with stringent regulations do not diminish profit margins to the point that drilling ceases (see eg., Chapter 2).

> Among the biggest oil companies, the vast majority of volume growth has come from state-controlled entities ... Saudi Aramco's breakeven costs are on the order of $10 per barrel. The very best American Shale plays breakeven at more like $40 (Helman and Christopher, 2015).

North American O&G producers are underperforming and growth is stagnant compared with the explosive growth of state-run foreign companies. In recent years, the growth and profitably of foreign, government-owned O&G producers have expanded significantly, while the growth of independent supermajors has been minimal or nonexistent. State-owned O&G producers are supported by governmental financing and protection and are not restricted by burdensome taxes, fees, bureaucratic delays, and environmentally motivated overregulation. The top 21 global O&G producers, which generate 50% of the world's supply, are given in Table 12.2.

> An era of state-driven capitalism has dawned, in which governments are again directing huge flows of capital – even across the borders of capitalist democracies – with profound implications for free markets and international politics. China and Russia are leading the way in the strategic deployment of state-owned enterprises, and other governments have begun to follow their lead... Collectively, multinational oil companies produce just 10% of the world's O&G reserves. State-owned companies now control more than 75% of all crude oil production (Bremmer and Ian, 2010).

HOW CAN US NATIONAL ENERGY POLICY BE MOST EFFECTIVE?

The energy policy of the US is established by a group of laws generated by various federal departments and agencies and state equivalents, including the Department of Interior, the Department of Transportation, the Department of Energy, and EPA.

The efficacy of state O&G conservation laws is limited by national policies. An effective national energy policy would pave the way for effective state regulation. Policies that adhere to the rule of law and promote regulatory efficiency result in increased O&G production and economic growth. Domestic production flourishes when operators have access to drill on federal land and offshore locations, when subsidies do not artificially favor one type of energy over another, and when regulatory burdens are not unduly restrictive. Approval of the Keystone XL pipeline, lifting the crude oil export, eliminating tariffs on energy technology, and limiting regulations that result in job losses would benefit domestic production. Congress (rather than the EPA and other federal regulatory agencies) should approve new federal regulations (thereby adhering to constitutional boundaries), and regulatory reform should give appropriate weight and consideration to the costs incurred by producers and end users. Compliance

Table 12.2 Global O&G Producers, Which Generate 50% of the World's Supply

S. No.	Producer	Barrel of oil Equivalent Production per day (Million)
1.	Saudi Aramco (Saudi Arabia)	12
2.	Gazprom (Russia)	8.3
3.	National Iranian Oil (Iran)	6
4.	Exxon Mobil (United States)	4.7
5.	Rosneft (Russia)	4.7
6.	PetroChina (China)	4
7.	BP (United Kingdom)	3.7
8.	Royal Dutch Shell (United Kingdom/ Netherlands)	3.7
9.	Petróleos Mexicanos (Mexico)	3.6
10.	Kuwait Petroleum (Kuwait)	3.4
11.	Chevron (United States)	3.3
12.	Abu Dhabi National Oil (United Arab Emirates)	3.1
13.	Total (France)	2.5
14.	Petrobras (Brazil)	2.4
15.	Qatar Petroleum (Qatar)	2.4
16.	Lukoil (Russia)	2.3
17.	Sonatrach (Algiers)	2.2
18.	Iraq Ministry of Oil (Iraq)	2
19.	PDVSA (Venezuela)	2
20.	ConocoPhillips (United States)	2
21.	Statoil (Norway)	2

Sources: Helman and Christopher, 2015.

costs should be balanced against the environmental benefit gained. Regulations resulting in costs greater than the environmental benefit gained should be retired.

> President Obama and his administration are embroiled in an all-out war on fossil fuels. Under the semblance of reducing global warming, the administration has issued rampant regulations trying to further reduce CO_2 emissions under the so-called Clean Power Plan. Not only have these regulations been shown as ineffective at actually impacting global temperatures or sea levels, it turns out that the rules will have serious

economic impacts on all Americans, especially low-income and minority families (Inhofe, 2015).

A study by the American Coalition for Clean Coal Electricity (ACCCE) found that the President's climate agenda would only reduce CO_2 concentration by less than one-half of a percent; reduce the average global temperature by less than 2/100th of a degree; and reduce the rise of sea levels by 1/100th of an inch – or the thickness of three sheets of paper. These paltry numbers make the president's agenda seem downright reckless in light of the $479 billion price tag that it comes with (Inhofe, 2015).

Under the Clean Power Plan, electricity prices will double in most states in the next 10 years, which will result in the loss of tens of thousands of well-paying jobs, which will move to overseas markets with fewer environmental restrictions, such as China. Expert testimony submitted to the Senate Environment and Public Works Committee, documents that approximately 59 million American households earn $50,000 or less ($22,732 after taxes, being less than $1,900 a month). Families in that earning bracket spend 17% of after-tax income on energy costs. If energy prices double, they would be spending 34% of their after-tax income on energy costs (Inhofe, 2015).

In an Environment and Public Works Committee hearing held on June 23, 2015, Harry Alford, president of the National Black Chamber of Commerce, testified that the administration's Clean Power Plan would, by the year 2035, increase poverty by 23% and result in 7 million jobs lost in the Black community and increase poverty by 26% and result in 12 million jobs lost in the Hispanic community. The Clean Power Plan mandates that Americans substitute affordable fossil fuels with high-cost wind energy and solar energy so that renewables generate 28% of electricity production by 2030. Currently, wind and solar energy constitute less than 5% of our electricity, after decades of effort to achieve that level (Inhofe, 2015).

The cost of doing business is a significant factor to the bulk of O&G operators active in the US (80% of producers are small companies, frequently employing fewer than 10 people). "Numbering in the thousands, these smaller companies usually operate the most marginal wells, and thus, are very sensitive to price and operating cost changes. A large number of proposed environmental regulations are in force or are under consideration and could affect the economic viability of many domestic operators" (American Petroleum Institute, 2010).

The data evidence that current US national energy policy is not effectively promoting or benefiting the domestic production of O&G, at a cost to the nation, to the O&G industry (which is a major driver of a healthy economy) and to lower income citizens, who will be required to spend an increasingly larger percentage of their income on energy costs.

References

American Petroleum Institute, 2010. Environmental Regulation of the Exploration and Production Industry. Available from: www.api.org/environment-health-and-safety/environmental-performance/environmental-stewardship/environmental-regulation-exploration-production-industry

Bremmer, I., 2010. The Long Shadow of the Visible Hand – Government-owned firms control most of the world's oil reserves. Why the power of the state is back. Wall Street J. Available from: http://www.wsj.com/articles/SB10001424052748704852004575258541875590852 (accessed 22.05.2010.).

Bronstein, S., Griffin, D., 2014. Self-funded and deep-rooted: How ISIS makes its millions. CNN. Available from: www.cnn.com/2014/10/06/world/meast/isis-funding

FracFocus: Hydraulic Fracturing Chemical Disclosure State-by State, 2015. Available from: fracfocus.org/welcome

Groundworks: Hydraulic Fracturing Regulations, 2015. Available from: groundwork.iogcc.ok.gov/topics-index/hydraulic-fracturing/hydraulic-fracturing-regulations

Helman, C., 2015. The World's Biggest Oil and Gas Companies–2015, Forbes. Available from: www.forbes.com/sites/christopherhelman/2015/03/19/the-worlds-biggest-oil-and-gas-companies (accessed 19.03.2015.).

Inhofe, J. Senator (2015). The Obama Administration's War on Fossil Fuels. Human Events. Available from: humanevents.com/2015/09/14/the-obama-administrations-war-on-fossil-fuels (accessed 14.09.2015.).

Joy, M.P., Thomas, E., Leggette, L.P., 2015. Cuomo Decision on HF Doesn't Appear to be Based in Science – or the Law. Energy in Depth. Available from: energyindepth.org/marcellus/cuomo-decision-on-hf-doesnt-appear-to-be-based-in-science-or-the-law

U.S. Energy Consumption by Energy Source, 2014. Available from: www.eia.gov/Energyexplained/index.cfm?page=us_energy_home

U.S. Environmental Protection Agency, 2015. Assessment of the Potential Impacts of Hydraulic Fracturing for Oil and Gas on Drinking Water Resources, Executive Summary. Available from: www2.epa.gov/sites/production/files/2015-06/documents/hf_es_erd_jun2015.pdf

CHAPTER 13

An International Perspective of Challenges and Constraints in Shale Gas Extraction

Katharine Blythe*, Robert Jeffries, Mark Travers†**
*Ramboll Environ UK Ltd, Edinburgh, United Kingdom;
**Ramboll Environ UK Ltd, London, United Kingdom; †Ramboll, Copenhagen, Denmark

Chapter Outline

INTRODUCTION

The chapter examines the global situation in terms of onshore shale gas extraction, focusing on non-United States reserves.[1] In doing so, we consider the projected reserves of shale gas resources globally, the current picture in terms of

[1] "Resources" refers to an estimate of the amounts of oil or gas believed to be physically contained in the geological formation. "Reserves" refers to an estimate of the amount of oil or gas that can technically and economically be expected to be produced from a geological formation.

Environmental and Health Issues in Unconventional Oil and Gashttp://dx.doi.org/10.1016/B978-0-12-804111-6.00014-5

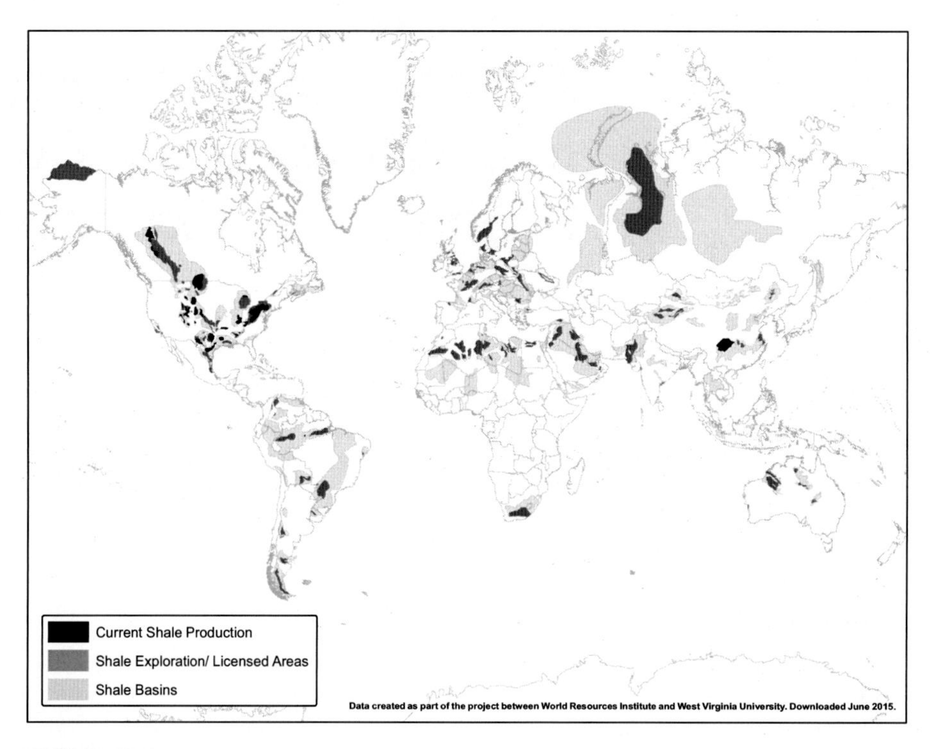

FIGURE 13.1
Map of global shale basins.

exploration and production in countries with sizable reserves, and, critically, the challenges and constraints affecting the extraction of shale gas.

For the purposes of this chapter we use the term "constraint" to mean a fixed environmental or other factor, which limits the potential for shale gas exploration to be successful and the term "challenge" to signify a more resolvable factor, generally more sociopolitical in nature, which could be removed through investment or political will. The absence of a particular constraint or challenge is considered to result in a "success factor."

GLOBAL BACKGROUND

The geographic extent of shale basins globally (for which estimates of natural gas in place have been provided and documented) is shown in Figure 13.1 (also see fig 1.4 in Chapter 1).[2] Although shale basins are present in many countries,

[2] Adapted using dataset from US Department of Energy & EDX (2015) and updated in line with references in Section 3: Current Status of Regional (and Country) Shale Gas Activity. Dataset consists of GIS data of major shale basins and plays in which unconventional production of gas and liquid hydrocarbon is occurring or has the potential to occur. The data collection procedure involved georeferencing, collecting, and attributing basins and plays present in publications by the Energy Information Administration, *Journal of Petroleum Technology*, China University of Geosciences, State geological surveys, *Oil and Gas Journal*, and other academic, private, and governmental sources.

Table 13.1 Countries Ranked by Estimated Shale Gas Reserves (wet and Dry Gas) (From EIA, 2013), indicating Current State of Shale Gas Activity

Country	Estimated TRR (TCF)[a]	Current Status[b]	Country	Estimated TRR (tcf)	Current Status[b]
China	1115	Production	Colombia	55	Exploration
Argentina	802	Production	Romania	51	On hold
Algeria	707	On hold	Chile	48	Exploration
Canada	573	Production	Indonesia	46	Exploration
United States	567	Production	Bolivia	36	Exploration
Mexico	545	Exploration	Denmark	32	On hold
Australia	437	Production (limited)	Netherlands	26	Moratorium
South Africa	390	On hold	United Kingdom	26	On hold
Russia	287	On hold	Turkey	24	Exploration
Brazil	245	Exploration	Tunisia	23	Exploration
Venezuela	167	Exploration	Bulgaria	17	Moratorium
Poland	148	Exploration	Germany	17	Moratorium
France	137	Ban	Morocco	12	Exploration
Ukraine	128	Exploration	Sweden	10	On hold
Libya	122	On hold	Spain	8	Exploration
Pakistan	105	On hold	Jordan	7	Exploration
Egypt	100	Exploration	Thailand	5	On hold
India	96	Exploration	Mongolia	4	Exploration
Paraguay	75	On hold	Uruguay	2	Exploration

[a]TRR: estimated (unproved) wet shale gas total recoverable reserves.
[b]Current status: "Production" – commercial production of shale gas either for electricity or transmitted gas use; "Exploration" – ongoing pilot tests or exploration drilling, including test extraction of gas; "On hold" – no official countrywide ban or moratorium in place or no or very limited active exploration (e.g., as a result of reassessment of geology, environmental regulations, sociopolitical opinion, or economics); "Moratorium" or "Ban" implies there is a political or legal instrument in place at the countrywide level to prevent activities associated with shale gas extraction on a temporary or permanent basis.

it can be seen that areas currently producing shale gas on a commercial scale (shaded in black) are confined to North America (United States [US] and Canada), with smaller areas in Argentina and China. Areas that have been either subject to, or licensed for, exploratory works are shaded in dark gray.

A listing of the countries ranked by size of shale gas reserves is provided in Table 13.1. This uses data from the US Energy Information Administration (EIA) (EIA, 2013). EIA estimates are based on (unproved) shale gas total recoverable reserves (TRR) and provide a reference point for global comparison. Certain shale basins were excluded from the EIA's assessment, as insufficient data were available to make an estimate of shale gas resources. These basins (primarily in Russia, Australasia, the Middle East, and to a lesser extent Indo-China, North Africa, and

Brazil) may have potentially productive shale, but no information gained from exploratory or desk-based studies was available.

In addition to the EIA data, Table 13.1 includes a brief summary of the current status of shale gas activity in each country gathered from recent publicly available briefings and media reports (as available in August 2015). It is acknowledged that these data are very time specific, and the situation in many countries is continually being reassessed and subject to change.

Overall, the EIA (2013) report estimated there are over 7000 trillion cubic ft. (tcf) of technically recoverable shale gas resources globally (including wet gas which may be produced alongside oil). Significant uncertainties are noted, due to relatively sparse geological data in many countries. More detailed country-specific shale gas resource assessments are proposed in future.

REGIONAL SUMMARY OF SHALE GAS ACTIVITY

This section discusses the current status of shale gas activity with reference to the countries listed in Table 13.1 on the following regional basis:

- Europe (Bulgaria, Denmark, France, Germany, Netherlands, Poland, Romania, Spain, Sweden, the United Kingdom);
- Russia and Ukraine;
- North America (Canada, Mexico, United States[3]);
- Asia and Pacific (Australia, China, Indonesia, Mongolia, Thailand);
- South Asia (India, Pakistan);
- Middle East and North Africa (Algeria, Egypt, Jordan, Libya, Morocco, Tunisia, Turkey);
- Sub-Saharan Africa (South Africa); and
- South America (Argentina, Bolivia, Brazil, Chile, Colombia, Paraguay, Uruguay, Venezuela).

Europe

Studies across Europe suggest there is potential for substantial amounts of extractable shale gas and oil (BGS, 2012; Natural Gas Europe, 2015; Perkins, 2015a; Vetter, 2015). Europe has a strong history in conventional gas production, with an associated good geological understanding, robust regulatory regimes, good infrastructure to transport gas to demand centers, and technical experience. Hydraulic fracturing has an established history of activity in Europe to improve productivity of conventional wells both onshore and offshore. For example, Germany has fractured 300 conventional wells since the 1950s (Vetter, 2015); the Netherlands over 200 (Van Leeuwen, 2015); and the United Kingdom (UK) has fractured approximately 200 onshore wells, of the 2000 drilled to date (Royal Society & The Royal Academy of Engineering, 2012).

[3] Although the United States is outside the scope of this chapter, reference to production levels have been included for comparative purposes.

Declining conventional reserves have meant European countries are now generally dependent on imports of gas. Germany, for example, currently imports around 90% of its gas requirements, which have increased following closure of its nuclear plants in 2011 (Vetter, 2015). Poland is highly dependent on gas from Russia and coal, which currently makes up approximately 56% of Poland's energy mix (PWC, 2014), and the United Kingdom (UK) imports approximately half its gas through European pipelines and as liquefied natural gas (LNG) from Qatar and the Middle East (May et al., 2015). Key exceptions are Denmark and the Netherlands, which still have significant conventional gas production, strong remaining reserves, and are also investing heavily in renewable energy (Bird and Bird, 2015b; Perkins, 2015b; Vinson and Elkins, 2015).

There is therefore potential for European shale gas to increase energy independence, reduce carbon emissions, and reduce gas prices, as well as providing job opportunities and tax benefits, particularly in areas with high unemployment and depressed economy (e.g., Eastern Europe and northern United Kingdom) (May et al., 2015) (see Chapter 2). European oil and gas (O&G) mineral rights are held by the State, so landowners have no rights to gas below their landholding; however, incentives are being introduced to encourage exploration. For example, the UK has reduced taxation on shale gas, and the UK and Spain have introduced benefits for local communities, including a tax of up to 5% of well profits to landowners and communities in Spain, and in the UK a one-off community payment of £100,000 (US$153,000) per fractured well (Shale Gas Europe, 2015b; May et al., 2015).

Extraction of shale gas in Europe is generally controlled by mining or hydrocarbon regulations (including European legislation offering a certain degree of consistency across countries). This typically involves several stages before a company is allowed to drill, with requirements for permits for land use change (planning) including environmental impact assessments as best practice, well construction and maintenance, establishment of surface and underground facilities, and environmental protection (including protection of land, surface water, groundwater, and air quality) (Shale Gas Europe, 2015b,d). Certain countries are also developing separate, shale-specific legislation, such as the UK Infrastructure Act (2015), which prohibits hydraulic fracturing at a depth of less than 1000 m and in certain "protected areas" (including groundwater source areas and national parks), requires monitoring of baseline methane levels in groundwater and air before hydraulic fracturing begins, and requires disclosure of chemicals in fracturing fluids (see Chapter 8). Such measures had previously been considered through the permitting regime rather than within legislation (May et al., 2015). Other countries, particularly in Eastern Europe have a more bureaucratic system than Western European countries, with slower moving regulations; for example, delays in establishing a robust and streamlined regulatory system and taxation framework, and volatile legal procedure in Poland and Romania have led to withdrawal by some investors (Mihalache, 2015; Shale Gas Europe, 2015a,d).

Despite potentially favorable geology, regulation, and political will, little exploration of shale gas has been undertaken across Europe in the last 5–10 years in contrast to the situation in the US. Poland, the most advanced country, has drilled over 60 shale wells since 2010, 25 of which have been hydraulically fractured (PWC, 2014). France hydraulically fractured 20 shale wells up until 2011 (Martor, 2015) and Germany also carried out test drilling and hydraulic fracturing in shale in 2008 (Vetter, 2015; Natural Gas Europe, 2015). In the UK, seven test wells have been drilled since 2010, only one of which was fractured in 2011 near Blackpool in northern England. This resulted in two earth tremors (as discussed in Chapter 11) and a temporary moratorium was put in place (lifted in 2012) while research into the cause of these tremors was undertaken (Green et al., 2012). For some countries such as Poland, Romania and Sweden, where exploratory drilling has been undertaken, there have been disappointing preliminary results and drilling has been uneconomic, given the relatively small reserves identified (see further discussion in Chapter 2) (Bird and Bird, 2015a; Shale Gas Europe, 2015a,d; Mihalache, 2015).

However, across Europe there has been substantial political and public objection over the hydraulic fracturing process. Concerns have focused on possible environmental impacts of the process such as contamination of groundwater, earthquakes, elevated greenhouse gas emissions, water consumption, and risks due to improper disposal of flowback water. There are also queries over the transparency and openness of companies, and whether existing regulatory frameworks are appropriate. These concerns are generally based on perceived problems with the process rather than reported incidents in the countries concerned (Martor, 2015; Vetter, 2015) (see discussion in Chapter 8).

As a result, several countries have introduced moratoria and bans, subject to further research, including Germany (since 2011), Belgium (since 2012), the Netherlands (since 2013 and extended for 5 years in 2015), Scotland and Wales (since 2015), and France (since 2011) (Perkins, 2015a; Vetter, 2015; Government of the Netherlands, 2015; Deans, 2015; Martor, 2015). These have been introduced contrary to expert advice that although health, safety, and environmental risks exist, they can generally be managed effectively by implementing best operational practice, potentially amending existing legislation, and increasing public participation in the process (Royal Society & Royal Academy of Engineering, 2012; Mackay and Stone, 2013; May et al., 2015; Martor, 2015; Vetter 2015). Even countries without shale gas bans have, to date, been unable to establish a diverse exploration program, largely due to public opposition. For example, in the UK, planning applications to drill and fracture eight shale wells over two sites were refused by the local council in June 2015 on the grounds of "industrialization of the countryside," and noise and traffic concerns (Beattie, 2015). Other applications for test wells and coring wells (for shale and coal-bed methane (CBM), and in some cases conventional onshore drilling) in the UK have led to widespread public opposition, including rejection of small-scale applications for temporary works (Beattie, 2015; Deans, 2015).

Several countries with moratoria have stated that they are still considering shale gas as a viable part of the future energy mix, and will permit hydraulic fracturing for shale gas for research in the future (AFP, 2015; Government of the

Netherlands, 2015; Martor, 2015; Perkins, 2015a). Delays in exploration and collection of geological data have meant knowledge of shale reserves in many countries is limited (Government of the Netherlands, 2015; Shale Gas Europe, 2015b), and it is unclear whether shales will be productive or economic. The issue of gaining social acceptance in Europe is likely to be the most challenging to address before further exploration to address these unknowns can commence.

Russia and Ukraine

Although Russia and Ukraine are geographically adjacent, their reasons for developing shale are distinct. Russia has the world's largest conventional natural gas resource – so has limited necessity to exploit its unconventional reserves. Only 3% of Russia's unconventional reserves comprise shale gas, and it is largely in remote, sparsely populated regions of northeastern Siberia, the Ural Mountains, and Arctic limiting accessibility. European and North American investment and transfer of expertise are prevented due to political sanctions in place. Therefore, Russian companies are strengthening ties with Asian economies, such as India, China, Indonesia, Vietnam, and South Korea, and investing in exploration elsewhere, such as Argentina. Very limited shale gas exploration has taken place in Russia itself (Nowak and Boczek, 2015; Shale Gas International, 2015c,d).

Ukraine, however, is more advanced in exploration, and wishes to reduce its dependence on Russian gas by diversifying supply. Political instability in recent years has highlighted the importance of energy security; for example, Russia cut off its supply of gas to Ukraine in January 2006 and January 2009, also impacting supply to the rest of Europe. Investment in Ukrainian gas is still being undertaken by the US and European companies (Shale Gas Europe, 2015c). However, both Chevron and Shell recently withdrew from the country due to *force majeure*, understood to be a combination of regional political instability and exploratory works not confirming significant presence of gas. Works on exploration, evaluation, and eventual extraction of shale gas in Ukraine are continuing through the state-owned company in cooperation with Poland (Shale Gas International, 2015c).

North America

Evidently, production in North America is dominated by that from the US, primarily from wells in Arkansas, Colorado, Louisiana, North Dakota, Pennsylvania, and Texas (Tran, 2014) (see eg., Chapter 1). Although certain cities and states have moratoria or bans in place, as a political response to local regulations and perceived shortcomings, in general, the US is progressing with exploration and production. This has affected the export relationships among the neighboring countries of Canada and Mexico. Both historically exported gas to the US; however, as a result of increased gas production from the US and declining costs, these two countries have had to modify their export market.

Approximately 5% of Canada's natural gas production is as shale gas from formations in British Columbia, Northwest Territories, and Alberta. Canada has a well-developed scheme to encourage foreign investment (EIA, 2013). Taxes from gas extraction are used to fund health, education, and infrastructure. However, the remote locations of the productive basins restrict transport of the gas. Therefore, currently it is largely fed into small-scale LNG facilities, for use as a transport fuel and as a fuel for offgrid properties and businesses. Large-scale LNG projects are also being developed on Canada's east and west coasts, though this has its challenges including transportation distances, construction on wild land, the availability of skilled resources, and complex regulatory process (EIA, 2013; Gomes, 2015) (see eg., Chapter 9).

Mexico is currently importing gas from the US through long-distance pipelines, and its conventional reserves are declining. Mexico's location adjacent to the shale gas and oil–producing region of Texas (Eagle Ford) suggests large potential for the country to also develop the same shale basin (called the Burgos Basin in Mexico), located in Mexico's industrial northeast. Approximately 20 exploratory wells have been drilled since 2014, largely by the state oil company Pemex (Burnett, 2015; Collins, 2015; Fisher, 2015; Seelke et al., 2015). Further investment is sought to cover the cost of commercial shale gas wells, and foreign investors have been allowed into the energy sector since 2013, with a reduction in the taxation regime, and increased autonomy and competition (Collins, 2015).

Constraints to development in Mexico include water scarcity, especially in northern Mexico, which is very arid. In addition, challenges in the region include security problems, pipelines have been tapped and tankers hijacked by "fuel pirates" leading to environmental problems such as oil spills, as well as a reluctance for skilled staff to work in the country. Local infrastructure to gather and send the gas and oil to market is also lacking, as are roads, rail, and housing in shale-rich areas (Burnett, 2015; Fisher, 2015).

Asia and Pacific

This is a diverse region comprising Australia, China, Indonesia, Mongolia, and Thailand, with very different experiences in hydrocarbon extraction, regulatory regimes, and economic systems. Countries here, as globally, are keen to develop shale gas reserves to increase energy independence, increase export potential, and, especially in the case of China, to reduce carbon and other pollutant emissions in energy production (Shale World, 2013).

Exploration is progressing with commercial production in Australia and China. In Australia, the state of Western Australia holds approximately two thirds of Australia's identified reserves of shale gas – the Cooper Basin has had commercial production since 2012 with associated infrastructure, pipelines, and treatment stations, and the Perth Basin is also being explored due to its proximity to the Perth City region (Allnutt and Yoon, 2015). Since 2013, China's national companies PetroChina and Sinopec have drilled over 200 exploration wells across the Sichuan Basin, in south-central China, with 74 of them taken into production by the middle of 2015. It has been estimated that production could be up to 219 billion cubic ft. (bcf) annually by the end of 2015, over 30 times production in 2013, though it is anticipated that the majority of gas will still be imported from Russia over the next 30 years (Economist, 2014; Aloulou, 2015).

Indonesia has potentially productive shales around the major islands of Sumatra, Borneo, and Papua, which are similar in geology to those in the US. However, exploration has been limited, with Indonesian investment being primarily in shale basins in the US, Canada, and South America. Thailand is also investing in the US and Canadian shale development, rather than developing domestically, and the military government, which took power in 2014 is slow to encourage further investment (Razavi and Hidayanto, 2013; Riaz, 2013; Jaipuriyar, 2015). Limited exploration is also ongoing in Mongolia, in association with shale oil exploration (Shale World, 2013).

The incentives for international companies to invest in shale extraction vary across the region, with regulation being very different due to the differing political backgrounds of the countries. Regulation in Australia is controlled at the national and state level – the Australian government has established a broad economic policy, providing a regulatory framework for petroleum operations, and promotes collection of geoscientific information. States can introduce their own regulations and variations. For example, Western Australia has recently introduced regulatory changes specifically directed at the shale gas industry, and is developing a more robust, enforceable, and transparent regulatory framework, whereas the state of Victoria introduced a moratorium on hydraulic fracturing and CBM in 2012 (Allnutt and Yoon, 2015). Indonesia also adopted a specific shale gas regulatory framework in 2012 based broadly on the O&G regulation for deep-water conventional reserves. This opened up exploration to international companies, in joint agreements with the state-owned Pertamina (Shale Gas International, 2015b). However, taxation, incentives, and the potential for sharing of infrastructure and marketing restrictions are deterring foreign investors, although knowledge and skills transfer is improving through Pertamina investing in foreign reserves (Razavi and Hidayanto, 2013). Similarly, most investment in China and Mongolia has been made by state-owned companies (Economist, 2014), though Mongolia in particular is hampered by the lack of any previous hydrocarbon extraction history and an associated lack of expertise, technology, and infrastructure (Shale World, 2013).

The key constraint across the region is the remote locations of the productive shale basins, which are generally far from demand centers, leading to challenges of developing infrastructure, and associated increased costs in drilling and production (including development of LNG facilities) (Economist, 2014; Jaipuriyar, 2015). Water resources are also a constraint across the region – including in Western Australia, China, and Mongolia (Shale World, 2013; Economist, 2014; Allnutt and Yoon, 2015).

Geological constraints such as the depth and quality of the shales are also a consideration. Shales are comparatively deep in the region, and there is a lack of equipment such as rigs to drill to the required depths and, as a consequence, wells are expensive (Allnutt and Yoon, 2015). However, at present, imported gas (LNG) costs in Indonesia, for example, are approximately four times those in the US, so there may still be potential to reduce overall gas cost (Razavi and Hidayanto, 2013). In China, shales are also less brittle, and not as easily fractured as in the US. Some Chinese and Mongolian shale regions are also in seismically active regions (Shale World, 2013).

South Asia

India and Pakistan both produce some gas from conventional sources but import around half (Batra, 2013). Both countries aim to become more energy independent through the use of shale reserves. India also hopes to cut petrol and liquefied petroleum gas consumption through the use of compressed natural gas and piped natural gas (EMIS, 2015).

Exploration in India is concentrated around six basins in the northwest of the country. A shale gas exploration policy was approved in September 2013, following public consultation, allowing international competitive bidding for blocks and developing a fiscal regime on royalty and production-linked payments increasing the state-controlled cap on gas prices (EMIS, 2015). Exploration drilling took place in 2012 and 2013 in Gujarat and West Bengal by the national companies ONGC and OIL, and further investment has been announced. India is also focusing on exploitation of reserves in other countries – providing experience of the technologies being used.

However, foreign investment is not encouraged, and low gas prices and high exploration and production costs further discourage investment. Although mineral rights are held with the government, shale exploration companies or state authorities have to buy the land needed for the projects, in a lengthy and slow-moving process. Shale deposits also largely occur in populated areas leading to concerns over earthquakes and the need to displace large masses of people.

Exploration in Pakistan is less well advanced, as governmental policy appears to favor importing piped gas, or LNG from Qatar, which are seen to be cheaper options (Bhutta, 2015). Although Pakistan has a good geological understanding of shale reserves there is limited political will to develop the resource (Abbasi, 2015). There is no formal shale gas policy and pressure from environmentalist groups appears to have prevented the move toward exploration. This is despite the potential for the industry to reduce unemployment, which is a major issue in rural areas, where shale would be developed (Abbasi, 2015).

Both India and Pakistan suffer from physical and economic water scarcity. Potential shale gas bearing areas are also areas that could experience severe water stress by 2030 (Batra, 2013). Existing water supply constraints mean that not all possible shale gas deposits can be explored, limiting the potential benefit that gas extraction has to offer (EMIS, 2015). Although environmental regulation in India recognizes the importance of treatment of water used in the extraction process before discharge, enforcement of environmental legislation can be problematic, and sporadic, and wastewater disposal in particular is often not adequately enforced (Batra, 2013).

Middle East and North Africa

Middle East and North African countries historically have experienced large amounts of O&G production, though some, such as Jordan, Morocco, and Turkey, generally import hydrocarbons from other countries in the region. However, as conventional reserves are declining, in part as a result of revolution and political unrest in the region, shale gas extraction is receiving increasing consideration (along with other options) to help maintain domestic supplies and export contracts, as well as the move to gas electricity production from coal and diesel to reduce carbon emissions (GlobalSecurity.org, 2014; Rebhi, 2014; Nakhle, 2015; O'Neill et al., 2015; Santos, 2015; Vladimirov, 2015). Importing countries are also keen to move toward greater energy dependence and reduce reliance on countries such as Russia and Iran. In Libya, a rapid deployment of shale wells is seen as a potential benefit by reducing unemployment in the country (Yee, 2015).

Various changes to regulatory regimes have been made. For example, in Algeria a hydrocarbon law was introduced in 2005 and subsequently amended to reduce the monopoly of Sonatrach, Algeria's national oil company, and provide tax incentives related to shale gas but still restrict imports and curbs on foreign investment. In Libya there has also been a move to allow foreign companies higher stakes in shale projects owned by the state oil company NOC, and the 1955 Petroleum Law is being amended to provide more clarity for foreign investors, though this is being hampered by lack of a national constitution (Yee, 2015).

Some agreements with international firms have been made to invest capital and expertise – such as in Algeria (Nakhle, 2015), along with three test wells in the Western Desert of Egypt (Chodkowski-Gyurics, 2014), four exploration wells in Tunisia (Rebhi, 2014), and exploration wells in Morocco's Lalla Mimouna Basin (Vladimirov, 2015). However, these agreements are limited. Jordan's reserves are primarily shale oil (USGS, 2014) – mostly in west-central Jordan where there is shale relatively near the surface and in the vicinity of infrastructure, and international companies from Estonia, Brazil, and the US are exploring these resources. Although some wells such as those drilled to date in Morocco have been promising, others, such as in Libya, for example, have had disappointing results (Yee, 2015).

In Algeria and Tunisia there has been much public protest over the decision by the government to invest in the process, primarily due to large water use, which was perceived as having the potential to decrease water supplies in regions that largely depend on agriculture (EJ Atlas, 2013; Nakhle, 2015). For example, more than 95% of Algeria's shale plays are covered by the Sahara Desert. Other countries in the region also suffer from water scarcity. In Tunisia there has also been concern that international companies were exploiting shale gas without proper permits and licenses (Rebhi, 2014).

Other challenges such as gaps in infrastructure, limited drilling rigs, slow government approval processes, difficulties attracting investment partners, and technical problems have discouraged investment in the region. In addition, there is perceived to be an unstable legislative and regulatory environment, protectionist policies, a tough tax regime, corruption, and high security risks. (Nakhle, 2015). Political instability leading to a reluctance for international companies to invest is a feature across the region (Goddard, 2013; O'Neill et al., 2015). In Libya, for example, there is still a state of civil unrest, with sporadic oil worker strikes and port seizures discouraging foreign investors (NARCO, 2015). Such risks are a discouragement for foreign investors – especially when coupled with low profits as a result of restrictive policies (including subsidized domestic gas prices), and disappointing preliminary results.

Sub-Saharan Africa

Only South Africa is understood to hold substantial shale reserves in sub-Saharan Africa (Figure 13.1). South Africa's government is promoting shale gas exploration to increase energy independence. Shale gas reserves are primarily located in the Karoo Basin in the Western Cape. In April 2011, a shale gas exploration moratorium was imposed as a result of environmental groups' concern over potential impact on the Karoo's biodiversity and groundwater resources, leading to pressurizing the South African government to refuse any exploration permits (Pinsent Masons, 2015). This moratorium was lifted in 2012, on expert advice that exploration took place under strict supervision by a monitoring committee, and could be halted if unacceptable issues arose. Geological field mapping and other data-gathering activities were recommended under the existing regulatory framework, while this was amended to allow for increased supervisory provisions. However, delays in finalizing this legislation have created uncertainty in the process, which has led to recent withdrawals from the country by international companies, such as the withdrawal of Shell from the Karoo Basin in March 2015 and Anadarko Petroleum in 2014 (Prinsloo, 2015).

South America

Many South American countries have a history of conventional gas production, though this is now declining – for example, Argentina (Platts, 2015), Bolivia (Hill, 2015; Shale Gas International, 2015a), and Colombia (Mares, 2013; Naziri, 2014). These countries wish to regain their export capacity and domestic supply, as well as develop domestic industry. Other countries, which are net importers, such as Brazil, Paraguay, and Venezuela, are aiming to reduce domestic gas costs and increase the supply for residential and commercial use (Mares, 2013; Cunha, 2014; Regester Larkin, 2014). Brazil is also interested in using revenue from shale extraction to invest in other areas such as "subsalt" offshore O&G fields (Platts, 2015). Venezuela is keen to develop natural gas production for industrial use and consequently is developing its gas infrastructure in support of this effort.

A high level of shale gas activity has been progressing in Argentina; for example, more than 275 wells since 2013 in the Vaca Muerta shale formation, located in west-central Argentina, producing approximately 25 bcf per annum of shale gas. The key Argentine operator is the state company YPF, which has entered into joint venture agreements with international companies. Colombia has also progressed with exploration in the La Luna formation (Naziri, 2014), with approximately 10 wells drilled in the country in 2014 (Arthur, 2014). Other countries such as Bolivia, Brazil, Chile (which shares the Argentinian shale basin), and Venezuela are less well advanced, but entering into agreements for exploration wells (Shale Gas International, 2014; Arthur, 2014). Uruguay shale contains primarily oil rather than gas, and exploration in 2013 focused on this aspect, resulting in potentially productive oil wells with associated gas (Proactive Investors, 2013).

Reforms in hydrocarbon legislative regimes in South America are increasing investments from international companies. These changes aim to reduce economic uncertainties, and restrictive policies, such as restrictions on sending profits out of the country and price controls, providing tax breaks, energy price rises, and other incentives (Naziri, 2014; Platts, 2015; Shale Gas International, 2015a). Countries are also developing regulations specifically for unconventional gas where exploration has primarily been carried out under broader hydrocarbon regulations and, in countries such as Brazil, where environmental regulation is currently not well defined (Regester Larkin, 2014), to help encourage international investment needed to provide the finance and expertise to develop shale gas in the region (Cunha, 2014; ITE Oil & Gas, 2015). Some countries such as Colombia have significant numbers of international companies working in conventional reserves (Naziri, 2014), though the falling oil price has made many of these divert investment to other countries such as Mexico where the geology and regulatory framework is perceived as more inviting (Platts, 2015). International investment in countries with little baseline knowledge and estimated small reserves of shale gas – such as Paraguay and Uruguay – is unlikely (Mares, 2013).

Public concern is high in Bolivia with concerns over water and chemical use, health impacts, and increasing carbon emissions. This is especially the case where remote regions, such as Bolivia's Chaco region, are considered environmentally vulnerable and are home to indigenous people (Hill, 2015); whereas in neighboring Brazil there is more public acceptance due to perceived economic benefits.

Further constraints include the remoteness of many basins and associated challenges including a lack of infrastructure including pipelines and drill rigs – especially in Brazil and Chile. Additional challenges include a lack of geological and technical knowledge and high operational costs with wells costing up to five times the equivalent US cost (Arthur, 2014). Some of these challenges are being addressed by cooperation with other countries (e.g., Bolivia is working with Argentina to train their workforces in extraction techniques – Hill, 2015 – and Argentina is developing resources in China and Russia – Aloulou, 2015). Challenges remain with corruption and security risk in the O&G industry, which is affecting the profitability of the industry at present, such as in Brazil (Platts, 2015) and Colombia, where over 250 pipeline bombings took place in 2013 (Naziri, 2014).

GLOBAL CHALLENGES AND CONSTRAINTS TO SHALE GAS PRODUCTION

A review of the current status of shale gas activity worldwide has identified a number of challenges and constraints to successful shale gas extraction. A summary of these is shown in Table 13.2. This table presents a "traffic light" system: white cells signify there is a success factor – or limited constraint or challenge; tan-shaded cells signify some moderate constraint or challenge (either varying in intensity around the country, or of moderate intensity overall); and red-shaded cells signify a significant constraint or challenge covering much of the country. Cells with a question mark indicate there is insufficient data available at present to establish whether a constraint is likely to exist.

The data used to compile this table were adapted from the references cited in Section "Regional Summary of Shale Gas Activity" of this chapter, and from global mapping of water stress and global seismic hazard (World Resources Institute, 2014; GSHAP, 1999). This is a qualitative matrix allowing approximate comparison of the relative status between countries.

Constraints

Constraints refer to environmental and other factors, which are part of the nature of the country and shale gas resource in question, which cannot be easily modified through investment or changes in the political or legislative regime. These include factors such as the nature of the shale and geology, its location, water availability in the area where shale is in place, and factors such as a country's constitution and past history in hydrocarbon extraction. Although constraints can be mitigated against, they cannot generally be removed altogether. If a constraint is absent (e.g., if there are no water restrictions in a country), it is considered to be a "success factor" (i.e., there is a potential supply of water for the hydraulic fracturing process). Mitigation against this constraint could include importing water from elsewhere, or developing techniques to minimize water use – but it would not remove the constraint.

GEOLOGY AND LOCATION

The geological characteristics of different shale deposits can vary significantly, in particular depth, temperature, and nature of the organic material. The shales also need to be "brittle" (e.g., with a low percentage of clays) in order for the current extraction technology of hydraulic fracturing to be successful. The nature of complex or unfavorable shale geology will affect the technical recoverability

Table 13.2 Summary of Challenges and Constraints to Shale Gas Production Globally

		Europe										Former Soviet Union		North Ameica			Asia & Pacific				
	Country	Bulgaria	Denmark	France	Germany	Netherlands	Poland	Romania	Spain	Sweden	UK	Russia	Ukraine	Canada	Mexico	USA	Australia	China	Indonesia	Mongolia	Thailand
	Current Status	M	O	B	M	M	E	O	E	O	O	O	E	P	E	P	E/P	P	E	E	O
Constraints	Water availability																				
	Deep seams / shale basins																				
	Poor initial results	?		?	?	?			?	?	?									?	?
	Seismically active																				
	Remote locations																				
	Private / govt ownership of resources																				
	Surplus existing conventional energy supplies																				
Challenges	Limited geology data																				
	Expensive wells																				
	Limited existing infrastructure																				
	Lack of onshore rigs & equipment																				
	Heightened public objection																				
	Uncertain/ lacking environmental regulations																				
	Political instability																				
	Complex regulatory process/ restrictive govt policies																				
	Security risks & corruption																				

		South Asia		Middle East & North Africa							Sub-Saharan Africa	South America							
	Country	Pakistan	India	Algeria	Egypt	Jordan	Libya	Morocco	Tunisia	Turkey	South Africa	Argentina	Bolivia	Brazil	Chile	Colombia	Paraguay	Uruguay	Venezuela
	Current Status	O	E	O	E	E	O	E	E	E	O	P	E	E	E	E	O	E	E
Constraints	Water availability																		
	Deep seams / shale basins																		
	Poor initial results	?	?						?	?	?				?		?		?
	Seismically active																		
	Remote locations																		
	Private / govt ownership of resources																		
	Surplus existing conventional energy supplies																		
Challenges	Limited geology data																		
	Expensive wells																		
	Limited existing infrastructure																		
	Lack of onshore rigs & equipment																		
	Heightened public objection																		
	Uncertain/ lacking environmental regulations																		
	Political instability																		
	Complex regulatory Process/ restrictive Govt policies																		
	Security risks & corruption																		

Key

P	Production		Success factor or limited constraint/ challenge
E	Exploration		Some constraint/ challenge
O	On-hold		Key constraint/ challenge
M	Moratorium	?	Unknown/ not enough information
B	Ban		

of the shale gas (i.e., the volumes of natural gas that could be produced with current technology regardless of natural gas prices and production costs).

At present, there is substantial uncertainty on the geological characteristics of shale deposits in many countries. For example, although some central European shales were expected to be productive, test wells in Poland, Romania, and Denmark have been less productive than hoped, whereas similar wells in North and South America were more promising. Other countries such as China appear to have high-clay shales, which are not brittle, and therefore not as productive as their US equivalents. More exploration and research is required to establish the nature of the geology in different countries not only through both desk-based studies and seismic surveys but also through coring wells and pilot test wells. This research requires substantial investment before any revenues can be established. Investor uncertainty may prevent this research being carried out.

The potential for induced seismic activity will also affect shale gas extraction in affected areas such as China and Mongolia. As discussed in Chapter 11, induced seismicity could result from wastewater injection or the hydraulic fracturing process itself (see Chapter 5 on wastewater treatment). Although it is unlikely that any induced seismicity from hydraulic fracturing will cause property damage, it could create nuisance issues, and there is some risk of reduction in well integrity.

Locational considerations will also be important. Where shale reserves are located in remote areas there will be associated challenges in developing infrastructure to transport and export gas, with additional cost and environmental impact.

WATER RESOURCES

Significant volumes of water are required for the hydraulic fracturing of shale using current technologies, and this can be a limiting factor for progressing shale gas activity in many countries where water resources are scarce and in countries where the demands for water resources are highly competitive (e.g., agricultural use in North Africa and South Asia) (see eg., Chapter 4).

Information provided by the World Resources Institute (2014) and summarized in Table 13.3 indicates that 8 of the top 20 countries with the largest shale gas resources face arid conditions or high to extremely high baseline water stress where the shale resources are located.

Current research into reduced water requirements, onsite technologies for treatment, and reuse of flowback/wastewater, using nonwater alternatives such as propane gels is ongoing and will be important for offering alternative strategies for drilling and hydraulic fracturing in water-stressed environments (see eg., Chapter 5 for technologies).

HISTORICAL FACTORS

Although most constraints identified are environmental, historical and economic factors such as past experience in hydrocarbon extraction and the financial ownership of gas reserves can also influence the success of shale gas extraction.

Table 13.3 Average Exposure to Baseline Water Stress Across Shale Plays in 20 Countries With the Largest Technically Recoverable Shale Gas Resources

Rank[a]	EIA Estimated Shale Gas TRR (tcf)	Country	Average Exposure to Baseline Water Stress Over Shale Play Area[b]
1	1115	China	High
2	802	Argentina	Low to medium
3	707	Algeria	Arid and low-water use
4	573	Canada	Low to medium
5	567	United States	Medium to high
6	545	Mexico	High
7	437	Australia	Low
8	390	South Africa	High
9	287	Russia	Low
10	245	Brazil	Low
11	167	Venezuela	Low
12	148	Poland	Low to medium
13	137	France	Low to medium
14	128	Ukraine	Low to medium
15	122	Libya	Arid and low-water use
16	105	Pakistan	Extremely high
17	100	Egypt	Arid and low-water use
18	96	India	High
19	75	Paraguay	Medium to high
20	55	Colombia	Low

Adapted from World Resources Institute, 2014

[a]*Based on size of estimated shale gas TRR, taken from EIA (2013).*

[b]*Water Resource Institute's tool "Aqueduct Water Risk Atlas."*

A key difference between the US and many other countries is the private ownership of mineral reserves in the US, meaning royalties can be paid to landowners, avoiding objections on the grounds of nuisance and perceived imbalance of costs and benefits (EIA, 2013) (see discussion on incentives in Chapter 8). Also, the necessity for a new source of natural gas will depend on the current production of conventional reserves. Countries such as Russia and the Netherlands have significant reserves that meet or exceed current domestic demands. They are therefore less concerned with establishing a new energy supply to increase energy security, or increase the mix of gas in the domestic electricity supply, than net importers such as much of Europe, Asia, and North Africa.

Challenges

Challenges refer to factors, which could feasibly be overcome or substantially minimized through current knowledge, investment, and political will. For example, the challenge of an unclear regulatory system or unfavorable taxation regime could be removed by modification of those regimes. Similarly, a lack of infrastructure or technical knowledge could be overcome by investment in those areas. Again, absence or overcoming of a particular challenge indicates a "success factor." Evidently, some challenges such as regional political instability and security risks are less controllable by simple legislative or structural changes than others, but for the purposes of this chapter they are considered together.

Although there is much overlap in identified challenges, establishing a balance of favorable political, economic, regulatory, and environmental factors will be important for future shale gas development. The exact nature of this balance will vary between countries, depending on the sensitivity of each environmental setting. Indicative challenges and how particular countries are addressing these are listed later. Further details are outlined in Section "Regional Summary of Shale Gas Activity."

LIMITED GEOLOGICAL DATA

Geological survey data and data from test shale gas wells are lacking in many countries, which could hinder international investment. This is particularly a problem in Europe, where bans and moratoria prevent the gathering of test data, and in countries with remote shale basins and limited technical expertise such as China, Russia, North Africa, and some South American countries such as Paraguay and Brazil.

LIMITED EXISTING INFRASTRUCTURE

Where shale basins are located in areas far from demand centers, there could be additional requirements for infrastructure such as pipelines, roads, and water treatment facilities. Areas where this is a key challenge include western China, Australia, and Russia. This also leads to wells being expensive to drill in comparison with countries with this infrastructure already in place. Rig availability and expertise can also be a challenge limiting the potential for extraction, such as in Europe, South America, and some of North Africa without a history of hydrocarbon extraction.

ENVIRONMENTAL REGULATORY FRAMEWORK

A clear, robust and comprehensive environmental framework is important in attracting international investment; insuring external investors understand relevant environmental risks and liabilities. Factors to be considered include:

- Planning ahead of developments and evaluating possible cumulative effects before granting licenses;
- Carefully assessing environmental impacts and risks;
- Ensuring that the integrity of wells is up to best practice standards;

- Checking the quality of local water, air, and soil before operations start, in order to monitor any changes and deal with emerging risks;
- Controlling air emissions, including greenhouse gas emissions, by capturing the gases;
- Informing the public about chemicals used in individual wells; and
- Insuring that operators apply best practices throughout projects.

Unclear and evolving environmental legislation is a particular challenge across Africa, the Middle East, Asia, and South America, but also in Europe, where public expectations are high and some countries are still developing regulations specifically to cover unconventional hydrocarbon extraction.

HEIGHTENED PUBLIC OBJECTION

This is generally a challenge relating to concerns over environmental impacts, and can be addressed through a variety of engagement techniques. Public objection is an issue worldwide and is a key reason for delaying shale gas exploration in many countries in Europe and to a lesser extent in South Asia and North Africa. Such objections can lead to bans or suspensions of hydraulic fracturing, as can be seen in Poland, Romania, and South Africa. In particular, the means to avoid public objection include the establishment of a clear environmental regulatory system, as indicated earlier, and ensuring transparent engagement with affected communities throughout the process. Public opposition can be pronounced where hydraulic fracturing is proposed in both densely populated areas (such as northern England) and in more remote "pristine" environments, such as Poland and Bolivia (see eg., Chapter 8).

POLITICAL AND ECONOMIC INSTABILITY AND SECURITY RISKS

Countries with political instability present a risk for investors, making even potentially productive shale basins less favorable. Such challenges were apparent during the recent conflict between Ukraine and Russia, and ongoing internal conflicts in Libya and other North African countries. In addition, countries such as Mexico and Colombia experience high levels of violence and infrastructure sabotage.

GOVERNMENT SUPPORT

The level of government support and commitment for shale gas activity at the national level is important strategically and financially (e.g., in establishing favorable taxation regimes). There is currently political support in countries such as Canada, Spain, and England (UK), contrasted with moratoria and bans in others such as Bulgaria, France, the Netherlands, and Scotland and Wales (UK).

SUMMARY

The successful exploitation of shale gas reserves is dependent on a combination of factors, and with limited information available it is not possible to predict, which countries will be successful. In general, successful exploitation of a shale reserve is dependent on a balance of maximizing success factors and minimizing and controlling remaining challenges and constraints. However, in some cases, very few constraints or challenges can prevent exploration, whereas in other cases successful exploration or even production can proceed in a highly challenging and constrained environment.

Countries with current programs of commercial shale gas production (the US, Canada, Argentina, China, and to a lesser extent Australia) would be anticipated to have fewer constraints and challenges than those with less well-developed industries. However, although the US, Canada, and Argentina have comparatively many "success factors" (as shown in Table 13.2), China and Australia show more constraints and challenges, even than less successful countries. This is likely to reflect the highly variable nature of shale resources across countries such that production is taking place in a less constrained area. Conversely, countries such as the UK, Sweden, and Denmark, which theoretically have few environmental constraints and several "success factors," are "on-hold" with little if any exploration proceeding – due largely to public opposition to the process.

In Europe, the "socio-political" challenges are limiting further exploration and resulting in the countries being more "cautious" than would be expected. Exploration is progressing in Africa, the Middle East, South America, and Asia, in spite of a high number of constraints and challenges remaining. In general, these countries explore in many cases through state-funded companies, even though political challenges remain, which could deter international investors.

Further research and exploration is needed globally to establish reserves, and where and how those reserves can be extracted. However, the combination of success factors and few constraints and challenges present in North America, in particular, are unlikely to be replicated in other countries in the short term.

References

Abbasi, A.H., 2015. Shale gas: a real game-changer – The International News. http://www.thenews.com.pk/Todays-News-9-321573-Shale-gas-a-real-game-changer (accessed 05.08.2015.).

AFP, 2015. Germany restricts fracking but doesn't ban it. http://phys.org/news/2015-04-germany-restricts-fracking-doesnt.html (accessed 14.08.2015.).

Allnutt, L., Yoon, J., 2015. Australia – Shale Gas Handbook – June 2015 Norton Rose Fulbright LLP. http://www.lexology.com/library/detail.aspx?g=314ab6de-8c0f-4e82-868b-77468e2b2b95 (accessed 05.08.2015.).

Aloulou, F., 2015. Argentina and China lead shale development outside North America in first-half 2015 –Today in Energy, US EIA. http://www.eia.gov/todayinenergy/detail.cfm?id=21832 (accessed 03.08.2015.).

Arthur, A., 2014. Shale Oil and Gas the Latest Energy Frontier for South America. http://oilprice.com/Energy/Energy-General/Shale-oil-and-gas-the-latest-energy-frontier-for-South-America.html (accessed 10.08.2015.).

Batra, R.K., 2013. Policy brief June 2013 Shale Gas in India: Look Before You Leap The Energy and Resources Institute (teri). http://www.teriin.org/policybrief/docs/Shale_gas.pdf. (accessed 05.08.2015.).

Beattie, J., 2015. Green campaigners rage as ministers plan to fast-track fracking applications. The Mirror. http://www.mirror.co.uk/news/uk-news/green-campaigners-rage-ministers-plan-6224884 (accessed 11.08.2015.).

BGS, 2012. Unconventional Hydrocarbon Resources of Britain's Onshore Basins – Shale Gas. https://www.og.decc.gov.uk/UKpromote/onshore_paper/UK_onshore_shalegas.pdf (accessed 10.08.2015.).

Bhutta, Z., 2015. Not a better choice: US pushes for expensive shale gas extraction, The Express Tribune. http://tribune.com.pk/story/891551/not-a-better-choice-us-pushes-for-expensive-shale-gas-extraction/ (accessed 05.08.2015.).

Bird & Bird, 2015a. Shale Gas in Sweden - http://www.twobirds.com/en/hot-topics/shale-gas/shale-gas-in-sweden (accessed 01.08. 2015).

Bird & Bird, 2015b. Shale Gas in Denmark http://www.twobirds.com/en/hot-topics/shale-gas/shale-gas-in-denmark (accessed 10.08.2015.).

Burnett, J., 2015. Excitement Over Mexico's Shale Fizzles As Reality Sets In – npr. http://www.npr.org/sections/parallels/2015/03/16/393334733/excitement-over-mexicos-shale-play-fizzles-as-reality-sets-in. (accessed 05.08.2015.).

Chodkowski-Gyurics, G., 2014. Egypt begins shale gas exploration. http://shalegas.cleantechpoland.com/?page=news&id=268&link=egypt-begins-shale-gas-exploration. (accessed 01.08.2015.).

Collins, G., 2015. Mexico Is Becoming the Single-Largest U.S. Shale Gas Export Customer.5 http://www.mondaq.com/mexico/x/365838/Oil+Gas+Electricity/Mexico+Is+Becoming+the+SingleLargest+US+Shale+Gas+Export+Customer(accessed 04.08.2015.).

Cunha, R., 2014. Outlook for Brazil's Oil and Gas Onshore Segment May 2014 http://buyusainfo.net/docs/x_4366673.pdf (accessed 05.08.2015.).

Deans, D., 2015. UK Government will not award any new fracking licences in Wales ahead of powers being devolved to the Assembly-Wales Online. http://www.walesonline.co.uk/news/wales-news/uk-government-not-award-any-9811668 (accessed 11.08.2015.).

The Economist, 2014. Natural Gas in China: Shale Game – The Economist. http://www.economist.com/news/business/21614187-china-drastically-reduces-its-ambitions-be-big-shale-gas-producer-shale-game (accessed 03.08.2015.).

EIA, 2013. Technically Recoverable Shale Oil and Shale Gas Resources: An Assessment of 137 Shale Formations in 41 Countries Outside the United States June 2013 – US Energy Information Administration and Advanced Resources International (ARI). http://www.eia.gov/analysis/studies/worldshalegas/ (accessed 12.08.2015.).

EMIS, 2015. India: Shelving the shale? http://www.securities.com/emis/insight/india-%E2%80%93-shelving-shale (accessed 05.08.2015.).

Environmental Justice Atlas, 2013. Fracking opposition Tunisia - https://ejatlas.org/conflict/fracking-opposition-tunisia (accessed 10.08.2015.).

Fisher, J., 2015. Mexico's Shales Present Opportunities, Challenges, Researchers Find Natural Gas Intel. http://www.naturalgasintel.com/articles/102487-mexicos-shales-present-opportunities-challenges-researchers-find (accessed 04.08.2015.).

Global Security.org, 2014. Libya – Oil and Gas. http://www.globalsecurity.org/military/world/libya/petrol.htm (accessed 01.08.2015.).

Goddard, P., 2013. Fracking: Saying yes to shale – Global Trader. http://www.gtglobaltrader.com/news/fracking-saying-yes-shale (accessed 16.08.2015.).

Gomes, I., 2015. Oxford Institute for Energy Studies (OIES) Paper NG98: Natural Gas in Canada: what are the options going forward? http://www.oxfordenergy.org/wpcms/wp-content/uploads/2015/05/NG-98.pdf (accessed 04.08.2015.).

Government of the Netherlands, 2015. No extraction of shale gas during the next five years. http://www.government.nl/news/2015/07/10/no-extraction-of-shale-gas-during-the-next-five-years.html (accessed 12.08.2015.).

Green, C.A., Styles, P., Baptie, B.J., 2012. Preese Hall Shale Gas Fracturing Review & Recommendations For Induced Seismic Mitigation. https://www.gov.uk/government/uploads/system/uploads/attachment_data/file/15745/5075-preese-hall-shale-gas-fracturing-review.pdf (accessed 16.08.2015.).

GSHAP, 1999. Global Seismic Hazard Action Programme. http://www.seismo.ethz.ch/static/GSHAP/ (accessed 15.08.2015.).

Hill, D., 2015. Is Bolivia going to frack "Mother Earth"? The Guardian. http://www.theguardian.com/environment/andes-to-the-amazon/2015/feb/23/bolivia-frack-mother-earth (accessed 11.08.2015.).

ITE Oil & Gas, 2015. Oil and gas news roundup: South America. http://www.oilgas-events.com/market-insights/sector-news/oil-and-gas-news-roundup-south-america/801791026 (accessed 16.08.2015.).

Jaipuriyar, M., 2015. At the Wellhead: Thailand's oil, gas exploration under pressure – The Barrel, Platts. http://blogs.platts.com/2015/03/09/thailand-oil-gas-exploration/ (accessed 16.08.2015.).

Mackay, D.J.C., Stone, T.J., 2013. Potential Greenhouse Gas Emissions Associated with Shale Gas Extraction and Use. https://www.gov.uk/government/uploads/system/uploads/attachment_data/file/237330/MacKay_Stone_shale_study_report_09092013.pdf (accessed 14.08.2015.).

Mares, D.R., 2013. Energy Policy Group Working Paper: Shale Gas in Latin America: Opportunities and Challenges. http://www.thedialogue.org/wp-content/uploads/2015/03/MaresShaleGasfor-Webposting.pdf (accessed 05.08.2015.).

Martor, B., 2015. France: Evolutions in the legal framework for shale oil and gas. http://www.shale-gas-information-platform.org/categories/legislation/expert-articles/martor-article.html#c3132 (accessed 22.08.2015.).

May, C., Field, D., Bruce-Jones, L. Baines, T., 2015. Shale Gas Handbook: United Kingdom – June 2015 - Norton Rose Fulbright. http://www.nortonrosefulbright.com/knowledge/publications/129593/united-kingdom (accessed 10.08.2015.).

Mihalache, A.E., 2015. No Shale Gas, After All – Implications of Chevron's Exit from Romania – natural Gas Europe. http://www.naturalgaseurope.com/chevron-romania-shale-gas-exit-22947 (accessed 03.08.2015).

Nakhle, C., 2015. Algeria's Shale Gas Experiment Article April 23, 2015 Carnegie Middle East Centre. http://carnegie-mec.org/2015/04/23/algeria-s-shale-gas-experiment (accessed 31.07.2015.).

Natural Gas Europe, 2015. Germany Delays Parliamentary Vote on Fracking. http://www.natural-gaseurope.com/germany-delays-parliamentary-vote-on-fracking-24436 (accessed 14.08.2015.).

Naziri, A., 2014. Colombia – The Next Shale Gas Revolution? Palantir Solutions. http://www.palan-tirsolutions.com/blog-research/blog/2014/july/2/colombia-next-shale-gas-revolution (accessed 05.08.2015.).

North Africa Risk Consulting Inc (NARCO), 2015. North Africa Round Up. http://northafricarisk.com/analysis/2015-02-02 (accessed 31.07.2015.).

Nowak, Z., Boczek, S., 2015. Will Equals Way: Unconventional Gas in Russia – The Polish Institute of International Affairs (PISM) Bulletin No. 61 (793). http://www.pism.pl/files/?id_plik=20015 (accessed 01.08.2015.).

O'Neill, L., Watson, A., Petilon, J.-C., 2015. Oil and gas in Egypt – Oil & Gas Financial J. http://www.ogfj.com/articles/print/volume-12/issue-6/features/oil-and-gas-in-egypt.html (accessed 01.08.2015.).

Perkins, J., 2015a. Bulgaria to keep fracking moratorium. http://www.shaleenergyinsider.com/2015/02/02/bulgaria-to-keep-fracking-moratorium/ (accessed 12.08.2015.).

Perkins, J., 2015b. Denmark suspends Total shale gas exploration - Shale Energy Insider. http://www.shaleenergyinsider.com/2015/05/07/denmark-suspends-total-shale-gas-exploration/ (accessed 11.08.2015.).

Pinsent Masons, 2015. Out-Law – South African legislators call for fracking report as ministers consider shale gas exploration bids.

http://www.out-law.com/en/articles/2015/june/south-african-legislators-call-for-fracking-report-as-ministers-consider-shale-gas-exploration-bids-/ (accessed 31.07.2015.).

Platts, 2015. Latin America Oil Outlook 2015. http://www.platts.com/news-feature/2015/oil/latin-america-oil-outlook/index. (accessed 05.08.2015.).

Prinsloo, L., 2015. Shell gets cold feet on SA shale-gas Business Times. http://www.timeslive.co.za/businesstimes/2015/03/15/shell-gets-cold-feet-on-sa-shale-gas-audio (accessed 31.07.2015.).

Proactive Investors, 2013. Petrel Energy in first deep onshore shale drilling in Uruguay for 30 years. http://www.proactiveinvestors.com.au/companies/news/45345/petrel-energy-in-first-deep-onshore-shale-drilling-in-uruguay-for-30-years-45345.html (accessed 16.08.2015.).

PWC, 2014. Know your way around Shale gas in Poland. http://www.pwc.com/en_GX/gx/oil-gas-energy/publications/assets/know-your-way-around-shale-gas-in-poland.pdf (accessed 22.07.2015.).

Razavi, S., Hidayanto, B., 2013. Shale Gas in Indonesia: an unconventional revolution? Int. Financial Law Rev, 11 December 2013. http://www.iflr.com/Article/3289114/Shale-gas-in-Indonesia-an-unconventional-revolution.html (accessed 10.08.2015.).

Rebhi, H., 2014. Governance: Tunisia presses ahead with fracking despite counter arguments – International Anti-Corruption Conference. http://16iacc.org/blog/2014/10/02/governance-tunisian-government-presses-ahead-with-fracking-despite-counter-arguments/ (accessed 12.08.2015.).

Regester Larkin, 2014. Brazil's shale prospects. http://www.regesterlarkin.com/news/rnl-shale-report-brazils-shale-prospects/ (accessed 05.08.2015.).

Riaz, S., 2013. Thailand's PTTEP wants to learn from North American shale projects – Shale Energy Insider. http://www.shaleenergyinsider.com/2013/09/30/thailands-pttep-wants-to-learn-from-north-american-shale-projects/ (accessed 16.08.2015.).

Royal Society and The Royal Academy of Engineering, 2012. Shale gas extraction in the UK: a review of hydraulic fracturing. http://www.raeng.org.uk/publications/reports/shale-gas-extraction-in-the-uk (accessed 12.06.2015.).

Santos, A.J., 2015. Turkey: Turkish Energy Market. http://www.mondaq.com/turkey/x/395770/Oil+Gas+Electricity/Turkish+Energy+Market+2015 (accessed 12.08.2015.).

Seelke, C.R., Ratner, M., Angeles Villarreal, M., Brown, P., 2015. Mexico's Oil and Gas Sector: Background, Reform Efforts, and Implications for the United States: Congressional Research Service. https://www.fas.org/sgp/crs/row/R43313.pdf (accessed 05.08.2015.).

Shale Gas Europe, 2015a. A pivotal year for Polish shale. http://shalegas-europe.eu/2015-pivotal-year-polish-shale/ (accessed 22.07.2015.).

Shale Gas Europe, 2015b. Spain, the forgotten shale gas country. http://shalegas-europe.eu/spain-the-forgotten-shale-gas-country-2/?lang=pl (accessed 10.08.2015.).

Shale Gas Europe, 2015c. Shale Gas in Ukraine: Opportunity to Reality? http://shalegas-europe.eu/shale-gas-in-ukraine-opportunity-to-reality-2/?lang=pl (accessed 22.07.2015.).

Shale Gas Europe, 2015d. European Shale Gas Round-up. http://shalegas-europe.eu/european-shale-gas-round-up-9/ (accessed 10.08.2015.).

Shale Gas International, 2014. Venezuela to start shale gas exploration. http://www.shalegas.international/2014/06/02/venezuela-to-start-shale-gas-exploration/ (accessed 05.08.2015.).

Shale Gas International, 2015a. Bolivia keen on fracking despite Green opposition. http://www.shalegas.international/2015/03/05/bolivia-keen-on-fracking-despite-green-opposition/ (accessed 11.08.2015.).

Shale Gas International, 2015b. Indonesia signs four contracts for unconventional oil and gas exploration. http://www.shalegas.international/2015/05/27/indonesia-signs-four-contracts-for-unconventional-oil-and-gas-exploration/ (accessed 10.08.2015.).

Shale Gas International, 2015c. Shell planning to quit shale exploration in Eastern Ukraine. http://www.shalegas.international/2015/06/12/shell-planning-to-quit-shale-exploration-in-eastern-ukraine/ (accessed 22.07.2015.).

Shale Gas International, 2015d. Could hydraulic fracturing play a part in Russia's future? http://www.shalegas.international/2015/06/25/could-hydraulic-fracturing-play-a-part-in-russias-future/ (accessed 01.08.2015.).

Shale World, 2013. Will Mongolia Become the Next Shale Gas Hotspot? Shale World. http://www.shale-world.com/2013/12/16/mongolia-shale-gas-hotspot/ (accessed 16.08.2015.).

Tran, T., 2014. Shale gas provides largest share of U.S. natural gas production in 2013 Today in Energy US EIA. http://www.eia.gov/todayinenergy/detail.cfm?id=18951 (accessed 03.08.2015.).

US Department of Energy & EDX, 2015. Shale Basins and Plays World. https://edx.netl.doe.gov/dataset/shale-basin-and-plays-world/revision_resource/ee967de8-9448-40dd-b111-65439aa22be0 (accessed 22.07.2015.).

USGS, 2014. Assessment of Potential Shale-Oil and Shale-Gas Resources in Silurian Shales of Jordan. http://pubs.usgs.gov/fs/2014/3082/pdf/fs2014-3082.pdf (accessed 16.08.2014.).

Van Leeuwen, M., 2015. Shale Gas Handbook: The Netherlands– June 2015 - Norton Rose Fulbright. http://www.nortonrosefulbright.com/knowledge/publications/129588/the-netherlands (accessed 04.08.2015.).

Vetter, A., 2015. Shale gas in Germany – the current status – April 2015. http://www.shale-gas-information-platform.org/areas/the-debate/shale-gas-in-germany-the-current-status.html (accessed 15.08.2015.).

Vinson & Elkins, 2015. Shale Development in Denmark. http://fracking.velaw.com/shale-development-in-denmark/ (accessed 10.08.2015.).

Vladimirov, M., 2015. Will Morocco Finally Realize Its Shale Dream? http://oilprice.com/Energy/Gas-Prices/Will-Morocco-Finally-Realize-Its-Shale-Dream.html (accessed 14.08.2015.).

World Resources Institute, 2014. Global Shale Gas Development – Water Availability and Business Risks. http://www.wri.org/publication/global-shale-gas-development-water-availability-business-risks (accessed 14.06.2015.).

Yee, A., 2015. Libya pins its hopes on foreign investors to develop shale gas – The National – Business. http://www.thenational.ae/business/industry-insights/energy/libya-pins-its-hopes-on-foreign-investors-to-develop-shale-gas (accessed 01.08.2015.).

Subject Index

C

D

E

F

G

I

J

K

L